336760041003372
W9-ATZ-555

DISCARD

Poison in the Well

Poison in the Well

Radioactive Waste in the Oceans at the Dawn of the Nuclear Age

JACOB DARWIN HAMBLIN

RUTGERS UNIVERSITY PRESS
New Brunswick, New Jersey, and London

Library of Congress Cataloging-in-Publication Data

Hamblin, Jacob Darwin.
Radioactive waste in the oceans at the dawn of the nuclear age / Jacob Darwin Hamblin.
p. cm.
Includes bibliographical references and index.
ISBN 978-0-8135-4220-1 (harcover : alk. paper)
1. Radioactive waste disposal in the ocean. I. Title.
TD898.H35 2008
363.72'8909162—dc22 2007015498

A British Cataloging-in-Publication record for this book is available from the British Library

Visit our Web site: http://rutgerspress.rutgers.edu

Manufactured in the United States of America

For my children

Contents

Acknowledgments

The first horror movie I ever saw was *Empire of the Ants*. It was on television when I was eight or nine years old. Drums of radioactive sludge had washed up onto a Florida beach, creating, well, an empire of giant ants. In retrospect it was not a very good movie, but it kept me awake at night. I mention this in honest acknowledgment of my first exposure to the subject of radioactive waste disposal at sea. I no longer harbor beliefs about the possibility of such monstrous mutations from human waste disposal practices. Still, I share most people's general concern about radioactive waste. Friends and colleagues expressed surprise when they discovered I was writing a book on the subject; "They didn't do that, did they?" was a common refrain. A few had vague recollections of hearing what the Soviets did; much fewer were aware that Western countries routinely dumped radioactive waste at sea in the 1950s and 1960s. It struck some people as so patently wrong that they supposed I was writing an exposé of a secret and sordid past. That is not what this book is about. Researching and writing this book has made me less concerned about the question "Was it wrong?" and more concerned to ask "Who made these decisions, and why?" I have tried to avoid condemnations and prescriptions for future action. The book is neither an environmental screed nor an apologia for nuclear power and the institutions of the cold war. It is a history: it attempts to see the past from more than one perspective, to explain decisions and actions, and to draw some conclusions about science, the environment, international relations, and democratic politics.

I began thinking about the subject seven years ago while in England finishing some research for an earlier book. At the Public Record Office I found some documents about accidents at sea in Britain's routine dumping operations back in the 1960s, and I soon began chatting with colleagues at the archives about writing an international history of radioactive waste disposal at sea. Historians Kurk Dorsey and Molly Girard were among them; I thank Kurk in particular for comments and encouragement over the years. Special thanks go also to Audra

Wolfe, who first took an interest in the manuscript at Rutgers University Press and helped bring the book to fruition. I would like to thank the other colleagues who have played some role in the evolution of this project, at conferences, over email, and by commenting on articles and papers. I owe many intellectual and professional debts to Lawrence Badash, Michael A. Osborne, and Fredrik Logevall. I also thank Kai-Henrik Barth, Stephen Bocking, Cathy Carson, John Krige, Morris Low, Mark Lytle, Pamela Mack, Roy MacLeod, Erika Milam, Edwin Moïse, Peter Neushul, Kenneth Osgood, Jahnavi Phalkey, Ronald Rainger, Ioanna Semendeferi, John Staudenmaier, J. Samuel Walker, Zuoyue Wang, and Benjamin C. Zulueta. I am also very grateful to Gordon Linsley and Tiberio Cabianca for chatting with me in Oxford about their experiences at the IAEA.

This book would not have been possible without an extraordinary number of archival documents from France, Britain, and the United States. I thank the archivists in all the institutions I have visited for allowing me to use a digital camera (Caltech was the sole exception). For their assistance on an individual level, I am grateful to Shelley Erwin, Bonnie Ludt, and Judith Goodstein at Caltech, Jens Boel and Mahmoud Ghander at UNESCO, Odile Frossard at the Commissariat à l'Énergie Atomique, Deborah Day at the Scripps Institution of Oceanography, Gary Lundell at the University of Washington, Janice Goldblum at the National Academies Archives, and the anonymous (to me) faces at other institutions. Research trips for this book were funded by the University of California's Institute on Global Conflict and Cooperation, the American Philosophical Society's Franklin Research Grant, and the History Department at California State University, Long Beach.

On a personal note, I would like to thank my mother and father, Sharon and Les Hamblin, who have passed to me their values, and my sister Sara and her daughter, Victoria, who remind me that new beginnings are always possible. Thank you to my in-laws Paul and Cathy Goldberg, who have made tremendous emotional sacrifices to support us. My beautiful daughter, Sophia, is almost three years old, and another child will arrive soon. This book is dedicated to them. They are my personal hope for the future, and it is a parent's hope: that they will learn not only to listen to what people say but also to think hard about why they are saying it. And, of course, I thank Sara Goldberg-Hamblin. We have shared many adventures, though sometimes we have lived with continents and oceans between us. I hope we never will again. I have never deserved unfailing support, but I have always received it: thank you.

Lastly, I thank my grandfather, Darwin Claude Hamblin—navy pilot, rancher, teacher, inspirational (if not very good) fisherman, and unrivaled cheater at cards: 1923–2007. Rest in peace.

Poison in the Well

Introduction

When Russian president Boris Yeltsin decided in the early 1990s to reveal some of the old Soviet regime's dark secrets, he dropped an environmental bombshell. A major report from his special advisor on the environment, Alexei Yablokov, unveiled the long history of dissimulations and lies by the Soviet government about dumping radioactive waste at sea. Despite decades of denials under communism, Yablokov now revealed that the Soviet Union had dumped large amounts of dangerous radioactive waste into rivers and seas, notably into the Arctic Ocean. Between 1959 and 1992, the Soviet Union routinely violated international norms and agreements, including the London Convention, which restricted marine pollution. In addition to effluent and packaged waste, it dumped sixteen nuclear reactors from submarines and icebreakers, some still with nuclear fuel, most of them in water less than one hundred meters deep.[1]

The international outrage that followed these disclosures seemed to suggest not merely corruption or incompetence, but also a problem of pathological proportions within the former Soviet Union that linked its decaying institutions with its blighted environment, already known to be marred by the poisons in Lake Baikal and the Chernobyl nuclear disaster of 1986. An appalled United States Congress put its Office of Technology Assessment to work gauging the consequences of the Soviets' Arctic pollution. Some scholars charged that Soviet political culture, lacking democratic openness and political accountability, invited such excesses. Others concluded that the Soviet regime concentrated on production at any cost and ignored its problems, assuming that scientists and engineers eventually would solve them. Exploring the links between communism and environmental degradation became a veritable cottage industry among academic writers.[2]

The irony was that the Soviet Union had used the issue of radioactive waste disposal at sea as a vehicle for waging a propaganda war against the West, thus putting the environment into its diplomatic arsenal during the 1950s and 1960s. During those years, most of the criticism for dumping radioactive waste at sea was directed at Western countries, especially the United States and Britain. When international debates raged over the issue in the 1960s, the Soviet Union raised the loudest voice of disapproval, calling Western countries "poisoners of wells" who dumped radioactive waste into the shared resource of the sea. Dumping radioactive waste at sea is banned by international treaty today, but the United States, Britain, and other countries practiced it for decades.[3]

Although it may be instructive to link environmental pollution from radioactivity to the corrupt institutions of a failed political ideology, as in the case of the Soviet Union, there is much to learn from the processes that shaped those same issues in the West, particularly the United States and Britain—the two countries that first did it and set standards that others followed. Their policy decisions, scientific conflicts, public relations strategies, not to mention mishaps and subsequent cover-ups, defy convenient generalizations about secrecy and openness in the East and West during the cold war era. Why did scientists and politicians choose the sea? What about it proved so attractive, and how did the negotiations about the uses of the sea change the way scientists, government officials, and ultimately the lay public envisioned the oceans? In the 1950s, leading oceanographers viewed the ocean as a sewer, using language that might have led to the professional ostracism of an aspiring marine scientist just a couple of decades later.

Poison in the Well is a history of the scientific, political, and diplomatic controversies connected to disposing radioactive waste at sea, told in the context of the democratic nuclear powers. It traces the development of the issue from the end of World War II to the blossoming of the environmental movement in the early 1970s, when old ideas about the oceans confronted new attitudes about protecting the seas, leading in 1972 to the first global treaty, known as the London Convention, limiting radioactive waste disposal and other forms of marine pollution. The United States and Britain, like the Soviet Union, promoted nuclear weapons and civilian nuclear power on a large scale during the 1950s. They rushed into atomic energy for a host of reasons—diplomatic, military, cultural, and economic. And simultaneously they faced an array of challenges to public health, including the fallout from nuclear tests, occupational safety, and radioactive waste.[4] By the 1960s, they were joined by France and other nations whose reasons for wanting nuclear reactors varied, but whose dilemmas were extraordinarily similar; the salient difficulty was that of radioactive wastes, the by-products of the nuclear age. Many controversial solutions were proposed over the years, and indeed the problem has yet to be solved. Even today scientists and politicians clash over American plans to house the nation's most dangerous materials in Nevada's Yucca Mountain.[5]

The term "radioactive waste" conjures an image of highly contaminated, dangerous material, the effects of which are limited only by the imagination. The links between acute radiation exposure and radiation sickness, exhibited in the victims of the 1945 atomic bombings of Hiroshima and Nagasaki, have been supplemented over the years by scientific reports connecting chronic exposure to radiation with various forms of cancer.[6] The genetic effects have captivated us most, with the possibility that radiation-induced mutations might produce outrageous changes in the gene pool. Exaggerations of these effects have been embedded in popular culture, beginning in 1954 with the monster-giant film *Godzilla*, produced in Japan after the American hydrogen bomb tests in the Pacific, and *Them!*, an American film about giant ants coming out of the New Mexico atomic test sites.[7] Hollywood suggested the implications of sea disposal with the 1977 film *Empire of the Ants*, with its poster tagline "It's no PICNIC!," about a washed-up canister of radioactive waste producing a colony of enormous ants in the Florida Everglades.

"Radioactive waste" actually was a broad term that described liquids and solids that had become radioactive through exposure to intense electromagnetic radiation or direct contact with radioactive elements. "Radioactivity," a term coined in France by Marie Curie at the turn of the twentieth century, described the property of emitting different kinds of radiation, classified as alpha, beta, and gamma activity. Wastes contained radioactive isotopes of a variety of elements—not just bomb materials like uranium and plutonium, but also polonium, cesium, strontium, and iodine, to name a few—and each radioactive isotope was unique, presenting different threats to humans. Probably the most well-known isotope, strontium-90, became infamous during the fallout controversy of the 1950s because it was known to concentrate in the bones—and in children's teeth.[8] It and other isotopes also concentrated in the sea's flora and fauna, presenting a pathway of radiation exposure to any community relying on the ocean for food.[9]

By the 1950s, the most problematic sources of waste were nuclear reactors, whose fuel rods were highly radioactive and needed to be disposed of safely or sent to chemical reprocessing plants. Such chemical reprocessing of "spent" nuclear fuel, to salvage its remaining fissionable material, produced additional large amounts of radioactive waste, mostly in the form of liquid effluent. These factories also separated out the plutonium for nuclear weapons, at sites like Hanford (United States), Windscale (Britain), and Kyshtym (Soviet Union). Hanford, Kyshtym, and other sites in the United States and

Soviet Union discharged their effluent into rivers, ultimately and often imperceptibly sending some of it to the sea. But until the late 1960s, when France began to operate a coastal reprocessing facility at La Hague, only Britain discharged such effluents directly into the sea in sizable quantities, putting that country at the center of the sea disposal controversies.[10] Britain also dumped radioactive waste from ships, though it was not the only country to do so. Institutions all over the world produced solid radioactive wastes, some from weapons and civilian power, but others from research activities associated with radioisotopes.[11] Surgical gloves, beakers from the laboratory, pieces of irradiated metal from reactors, carcasses of experimental animals, and other assorted objects—all were radioactive wastes. Beginning in the 1940s, both the United States and Britain packaged up solid wastes into drums, put them onto ships, and dumped them in the ocean.

For many years, the sea seemed to be the answer to the waste disposal problem. But atomic energy establishments increasingly met with fierce resistance, from the politicians, laypersons, and scientists who cast doubt on sea disposal by the 1960s. Distrust of atomic energy, combined with an ascendant concern for saving the oceans from pollution, accompanied bitter salvos across cold war lines about the effects of nuclear arsenals upon the environment. Ultimately, diplomats agreed to ban the practice. However, as this book shows, this did not significantly change what nations were doing—not in the case of the United States, which already had given it up, and not in the case of Britain, which continued to rely on the sea more than any country. In fact, the celebrated international agreement, certainly a product of the rise of environmentalism, had a curiously small effect upon the introduction of radioactivity into the sea, which continued long after the global dumping regime went into force. The history of ocean disposal of radioactive waste is less about the triumph of environmentalism and more about science, politics, and diplomacy.

Until now, sea disposal of radioactive waste has been one of the untold stories of the cold war, and this book begins to address a major deficiency in our knowledge of the recent past. It seeks to explain why it was practiced, why it was banned, and what the story tells us about science and the environment in the democratic nuclear countries. The narrative is guided by four broad and interrelated themes, explained in detail below, each highlighting the fact that sea disposal policies hinged upon struggles for scientific authority—between governments, between institutions, and between groups of scientists vying for influence, patronage, or decision-making power. Despite efforts at objective analysis, policies on sea disposal required subjective choices reflecting a range of priorities. The virulence of the controversy resulted from a series of political contests in which the battleground was the opinion of the lay public, at national and international levels. Waste disposal started as a challenging but apparently not insuperable technical problem, but its delayed solution made it atomic energy's Achilles' heel. This was not a product of an inherent nuclear fear in an unreasoning general public, but was the result of fractious contests for authority at several levels: national policymaking, international rhetoric and negotiations, and even scientific assessment.

The first theme is the power of threshold values in setting policies and justifying them to the lay public. It is tempting to ask fundamental questions about radioactive waste at sea: Was it unsafe? Was it wrong for these countries to practice it? Was it right to ban it? The answers to these questions depend upon highly contested notions of danger, risk, and acceptable levels of biological damage. Threshold values, employed routinely as benchmarks of safety, set the limits of acceptable levels of exposure. This concept was embedded in such terms as "tolerance dose," "permissible dose," and "safe capacity." Atomic energy establishments usually respected them in order to show the lay public that their practices were safe. Indeed the issue of setting permissible doses stands at the heart of most studies of radiation protection during the cold war.[12] Instead of asking whether sea disposal was unsafe, one more properly should ask, "How were threshold values chosen?" In doing so a new slate of concerns arises over who chose the thresholds and what motivated them.

How did atomic energy establishments decide how much radioactive waste they could put safely into the sea? Most policies were based on the recommendations of bodies such as the International Commission on Radiological Protection and their national counterparts; these bodies, however, were primarily concerned with setting limits on direct exposure to humans, without particular reference to the effects of releasing radioactive material to the environment.[13] This opened a wide avenue for atomic energy establishments' interpretations and also for scientists to create research projects. How best to understand how radioactivity reached humans through the environment was a major point of contention between scientists, atomic energy officials, and diplomats well into the mid-1970s. There was the possibility that thresholds might themselves be illusory, and that the sea did not have a "capacity" to wash away quantifiable amounts of radioactive waste every year. Perhaps the nations of the world were altering the composition of the seas, consigning the "normal" ocean to an irretrievably lost past. Despite this nagging concern, scientists did identify threshold levels and sea disposal did occur. Most of the controversies arose over how thresholds should be identified and what group of scientists should identify them.

The second theme is the struggle for scientific authority between health physicists and oceanographers. "Health physicists" is a broad term describing the scientists who first studied radiation safety during the wartime bomb projects and effectively created the first threshold values for occupational exposure. In subsequent years the field broadened further to include not just worker safety but also public health policy; in this book, radiobiologists, biophysicists, and sanitary engineers are often considered part of this group. They dominated thinking about the biological effects of radiation in both the United States and Britain; they forged a unique community of scientists who straddled the line between academic and government science, often working in national laboratories or directly for atomic energy establishments. The United States Atomic Energy Commission (AEC) and United Kingdom Atomic Energy Authority (AEA) supported health physicists' research and were the powerful epicenters of the well-funded world of atomic science, atomic bombs, and future atomic power. But there were ancillary issues, such as the environmental effects of nuclear weapons and the need for permanent waste disposal, which soon attracted scientists from other disciplines. In the United States, oceanographers were not the original "atomic scientists," but they yearned to be recognized for—and to benefit from—the important role they could play in the nuclear arena. Leading oceanographers worked hard to convince the Atomic Energy Commission, for example, to fund studies that might help it to understand the effects of waste disposal on the sea.

Over time the oceanographers' activities created a rift between them and the existing experts within atomic energy establishments in the United States and Britain. The health physicists felt that opportunistic oceanographers were turning the public against them, needlessly questioning their practices and counterproductively playing up the potential dangers of sea disposal—not because of their genuine environmental or public health concerns, but simply to assert a place at the nuclear table. The AEC felt this most acutely from about 1959 as several reports by the National Academy of Sciences began to question AEC policies. The British were particularly vehement in this view, and British oceanographers made no strides that could compare to their American counterparts; British atomic scientists viewed with antipathy the rise in stature of American oceanographers and particularly the French oceanographer Jacques-Yves Cousteau. While this may have been an overreaction, it poisoned relations between scientists inside and outside government, and it widened the gap of mutual distrust between the lay public and atomic energy establishments. Ironically, many of the acts that defied integrity, such as disingenuous monitoring programs, suppressed information about dumping, and the like, were the result of government scientists and officials striving hard to protect their scientific integrity against those they considered opportunistic interlopers preying on the ignorance of the lay public. Health physicists cynically observed how often the oceanographers used the uncertainty of knowledge surrounding ocean disposal as a means to ask for more sponsored research. Academic scientists were sometimes given credit for being able to "speak truth to power," but this stood in sharp contrast to the perceptions of atomic energy establishments.[14] To them, oceanographers were pesky meddlers, not because they exposed inconvenient truths but because they called government scientists' competence into question simply because they wanted a piece of the pie.

Public participation in policymaking, the principal feature of democratic politics, was a specter that haunted the relationships between atomic energy establishments and some academic scientists. Some claimed that the debates about the biological effects of atomic radiation were too emotionally charged to be discussed rationally. In fact, the notion of public irrationality, particularly the public's visceral fears of all things connected to atomic energy, was a common thread in discussions of the issue in newspapers, official statements, and even scientific reports. But the reverse was also true: atomic establishments developed their own fears of the lay public, often leading them to keep secrets, avoid opening their policies to public scrutiny, hide accidents, play down dangers, and deny the possibility of harm from radiation. The defensive posture adopted by atomic energy advocates reflected long-standing battles for authority and influence about the scientific underpinnings of ocean disposal. The effort to define safe practices became a series of contentious skirmishes to assert roles as legitimate, authoritative sources of knowledge.

The third theme is the role of radioactive waste in cold war international relations. Conflicts in expertise made atomic energy establishments vulnerable to criticism, not only at home but also in international venues. With President Dwight Eisenhower's "Atoms for Peace" initiative, the entire spectrum of nuclear weapons and civilian nuclear power came under international scrutiny, including waste disposal. The Soviet government began to use sea disposal as a way to criticize the United States and Britain, beginning in the late 1950s. Despite the hypocrisy of it, made clear after the collapse of the Soviet Union, labeling these two countries as the modern-day poisoners of wells gave the Soviet government a common cause with the rest of the world. Like its criticism of nuclear testing, the Soviet Union's claims about radioactive waste disposal fused cold war politics with what soon would be seen as an environmental issue. Their propaganda suggested that putting the dangerous by-products of the nuclear age into the oceans was like poisoning a village well, the shared source of life for all.

American and British scientists met frequently in the 1950s to negotiate common positions on scientific issues before airing their views publicly. They also exercised considerable influence in the few international bodies that set policy standards, such as the International Commission on Radiological Protection. In the late 1950s, attempts by the specialized agencies of the United Nations to assess sea disposal independently of Anglo-American views were not successful, and they conformed to existing Anglo-American practices. Scientists and officials of the two countries worked closely to ensure their scientific credibility and to shepherd international opinion in favor of their national policies. Western countries, including France, showed remarkable solidarity against the propaganda broadsides from the Soviet Union. Atomic energy establishments liaised with each other to find ways of educating the average citizen away from oceanographers' constant focus on uncertainty. Policy makers in the United States and Britain struggled to find ways to combat Soviet propaganda without compromising sources of nuclear intelligence. The United States supported its allies by making the scientific integrity of nuclear waste disposal policies an issue for solidarity in the North Atlantic Treaty Organization (NATO).

The fourth and final theme is the relationship between radioactive waste and environmental policy making. The Anglo-American solidarity began to disintegrate during the rise of the environmental movement, as American politicians began to condemn the practice. Despite the apparent break, however, neither country overhauled its practices to accommodate environmental concerns. The United States began to pass pro-environment legislation that banned actions no longer practiced. For its part, Britain continued to defend sea dumping vigorously. By the end of the 1960s, it turned toward continental Europe, promoting joint dumping operations within the Organization for Economic Cooperation and Development's European Nuclear Energy Agency. By acting only with other European countries through this body, Britain tried to avoid the impression that it dumped unilaterally and irresponsibly. It paid lip service to international environmental initiatives but worked to overcome the actions of what its health physicists called the "environmental lobby," in order to protect practices seen as indispensable to the entire nuclear program.

How, then, did the 1972 London Convention come about, and what did it really accomplish? The convention reflected new attitudes toward the environment, and radioactive wastes were included as one form of marine pollution to regulate. Some scholars, mainly political scientists, have used the idea of scientific consensus-building to understand the formation of such environmental protection regimes.[15] Others emphasize the importance of political activism in mobilizing public opinion.[16] In the case of the London Convention, the new environmental regime was based more on political than scientific consensus, a fact that is borne out by the lack of substantive change in policy under the new rules. Many leading scientists, particularly health physicists, fought hard behind the scenes to preserve the essence of existing practices while using language that appeared to make environmental concessions. For them, if the political struggle seemed lost to the environmentalists, the scientific struggle was not. Despite appearances, the rules set forth in the convention itself did not significantly alter the practices of the signatories. Although the rise of environmentalism was a powerful catalyst for new legislation and international agreements, it was not always a catalyst for genuine change. For radioactive waste, the convention itself was less meaningful than it seemed. The beliefs, policies, and practices of radioactive waste disposal at sea already had been constructed by the contentious battles for scientific authority during the first decades of the cold war.

Chapter 1 Threshold Illusions

As oceanographer Richard Fleming went home to the University of Washington in the fall of 1952, he was hoping to have lifted a heavy burden from his shoulders. He had just put together a draft statement about the disposal of radioactive waste at sea, and he admitted, "I am returning to Seattle feeling years younger and six inches taller with this load off my neck."[1] There was a new faculty position waiting for him there, and he was eager to devote his energy to it. But over the years, he had been part of a committee sponsored by the National Bureau of Standards to assess ways to get rid of contaminated material. It was particularly burdensome for Fleming because, as everyone began looking to the sea as the final solution to the problem, he was the only oceanographer on the committee. It should not have been a difficult task; after all, the bureau was not expecting any restrictive recommendations, and in fact the culture of the bureau discouraged them. But the more he looked into the problem, the bigger it became, revealing to Fleming the profound state of ignorance about ocean chemistry and dynamics, the biological dangers of particular isotopes, and the intricate details of waste disposal engineering.

His initial report did not go over well with colleagues at the bureau and at the Atomic Energy Commission. So the project continued to sap Fleming's time and energy as he reworked the draft over the next year. His colleagues at major oceanographic institutions waited for him to make what would be the authoritative statements about the risks of ocean contamination, and his colleagues in government waited for him to draft something that would not condemn present practices. But he struggled with how to set permissible levels of waste disposal when there was so little known about the effects, and he was uncomfortable with the idea that radioactive waste could safely disappear into the oceans without significant consequences. But as time passed and his competing responsibilities drew him away from the project, he settled for a document that reinforced the assumption of both the AEC and the bureau: there were thresholds of safety, below which one could expect no harm at all. Fleming wrote to his colleague Allyn Vine in 1953 that ultimately his job became one of identifying maximum tolerable quantities based on the accumulation of isotopes in the flesh of food fish.[2] Perhaps this was reasonable; after all, he reasoned, surely some radioactive waste could be put into the sea.

Despite the role he played in authoring the first major national guideline about radioactive waste at sea, the experience did not sit well with Fleming. Perhaps he came to realize that one man could not change institutional culture, and he was happy to flee to the comforts of academe, with its culture of individuality and intellectual freedom. His suggestions that the AEC should support more research on ocean disposal had been ignored, and the bureau had cast aside his cautious reservations about allowing indiscriminate dumping. When in 1955 he was asked by the bureau to participate in a reassessment of the problem along with other oceanographers, he declined. "I have 'had my say,'" he wrote. "I feel rather frustrated in my dealings with the AEC," he added, "and unless there is a change of heart I do not intend to press the matter any further."[3]

Despite Fleming's frustrations, neither he nor any other prominent oceanographers played a combative role in the virulent public controversies over radiation effects between academic scientists (often geneticists) and the AEC throughout the 1950s. Instead, the way Fleming framed his recommendations—with the goal of assigning maximum tolerable quantities—actually conformed to a manner of thinking about radiation that geneticists already had been critiquing and trying to revise. Maximum quantities implied threshold values below which no harm could be expected. According to geneticists, such thresholds were illusory for radiation effects. Eventually oceanographers would come to the similar conclusion about the seas, but it would be a long process. This was partly because thresholds conformed to marine science culture; for example, this was precisely the way scientists approached problems of overfishing, by assigning maximum sustainable yields.[4] It was no stretch to assume that, like fish stocks, the ocean could rejuvenate itself through dilution, and that it had a definable annual capacity to do so for radioactive waste.

At the dawn of the nuclear age, there was only a loose understanding of these concepts and there was no consensus about them. By the time the Atomic Energy Commission was created in the United States in 1946, radiation protection already had been addressed in widely divergent ways, and most of the people whose work was connected to atomic energy resisted strict standards being enforced by a new agency. And yet the radioactive wastes were piling up quickly, with no agreement about what to do with them. Many solutions were proffered, such as land burial, rocketing into space, and even dumping them onto enemy territory during the Korean War. Ultimately the sea proved attractive, not only to the United States but also to Britain. But the specifics were elusive—what kinds of waste could be put in the sea and how much of them? In an effort to avoid the appearance of the AEC regulating itself, and to placate those in the private sector who did not want to be controlled by the AEC, the AEC asked the National Bureau of Standards to step in and assess the problem. The result was a report written by a lone oceanographer under immense pressure and a precedent for incorporating the sea into the world of thresholds.

The Invisible Threat

In 1927 an article appeared in the journal *Science* confirming, for the first time, that it was possible for human beings to enhance the natural mutation rates of genes in organisms. Because genetic mutation was viewed as the principal vehicle for evolutionary change, this was exciting news. It meant that man no longer had to conform to the sluggish pace of nature to investigate the processes of biological inheritance. The author of the article, American geneticist Hermann J. Muller, had increased the rate of mutations in fruit flies by using X-rays. He traced these mutant genes down through three, four, and sometimes even more generations of flies. With X-rays, Muller predicted, one could produce "on order" enough mutations to experiment with artificial races and produce genetic maps, and, Muller wryly hinted, there might be interesting possibilities for the future of the human species.[5]

Almost two decades later, in 1946, Muller won the Nobel Prize (in Physiology or Medicine) for this work. But by then the societal landscape had changed, for Muller and for the world. He was distressed by the overproduction of mutations in human genes by the uncontrolled, indiscriminate use of X-rays, not only in research laboratories but also in hospitals and clinics, for purposes ranging from examining broken bones to clearing up pimples with radiation therapy. X-ray machines could be found even in shoe stores, where shopkeepers used them as gimmicks to help size feet. And, of course, the atomic bombings of Hiroshima and Nagasaki in 1945 introduced not only a new explosive power but also another biological threat. Atomic explosions unleashed extraordinary amounts of radiation and left radioactive debris that continued to produce it. Upon accepting the Nobel Prize in Stockholm in December 1946, Muller reminded his listeners that mutations, despite their importance for evolution, were very rarely beneficial. Mutations were random changes that, like any blind alteration to a complicated apparatus, typically proved damaging or lethal. Thus, adding such mutations by exposure to radiation should be avoided whenever possible.[6]

Most of the controversies about radiation exposure after World War II—including nuclear fallout, radioactive waste disposal, and the medical uses of radiation—hinged upon the powerful notion of "safe" levels of radiation. What was the safe amount for workers? What was the safe amount for the general population? What was the safe amount that could be dumped into the ocean without doing harm to it (and to the people who used it)? All of these questions implicitly presumed a threshold that, if respected, would prevent damage. Although precise values were debated and revised often, the notion that thresholds of safety *did exist* had a tremendous hold over scientists, politicians, and the lay public. Yet there certainly was evidence to the contrary. Geneticists on the whole did not believe there was a threshold value for radiation exposure below which no mutations would be produced. Muller's and others' experiments with fruit flies suggested that the number of mutations was proportional to the level of radiation, with even the smallest levels producing some.

Before and during the Second World War, one prevalent view was that X-rays or radioactive substances such as radium could be used until clear evidence of harm manifested itself visibly. There was no doubt that large doses of X-rays at one sitting—acute exposure—led to radiation burns. One index of overdosage was the appearance of erythema, or reddening of the skin. Those who wished to limit damage from X-rays did so by ensuring that exposure levels remained below that which produced erythema. At Memorial Hospital in New York City, for example, researchers Gioacchino Failla and Edith Quimby conducted experiments on humans, subjecting their skin to X-rays until it started to redden. In 1940 they warned about using particular kinds of apparatus that might inadvertently intensify the rays and produce erythema.[7] Their research, and that of others, reinforced the view that the erythema itself was a threshold indicator. This was an outlook that emphasized the immediate somatic, or bodily, effects of radiation.

But there were several widely reported examples of people who did not show evidence of harm until long after being exposed. Probably the most well known were the young women working as radium dial painters who licked their radium paint brushes in order to sharpen the points—many of them developed cancers in the mouth and throat in the 1920s.[8] Another serious ailment was the development of leukemia, which occurred to a number of patients whose doctors had treated their illnesses with radiation therapy, and to radiologists who spent years overexposing themselves. A further example was Marie Curie, the discoverer of radium and the person who coined the phrase "radioactivity," whose fingers were scarred permanently by radiation burns and who developed debilitating problems in the eyes and ears. The conclusion typically drawn was that Curie and these others, over the years, routinely had been exposed to acute doses of radiation that exceeded safe levels.[9]

Even those who recognized these long-term pathological dangers emphasized the importance of repeated acute (intense and brief) exposures. A huge dose of radiation at once could lead to debilitating illness or death—no one contested that. The pathologist's goal was to decide what acute levels of exposure could be considered safe, even over the long term. The genetic view, by contrast, emphasized that genetic harm had little to do with acute exposure. A massive acute exposure would have the same effect on the production of mutations as small exposures spread over a long period. Thus, geneticists worried about the cumulative effects of chronic exposure to small amounts of radiation. From the geneticists' point of view, thresholds were illusory, as were efforts to identify any safe level of exposure. This would be the primary reason that geneticists, including Muller, later became the most vocal scientific critics of American atomic energy policies, particularly of the denials of harm made by the Atomic Energy Commission through its chairmen or its Division of Biology and Medicine.

In the 1940s these safety questions were addressed in-house by the scientists of the U.S. Army's Manhattan Engineering District (the atomic bomb project), later by the AEC, and in Britain by the Ministry of Supply and the Medical Research Council. After the first successful atomic pile was built at the University of Chicago's Metallurgical Laboratory (Met Lab) in late 1942, demonstrating the feasibility of sustaining a chain reaction in atomic fission, the question still lingered whether the bomb could be built without threatening human life in the process. Radiologist Robert S. Stone oversaw the health aspects of the project. He and other scientists of the Manhattan Project studied the pathologic effects of atomic radiation in a number of ways, including the injection of plutonium into human patients without written consent—they believed, mistakenly in some cases, that these patients already were mortally ill. When Congress created the Atomic Energy Commission in 1946, the new body continued such human experiments.[10]

The scientists involved in studying the biological effects of radiation and devising methods and standards for radiation protection became known as "health physicists." Health physics sections were established during the war at three of the major centers of activity connected to the bomb: at the University of Chicago; at the uranium production facilities in Oak Ridge, Tennessee; and at the Hanford plutonium production facilities in Richland,

Washington. These sections were well aware of a discouraging history of radiation protection since the last years of the nineteenth century, as victims of overexposure to X-rays or radioactivity should have been better educated by officials who had some idea of the harmful effects. Karl Z. Morgan, who headed the health physics section at Oak Ridge, wrote in 1946: "Past records indicated that thousands of people, including dial painters, physicians, dentists, physicists, technicians, manufacturers, and engineers, had been injured by X-rays and radiation from radioactive substances and that many had died from these effects. The evidence was that all these radiation injuries and deaths were unnecessary and could have been avoided." The problem in avoiding such deaths, Morgan observed, was that "penetrating radiations produce their damage so insidiously and inconspicuously at first that men had seldom been aware of receiving excessive radiation" until they developed burns or even, possibly years later, developed cancer.[11] In other words, it was often impossible to know by physical evidence if a person had been overexposed until it was too late.

To prevent this gloomy past from repeating itself, Morgan and other health physicists developed the notion of "tolerance dose." This term described what American health physicists determined to be the threshold of radiation below which no damage would occur. To quote Morgan more precisely: "By 'tolerance' we mean the amount of radiation that is considered not to produce any biological damage."[12] Determining this tolerance was a tricky matter because each radioactive element posed different risks because of differing kinds of radiation and differing half-lives. In addition to these, nonradioactive elements put in a reactor resulted in artificially radioactive isotopes of those elements. The result was that there were hundreds of potentially radioactive substances, and each of them needed to be analyzed by health physicists to determine potential biological damage.[13]

Most evidence of biological effects on humans came not from new experiments by health physicists but through studies of the survivors of Hiroshima and Nagasaki. After Japan's surrender, teams of scientists and physicians from the United States—some from the Army's Manhattan District and others from the Naval Technical Mission in Japan—surveyed about fourteen thousand of the survivors of Hiroshima and Nagasaki to study the biological effects of radiation. A few weeks before Muller accepted his Nobel Prize, President Truman directed the National Academy of Sciences (NAS) to continue these studies on a long-range basis, in what became known as the Atomic Bomb Casualty Commission (ABCC).[14] From the outset the scientists involved in these studies doubted the existence of thresholds, but they also recognized that the ABCC was unlikely to provide conclusive evidence one way or another.[15]

While the NAS took over the study of Japanese survivors, the AEC prepared for an era of expanding use of fission products. This necessitated a serious effort to study and manage the biological effects of radiation. In 1947, the AEC convened a group of physicians, academic scientists, and military figures into an Advisory Committee for Biology and Medicine, with Alan Gregg as the chairman. Gregg, the director of the Rockefeller Foundation's Divisions of Medicine, had been a strong supporter of genetics over the years.[16] The committee's main task was to assess the initial problems of the AEC's future Division of Biology and Medicine and to pick its director.[17] This post would carry the colossal responsibility for radioactive materials not only in production of bombs or possibly commercial power, but also of the medical and laboratory uses of fission materials. Initially the committee offered it to Cecil J. Watson, of the University of Minnesota's Medical School, but he refused. Perhaps he was intimidated by the weight of responsibility. As he wrote, he knew it would be almost entirely administrative, allowing no time for research, and his role would be to preside over "what is virtually an empire of industrial health."[18] Watson had been at the top of the list of twenty-five candidates.[19] The job ultimately went to the third man, Shields Warren, a pathologist at the New England Deaconess Hospital who had little hand in the wartime work. Warren held the position until 1952, but his influence endured much longer.

Not only did the AEC establish safe threshold values for exposure ("tolerance doses"), but it also tried to enforce them. The Division of Biology and Medicine failed to change the entrepreneurial spirit among users of radioactive materials inherited from the prewar years. University and hospital researchers, for example, had been relatively free to do what they wanted, and they resisted change. When Robert S. Stone—the one who directed biological research for the atomic bomb project—began whole-body irradiation experiments on his patients with rheumatoid arthritis at the University of California Hospital, AEC scientists threatened to withdraw financial support and shipments because they lacked confidence in his judgment. Stone wrote to Gregg accusing him, Warren, and others at the AEC of assailing the tradition of academic freedom.[20]

The empire of industrial health described by Watson was resisted tooth and nail by industrialists, too, who wanted the AEC to relax its control. Probusiness politicians accused the AEC of being un-American, trying to exercise even more state control than the Soviet Union did. Even the first AEC chairman, David Lilienthal, doubted whether the strict control exercised by the commission really reflected the values of the American system. Later, when the Democrats lost control of the White House in the 1952 election, one of the first items on the new president's agenda would be to reform atomic energy legislation to enable easier use of radioactive materials.[21]

With deep-seated resistance to its authority, the AEC enlisted the help of scientists to buttress its credibility and to ensure its practices and recommendations were taken seriously. James B. Fisk, the director of the AEC's Division of Research, was concerned deeply about the entire new field of health physics and the broad discrepancies in safety precautions taken. The expanding operations of the AEC, through reactors and research laboratories, would necessitate a far better understanding of all the attendant medical problems. As these operations grew more intensive, it would be crucial, he and the others agreed, to establish some acceptable level of the "maximum permissible tolerance," to ensure the protection of employees who worked with radioactive materials. The major decision to be made was this: Should the AEC formulate the recommendations or ask an outside body to do it? Those present unanimously agreed that it would be better to have some other body do it, to let everyone know that it was drawing on the advice of acknowledged experts in the field. They chose the National Committee on Radiation Protection (NCRP).[22]

The NCRP suited the AEC's purposes for a couple of key reasons. One was that the NCRP was not a body with any regulatory teeth. It could make recommendations, but neither the AEC nor anyone in the private sector was obligated to follow them. The second was that the NCRP was not a research body. The AEC never saw the NCRP scientists as a group that should make a separate, independent study. They simply would use the data and experiments compiled over the years, mainly from the Manhattan Project, Atomic Bomb Casualty Commission, and the AEC. Gregg's committee was very specific about this—the decision to ask the NCRP for help should not entail any research program.[23]

Thus the first major American body to assess the entire problem of biological effects of atomic radiation, including waste disposal, was the NCRP. This committee existed prior to the atomic bomb project, though its primary focus had been the risks of exposure to X-rays and to radium. It came into existence because at the Second International Congress of Radiology, in 1928, attendees had been embarrassed that some countries' delegates disagreed amongst themselves at international meetings. They resolved to create committees in each country that would be responsible for negotiating a national consensus. In the United States, the natural body to do this was the Department of Commerce; its National Bureau of Standards, created in 1901, already determined the country's standard chemical and physical units, such as weight and electrical quantities.[24]

In 1929 the bureau tried to set standards for radiation protection by creating the Advisory Committee on X-ray and Radium Protection, but it chose its name carefully to reflect that it did not assume any regulatory responsibility or power—its purpose was to advise only. Until 1948 the bureau bore the cost of the committee's work, but in that year the AEC began shouldering some of it. Renamed the National Committee on Radiation Protection in 1946, the group tried to accommodate expected changes from the advent of atomic energy. No longer would medical problems be the sole important issue. Now the NCRP included representatives from a variety of organizations in government and business, including the Public Health Service, the Atomic Energy Commission, the Department of Defense, the American Medical Association, the National Electrical Manufacturers Association, and other bodies.[25]

The NCRP took a laissez-faire attitude toward regulations; in other words, it did not want to create any. Following the practice of the National Bureau of Standards, the committee used handbooks—essentially reports with recommendations—as its primary vehicles of influence. But its members made no particular effort to ensure these handbooks were disseminated or the ideas in them enforced. Instead, it was the responsibility of any business, university, hospital, or government body to seek them out. The NCRP's chairman, physicist Lauriston S. Taylor, wrote in 1949 that "it has been held from the outset, and does today, that the formulation of radiation safety codes into legal statutes by national or state bodies is an unwise procedure." Such laws, he asserted, would limit flexibility and make it more difficult to take advantage of developments in the field. Their work was not totally pointless, however; most court cases over the past couple of decades had been decided on the basis of guidelines set forth by the NCRP. Because these handbooks reflected the expertise of some of the country's leading scientists, they carried considerable authoritative weight. It might become necessary, Taylor acknowledged, to include representatives of labor unions on the NCRP, not because of any contribution to technical knowledge they might make, but to let them in on the negotiation in order to bring union practices and scientific standards closer together. "It is essential," Taylor argued, "to produce codes which on the one hand are acceptable to unions and on the other hand will not provide them with unreasonable levers with which to obtain undeserved fringe benefits." Under Taylor's leadership, the NCRP was a guide, and under no circumstances did it want to stand in the way of business or scientific research.[26]

The Atomic Energy Commission saw in the NCRP an opportunity—one that it also would pursue with the National Academy of Sciences—to create the appearance of its own practices being formulated by a body other than itself. At a time when the AEC had yet to establish any reputation, and some users of radioactive materials resented its interference, the NCRP provided much-needed credibility. And the NCRP scientists recognized they were being used in this way. As Taylor assessed the situation, "The [Atomic Energy] Commission decided, and I think wisely so, that they would be in an untenable position if, as the largest single user of radioactive materials, they also established their own safety rules. They could have subjected themselves to continued attack on the basis that they were setting their radiation levels to suit their operating needs rather than the safety of the worker." This was the reason, according to Taylor, that the AEC asked the NCRP to prepare recommendations of basic codes and practices. But Taylor added, "It should be noted that the Atomic Energy Commission never stated that they would accept the recommendations of the National Committee on Radiation Protection."[27]

The two bodies avoided the embarrassment of any discrepancies in their views by having plenty of AEC scientists on the NCRP, despite Taylor's acknowledgment that "it was important not to give the appearance of 'loading' the committee with Atomic Energy Commission interested members." But because many of the most experienced people in the field made their living with the AEC, the overlap seemed unavoidable. Influential figures in the AEC, including the director of the Division of Biology and Medicine, Shields Warren, were members of the NCRP or its subcommittees. Because the AEC often objected to any change in recommendations that might affect its operating practices, it was difficult for the NCRP to strike the right balance between getting along with the AEC and becoming merely a branch of it.[28]

It was through the NCRP that geneticists' attitudes about radiation saw their first serious policy effect. The Advisory Committee for Biology and Medicine included a geneticist, George Beadle of the California Institute of Technology, who recommended that geneticists' viewpoints be taken into account in any recommendations. Beadle was quite familiar with Muller's work on producing genetic mutations through exposure to radiation, and he recommended that Muller and other geneticists be enlisted to help the NCRP, to ensure recognition of long-term, chronic effects of radiation.[29] They were in fact asked, but in the subsequent years Muller complained that the NCRP refused to allow effects on future generations to affect the recommendations for permissible doses.[30]

Despite Muller's complaint, geneticists guided the NCRP scientists to recognize that health physicists at the AEC, predominantly worried about pathological effects, operated under a misleading assumption about thresholds of safety. From geneticists' point of view, there was no evidence for any threshold whatsoever if the findings in fruit flies held true for human beings. To the contrary, the consensus of the geneticists was that any amount of radiation would produce mutations to be passed down to one's children and that the vast majority of these mutations were harmful. According to this view, the demand for a tolerance dose was based on the false notion that an exposure limit could be set and that a definitive, quantitative figure of safety could be identified. In other words, any figure would be a false one, and, worse, it might mislead people into believing that it represented a level of safety.

To solve this problem, the NCRP made a subtle change in terminology that spelled a dramatic but unheralded shift in public health practice. Its scientists were uncomfortable with the phrase "tolerance dose" because of its obvious connotation that there was a true level of safety. Instead, it proposed "permissible dose," which the NCRP hoped would reflect better the findings by geneticists and others denying that there was any threshold of safety.[31] "Permissible" acknowledged the harm but implied that it was deemed acceptable. This immediately raised two serious issues. First, deemed acceptable by whom? By assigning a "permissible dose," the NCRP and the AEC dictated these levels. And, second, the phrase, despite being more correct, still retained the air of a safety level. Thus the appearance of a threshold remained.

With threshold values in place, by 1950 the health physics community in the AEC could give itself a clean bill of health, claiming that its safety precautions had prevented radiation from causing any harm. Ionizing radiation, it announced in a semiannual report, could be controlled like any other industrial hazard with appropriate safety measures. "Tens of thousands of workers are engaged," the AEC claimed of its operations. "Yet comprehensive records of radiation exposure show that no injuries have occurred in production operations." In fact, it said, the commission did not even need to offer workers hazard pay. Describing the plutonium production facility at Hanford, the report observed that most workers receive less radiation there in a year than they would by receiving a chest X-ray. "No Hanford worker has ever been injured by radiation," the AEC reported firmly.[32]

Wanted: Atomic Graveyards

Occupational exposure was not the only source of artificial harmful radiation. Another menace was the growing stockpile of radioactive waste. Scientists from the Atomic Energy Commission admitted as much in April 1948, at the annual meeting of the Greater New York Safety Council. They pointed out two possible dangers regarding such wastes. One was the problem of disposal of wastes from both weapons production and eventually peaceful power reactors. The other was the possibility of a radioisotope bomb being built—this was not a fission weapon, but rather a conventional explosive that contained radioactive materials. This kind of bomb would pose risks long after the initial blast. As for disposal, no proposals seemed likely to provide for the long-term safe disposal of wastes whose dangers would last for generations or longer. One promising solution appeared to be the sea: if wastes could be mixed with other materials into concrete, they could be deposited into remote areas of the ocean, where they might fall to the bottom and remain isolated from population centers.

The choice of forum indicated that the AEC considered, and hoped others would consider, the dangers of radioactivity as part of an array of risks posed in industrial processes, rather than as a unique danger. The meeting included several of the AEC's safety experts, including Merrill Eisenbud, the chief of the Industrial Hygiene and Safety Branch, and Forrest Western, assistant director of the Health Physics Division. Western warned that, in the event of an atomic war, blast effects would be only part of the problem. "The populations of large areas," he said, "may find the air that they breathe, the food they eat, the water they drink, and perhaps anything that they touch contaminated with harmful quantities of radioactive materials."

In times of peace, Western said, the greatest dangers would come from reactors. Yet these problems could be compared to the extant difficulties faced by industrial plants of other dangerous chemicals. The only difference was the possibility that dilution, so often used to render toxins less harmful, might not be effective for radioactive materials. Even given sufficient water dilution that would reduce radioactivity to levels acceptable for drinking, there always would be a risk that certain minerals or algae in the water would concentrate the radioactive isotopes and present further dangers.[33] Thus already in 1948, experts within the AEC recognized publicly that the possibility of organic or mineral concentration made dilution itself untenable as a panacea for radioactive wastes.

The AEC held a seminar on radioactive waste in late January 1949, inviting professionals connected to sanitary and waterworks engineering. Addressing the seminar, AEC chairman David Lilienthal cast the problem of waste disposal as part of "learning how to live with radiation," a process comparable with learning how to live with anything else unfamiliar. The important thing was to learn as much as possible about it and "to keep our shirts on" rather than give in to emotional and hysterical outbursts. He put a great deal of faith in human intelligence, saying that the problems of radiation would soon be conquered like any other industrial problem.[34]

At the seminar, AEC scientists Roger S. Warner and Arthur V. Peterson revealed how American atomic energy facilities disposed of their wastes. The Argonne, Brookhaven, and Oak Ridge national laboratories produced large quantities of radioactive materials, and these facilities were compelled to release them into the soil, air, or surrounding waters, decontaminating them first by filtering, scrubbing, dispersing, and diluting them sufficiently enough to meet established permissible limits. A much greater amount of waste was produced at the production centers, particularly at the Hanford site in Washington. The processes for decontamination were similar, except there was also some much higher level wastes that were held back to allow time for some radioactive decay, to be reprocessed, or to be stored indefinitely.[35] A later AEC report described the fate of the lower-level "wash" wastes: they were put into tanks and then, after given time to decay, discharged locally. This was a practice that began during World War II, the AEC acknowledged, "but it is being replaced by better methods as fast as they can be developed."[36]

At a press conference following the seminar, Lilienthal told reporters that the process of condensing the radioactive wastes was "moving along" and that this was probably the greatest contribution they could make. He acknowledged that the subject could become "a subject of emotion and hysteria and fear, and we do not believe those fears are justified, provided technology applies itself to eliminating the troubles." He added, "But it is a tough one, there is no doubt about it."[37]

Behind the scenes, however, some held more gloomy views. In a March 1949 secret report of the plutonium production facility at Hanford, E. I. Du Pont de Nemours & Company advised the AEC that the radioactive waste problem was "potentially one of the most serious," since it occurred not just with plutonium production but with any application using fissionable materials. Du Pont hoped that the future would hold "more elegant solutions" than just solidifying the waste and trying to reduce them in bulk. Noticing that waste disposal had a low priority in the AEC, the report pointed out that this could only be justified because of relatively favorable local situation at Hanford; this was unlikely to be the case in all facilities. Du Pont envisioned a future with serious cost penalties due to the problem of radioactive waste. The ideal, admittedly unachievable due to technological ignorance and presumably soaring costs, would be the total confinement of all radioactive material in solid form. Though the AEC was unlikely to attain that goal, Du Pont noted, it ought to set high goals; quite apart from safety considerations, the political consequences of doing otherwise "are frightening and demand that high goals be set and that the problem be attacked diligently."[38]

The problem of exposure to radioactive materials was exacerbated by the use of radioisotopes outside the commission's immediate control. The AEC distributed radioisotopes to medical centers and other bodies for research. Would these non-AEC users dispose of wastes wisely, without government regulations?[39] In October 1949, the Veterans Administration convened a conference for all of its radioisotope units to address the potential of research to combat disease by using radioactive tracers (to investigate metabolism, for instance) and to discuss the possible dangers of the by-products of research. The attendees emphasized the growing need for safe places to store all of the human clothing and rat carcasses, now highly radioactive, used in experiments. These had to be buried deeply, as the *Los Angeles Times* put it, in atomic "graveyards." For liquid wastes, they discussed dilution and subsequent introduction into normal sewage systems. In fact, the problem was larger than the *Los Angeles Times* acknowledged. Some hospitals and universities were sites of human experimentation, too, usually on patients with debilitating diseases. When these patients died, their bodies became radioactive waste.[40]

In 1950, atomic energy officials and concerned citizens began to take the problem of waste seriously. The AEC's facilities were fast becoming the nation's largest industrial units, with more expansion to come. This was a period of major growth in the United States' system of national laboratories.[41] Some argued that other agencies besides the AEC ought to take a more active role in ensuring that the wastes were properly taken care of. Should local, state, and federal authorities supplement the AEC's role? In January 1950, Abel Wolman and Arthur E. Gorman (the former of Johns Hopkins University and the latter of the AEC) spoke to a conference on industrial nuclear technology at New York University, noting it was "highly questionable" whether this important task should be left entirely to the AEC. They acknowledged that the commission had been diligent in setting high standards of safety, based on the latest knowledge about the dangers posed to humans. Still, scientific knowledge was meager, especially regarding the long-range effects of exposure on a large population. This was compounded by the fact that one had to rely on extrapolation of data from experiments on other animals, and human experimentation "is, of course, difficult, particularly where long-range effects are of significant importance." They pointed out that dilution did not necessarily lead to safety, because radioactive material can reaccumulate and still pose risks.[42]

Needs for atomic graveyards and cemeteries were reported frequently in major American newspapers. The University of California, Los Angeles, assembled its own conference on how to dispose of wastes safely in September 1950, and it included not only AEC and Public Health Service figures but also local advocates for fighting air pollution.[43] Walton A. Rodger, a scientist at the Argonne National Laboratory, put the problem in economic terms while giving a lecture at the Illinois Institute of Technology in 1950. The government spent ten dollars per cubic foot to store radioactive waste in specially designed canisters that were four feet long, six feet wide, and eight feet high, and these were housed in earth-covered concrete pits. All tolled, each of them cost the AEC $1,920. Rodger remarked, "and that's a lot of the taxpayers' money."[44]

Everyone seemed to have his own idea about how to get rid of the wastes. For example, a Bureau of Mines official told an audience at the Industrial Health Conference in 1951 that storing wastes in tanks was unnecessary. Under contract from the AEC, the bureau just burned all the waste to reduce it to more manageable ash. But this kind of discretionary freedom exercised by contractors disturbed some organizations, such as the Association of Casualty and Surety Companies. It claimed that hospitals, laboratories, and private industry all appeared to be more lax in dealing with radioactive wastes because they were not under the direct supervision of the AEC. It might prove too tempting for the AEC to outsource its problem and then turn a blind eye to the questionable practices of contractors. Without supervision, the association pointed out, Americans could expect major contamination on beaches, in drinking water, and in the air. This was especially true if the problem was cast as an economic one, because private firms inevitably would opt for the cheapest methods. The only viable solution would be government regulation, it argued, such as requiring that all waste be packaged into oil drums and dumped a hundred miles out to sea, or to insist they be buried in regularly monitored atomic cemeteries.[45]

Although waste disposal appeared to be a mounting problem, there were plenty of ideas about how to make efficient use of the wastes rather than simply discard them. Radioactive tracers were used in medicine, and they could be used in other realms as well. In Massachusetts, F. C. Henriques had been using radioisotopes to do medical experiments through the AEC's distribution program. He noted that tracers from waste could be used in industry; for example, petroleum prospectors could use tracers while drilling to detect possible leaks and to follow oil movements. Radioactive tracers would act as signals with which anyone with detection equipment (such as Geiger counters) could track movements of any substance. Henriques also suggested using such materials for the sterilization of food or for the development of pharmaceuticals.[46] However, Henriques was exactly the kind of overzealous user that provoked criticism about the safety standards of contractors. He had already been reprimanded by his company, Tracerlab Inc., in 1949, for using radioactive iodine (Iodine-131) on human subjects to study the effects of a substance called Krim-Ko gel, without gaining prior approval from the AEC. There were no consequences to this action, though the AEC sent letters to Tracerlab threatening to stop shipping radioisotopes if further violations occurred.[47]

Military commanders had very different solutions in mind. In 1947, Major General J. L. Homer openly suggested that guided missiles of the future might carry radioactive waste to the enemy. Homer was envisioning what later became cruise missile technology, but instead of affixing an atomic warhead to the missile, "other potential means of destruction carried by this weapon could be radioactive waste, worrisome by-products in the manufacture of fissionable material." At present, such wastes were kept in temporary storage tanks, but the tanks were piling up fast and they would need to find a permanent home or somehow be recycled for other ends, like a radioactive waste bomb.[48] Behind closed doors, the chairman of the Military Liaison Committee to the AEC, General Lewis Hyde Brereton, mentioned to the commission in 1948 that he and his committee members saw a significant connection between radioactive waste and "the solution of the materials problem for radiological warfare." By that time the AEC already had begun discussions with Ernest O. Lawrence, the director of Radiation Laboratory at the University of California, Berkeley, about using the lab to study the problem.[49]

Secretary of Defense Louis Johnson mentioned the possibility of radiological weapons in his March 1950 report to President Truman. A few months later, the AEC appeared to like the idea of radiological warfare so much that it reported to Congress its plan to devise a serious study of it. At a press conference, intrigued reporters teased out little information but got E. R. Trapnell of the AEC's Public and Technical Information Service to acknowledge that this was the first time the AEC had ever admitted to having such a program. Others at the press conference, such as Brigadier General James P. Cooney, assured the press corps that the project was too much in its infancy to give any technical details. But the report mentioned that the wastes, which had given the AEC such difficulty thus far, might conceivably have industrial uses, or even uses in warfare. Radiological weapons might not have a dramatic death toll, the AEC admitted, but they would have a powerful psychological impact, comparable to the first use of poison gas during the First World War.[50]

There were several reasons that radiological weapons, despite an initial rush of enthusiasm, seemed impractical. For one, it would have been difficult to spread the waste over a large area and maintain its lethality. The effects would diminish over time as the radioactive elements decayed. Also, most of the harm would be long term, which military commanders might not care about in a tactical environment. Conceivably, one U.S. Army publication reasoned, dropping such wastes on cities could make it possible to capture them without fighting at all, because the people would voluntarily evacuate, fearing the long-term damage. In such cases, however, one would have to use wastes of very short decay periods, so that American troops would not be affected.[51]

Despite such limitations, radiological weapons appeared to be an ideal way to make use of radioactive wastes—if they were harmful to people, why not adapt them to situations in which harm was desirable? Although the Defense Department and the AEC spoke in hypothetical terms, there was a real-world opportunity: the Korean War. Inspired by the AEC's report, Tennessee Senator Albert Gore advised President Truman to draw a line across the Korean peninsula and saturate it with radioactive wastes. This could, he reasoned, prevent the North Koreans or Chinese from crossing over into the southern part of the peninsula. Because radioactive wastes had to be stored at great expense to American taxpayers, why not get rid of them by dumping them on Korea as a deterrent? Gore advised Truman to evacuate all the Koreans from predetermined areas and then to "dehumanize a belt across the Korean peninsula by surface radiological contamination," as a drastic step to prevent the situation from worsening.

According to Gore, such a widespread use of radioactive waste would be "morally justifiable," because it was a deterrent rather than an attack on civilians. He wrote to the president: "Just before this is accomplished, broadcast the fact to the enemy, with ample and particular notice that entrance into the belt would mean certain death or slow deformity to all foot soldiers; that all vehicles, weapons, food and apparel entering the belt would become poisoned with radioactivity." Gore added that the belt would need to be replenished periodically with new waste until a "satisfactory solution to the whole Korean problem" was reached.[52] Skeptics typically noted the original limitations outlined by the AEC's report. Senator Brien McMahon, chairman of the Joint Committee on Atomic Energy, played down the possibility, remarking that a radioactive belt probably was not tactically feasible and would have at best a psychological effect.[53]

Despite obvious drawbacks, the increasing supply of dangerous wastes simply increased interest in the development of radiological weapons. In a 1954 article entitled "An Atomic Maginot Line," journalist Hanson W. Baldwin asserted in the *New York Times* that "all the general staffs of the world have recognized that if enough of these radioactive by-products could be retained, stored, and handled, they might offer an interdiction weapon of great value." Calling it a Maginot Line referred to the supposedly impenetrable line of fortifications built by France to deter German aggression prior to the Second World War. Baldwin noted that an effective radiological weapon—"one of the dreams of the atomic age"—would have more long-term and damaging effects than mustard or nerve gas. Such weapons were preferable to fission or fusion weapons, which produced the same kind of debris but on a large scale and in an unpredictable way.[54]

Enthusiasm for radiological weapons was an extreme manifestation of the tendency to ignore the mounting waste disposal problem and assume that new uses or technological solutions would be found before the situation ever crossed a threshold of safety. But even radiological weapons would utilize only a small amount of the heaps of contaminated garbage and effluent expected if ever the decision was made to develop nuclear power on a large scale. The problem of waste disposal seemed destined to limit the building of reactors, particularly reactors not deemed necessary for national security. It was becoming clear that the dream of cheap energy from atomic energy was just that—a dream—if the problem of waste disposal could not be solved.

Meanwhile, the sea was open for business. Although the United States developed the atomic bomb, pioneered studies of biological effects, and began dumping at sea in 1946, Britain would rely even more on the ocean as a radioactive waste dump. Its Atomic Energy Research Establishment (AERE) was established in 1946 on the grounds of a Royal Air Force base at Harwell, in Berkshire. These former aircraft hangars housed a variety of waste producers—the atomic piles (reactors), experimental equipment such as the cyclotron and linear accelerator, and equipment to enrich uranium through electromagnetic separation of isotopes 235 and 238. The next year, Britain decided to build an atomic bomb of its own, and by 1950 the Aldermaston airfield, also in Berkshire, was taken over for weapons research. The year 1952 would see not only Britain's first successful atomic bomb test, but also the official establishment of the Atomic Weapons Research Establishment (AWRE) at Aldermaston. Other sites included Risley, in Lancashire, for building a permanent reactor, and Windscale, in Sellafield, Cumberland, for plutonium production and reprocessing. Over the years, these sites produced most of Britain's radioactive wastes.[55]

The British thought of the ocean in terms of its capacity. How much could they safely dump while expecting it to be diluted or properly isolated? Public statements gave the impression of a threshold, using firm language that left little room for any thought of harm. For example, in preparing to dump a consignment into the Atlantic in April 1949, the Ministry of Supply reassured the press that the waste "is much too small to have any harmful effects on fish or on human life." These wastes were taken from the AERE and dumped in May. According to the government, the wastes were sealed into drums and dropped in waters two thousand fathoms deep, hundreds of miles southwest of Britain. These were solid wastes, composed of glass, paper, cloth, and other materials.[56]

The same year, the British atomic energy establishment was busy building a pipeline to carry liquid wastes, on a continuing basis, from its Windscale facility. Covering the story, the *New York Times* noted that the British Ministry of Supply said the pipeline would carry wastes into the Atlantic.

More accurately, the pipeline would discharge its waste into the Irish Sea. Windscale, later called simply Sellafield, would become the controversial site not only of a major fire in 1957 but also of a long-standing environmental controversy over its daily discharges into this body of water.[57]

In Britain's case, ocean disposal was politically less problematic than land disposal, because Britain had no vast, relatively unpopulated areas like those in the United States and Soviet Union. Still, there were other possible solutions afoot. Harwell scientists reportedly told London's *Sunday Chronicle* that the problem of neutralizing such wastes might soon be solved with rocket technology. The paper reported Harwell scientists as suggesting that perhaps the most dangerous wastes could be launched into space and dropped on the moon. In the United States, Rutgers University scientist Ira M. Freeman suggested the existing options—burial, ocean disposal, and pumping into dry oil wells—seemed undesirable compared to the possibility of shipping wastes all the way to another planet. Tanker rockets could be built to carry such wastes to Mars or Venus. If concern about possible life forms on these planets should preclude this, some uninhabited asteroid might be used. Otherwise, one could put the garbage between the earth and moon, and let atomic waste dumps orbit the earth at a safe distance.[58]

When the AEC released its semiannual report in July 1950, it was clear that the United States joined Britain in seeing the ocean as the obvious choice for some of its radioactive waste. Walter Claus, the acting head of the biophysics branch in the AEC's Division of Biology and Medicine, fielded questions in the press conference about the report. Explaining the practices in the United States and Britain, he said that both countries were disposing of wastes at sea in order to get rid of those wastes that could not be stored safely or be effectively decontaminated. These wastes were dropped in very deep water, where they would be isolated. The erosion of the containers would occur very slowly, so that no danger to humans would be posed.

Journalists received the impression, reported faithfully in the *New York Times*, that the oceans were to be used for the most hazardous of all wastes and that highly radioactive waste would rest at the bottom of the sea. There they would stay isolated, with most of the substances neutralized before saltwater eroded the containers enough to release the material into the waters. When asked if the United States dumped waste packed in concrete into the sea, Claus made no secret of it. "Quite a quantity of material has been handled in that way," he said.[59]

Scientists and engineers were trying to find ways to make sea disposal as safe as possible. At Brookhaven National Laboratory, for example, researchers under L. P. Hatch developed a method to incorporate special clay into radioactive waste. This clay, called montmorillonite, seemed capable of capturing radioactive substances, preventing their mixing with outside substances. This might solve the waste problem, some believed, adding feasibility to sea disposal and even, AEC chairman Lewis Strauss suggested in 1954, possibly providing for a safe long-term packing solution for radioactive material in general. But potential solutions such as these were only in their preliminary phases.[60]

Although recognizing the rudimentary state of knowledge, officials were confident that the sea was likely the safest place to put wastes. Producers of radioactive waste continued to dump at sea, periodically notifying the press about their activities. The Public Health Service announced its practices in 1953: it sealed the wastes from its experiments in canisters mixed with concrete, then dumped them in deep water about a hundred miles off the coast of Virginia. The canisters were supposed to last about a hundred years, after which time the radioactivity would have diminished considerably. The effect on the oceans, one hoped, would be minimal.[61]

Bringing Sea Disposal into Focus

The threshold concept permeated all facets of atomic energy, including waste disposal. While "tolerance" implied a threshold for human exposure, "capacity" implied a threshold for dumping waste into the sea. Most of the work on waste disposal was designed to establish a figure, comparable to a threshold value, to define what could be put safely into the oceans. And just as the NCRP suspected the falsity of threshold values for human exposure, it was the NCRP—and specifically its oceanographer, Richard Fleming—that began to question envisioning the ocean as a giant sewer with a specific capacity to accept radioactive wastes.

The NCRP had several subcommittees charged with specific technical aspects of radiation, such as the permissible external dose, permissible internal dose, and several other subjects. One of these subcommittees was devoted to the problem of waste disposal and decontamination, chaired by James H. Jensen, the head of plant pathology at North Carolina State College. Jensen had been on a leave of absence from his university post during 1948 and 1949 while serving as head of the Biology Branch of the AEC's Division of Biology and Medicine, so he epitomized what at times seemed the purely nominal detachment of the NCRP from the AEC.

Jensen's subcommittee began its work in 1949, initially producing handbooks on control and removal of contamination from laboratories and on the disposal of radioisotopes (such as phosphorus-32 and iodine-131) from medical facilities. Some of its work focused on specific materials like these and other radioisotopes, while still other of its work focused on the long-range possibilities of the final destination of radioactive material, whether it be on land or in the sea.[62]

The NCRP members began to rethink their waste disposal recommendations when the prospect arose of their individual views being set to print. In 1949 the editor of *Nucleonics*, Norman R. Beers, along with other representatives from the McGraw Hill publishing house, arranged a meeting with scientists from the AEC, the Public Health Service, the National Bureau of Standards, and other bodies. One of the subjects of discussion was radioactive waste, and Beers wanted to learn about the NCRP's work in this area. He and the others wondered if all of this work on radiation protection had any legal force. The answer, of course, was no. The reply, as Ross Peavey (the NCRP secretary) reported himself saying, was that "they were principally recommendations which carried very considerable authoritative weight, and had in the past been instrumental in court cases." But although the NCRP drew upon the expertise of leading individuals concerned with all aspects of radiation, ultimately its conclusions had no teeth from a policy point of view.[63]

The presence of publishers provided an opportune moment for some scientists to break ranks and criticize the NCRP's hands-off approach. Two of the participants, Abel Wolman of Johns Hopkins University and E. G. Williams of the Public Health Service, said that they needed a definite decision about procedures in using any radioactive materials, particularly for use in hospitals and in universities. These were not as closely monitored as the government laboratories, and in fact they seemed to be free to use, and to dispose of, radioactive materials at their own discretion. Although doctors and researchers tended to resist this kind of supervision, Wolman stated, there were a few cases of malpractice already in place. The most strident opposition to this view came from Edith Quimby (Columbia University) and Sergei Feitelberg (Mount Sinai Hospital), both working in large university or hospital laboratories. Government regulations, they felt, would stifle their research. But Wolman argued that their hospitals were large, probably with adequate monitoring. What about the smaller organizations with few resources to spare for close monitoring? These needed to be controlled better.[64]

This was a festering question among users of radiation and radioisotopes, particularly those who feared not only an era of top-down control by the AEC, but also the array of financial responsibilities that would come with regulations. The need to ensure adequate disposal practices raised a couple of practical problems. First, no instruments were available that were simple enough to allow laypersons to do adequate monitoring. Janitors, for example, would not be in a position to know if drains were contaminated with radioactive waste or not. More important, both Wolman and Williams "expressed considerable disquietude" about using the roentgen as a measure of biological effects of all types of radiation. The roentgen (named for X-ray discoverer Wilhelm Roentgen), after all, was a measure of physical energy, not a measure of the amount of radiation absorbed; units more specifically attuned to biological effects were needed. Such complaints, aired in the presence of publishers, were timed precisely right to bring pressure on those who resisted revisions of the prevailing practices.[65]

After reading a report about the meeting, Feitelberg and Quimby backtracked. They did not want to be understood as standing against the development of safety procedures. They agreed that NCRP recommendations would make training, protection, and discipline easier, because there would be standards by which to judge transgressions. They also concurred that the present practice of letting each laboratory develop its own rules probably was unwise and that pooling experiences for a unified outlook would be the best course in the long run. Feitelberg in particular, who wrote to Ross Peavey about it, was eager to change the impression that he and Quimby stood on one side of the issue, whereas Wolman and Williams stood on another, especially if the differences in views were to be set to print on the pages of *Nucleonics*.[66]

The only oceanographer on the NCRP's subcommittee on waste disposal and decontamination was Richard Fleming, then at the Hydrographic Office of the Navy. Fleming had earned some notoriety among oceanographers as one of the authors of the influential textbook *The Oceans: Their Physics, Chemistry, and General Biology*, published in 1942. But there was nothing on waste disposal in *The Oceans*. What little there was on radioactivity was drawn from the work of Roger Revelle, who (like most scientists) envisioned radioisotopes as important experimental tools. For example, *The Oceans* mentioned that because of the concentration of radioisotopes in marine sedimentation, their radium content could used to understand rates of deposition.[67]

For the NCRP, Fleming tried to pinpoint the factors that might govern the choice of areas in the ocean for disposal. His job was to provide "guesses as to permanency, lack of diffusion, etc.," as one NCRP meeting report phrased it. In a practical vein, Fleming had to choose the best (and the worst) ocean areas for disposal of waste in containers, as well as for uncontained waste dumped at the surface. Fleming's task was one of compilation and assessment, not one of new research projects. He was helped in this task by G. W. Morgan of Oak Ridge's Isotopes Division, whose committee assignment in late 1949 was to locate and assess extant British work on the role of dilution at sea, with a view to using it as a basis for American recommendations.[68]

The NCRP's work went beyond waste disposal at sea. It also explored geological disposal and touched on all aspects of radiation protection. But by the end of the year the problem of disposal proved more intractable than expected, and it took up everyone's time and attention, not just Fleming's. The committee tried to fragment the problem into separate issues—garbage, drains, digestion, incineration (including atmospheric problems), ground burial, sea dumping, and the possibility of simply returning such materials to the AEC for safekeeping. In addition, they needed to understand animal and human physiology and explore the possibility of naming tolerance levels—or rather, permissible doses. They developed priorities for each isotope, because the biological threat of each would differ. For example, chemical separation was discussed for cobalt-60, but not for carbon-14. These were the first two priorities for study, with other important isotopes to follow: strontium-90, gold-198, iron-55, iron-54, cerium-137, sulphur-35, and calcium-45. All of these were waste products needing special attention before adequate disposal methods could be developed, possibly with different policies for each.[69]

Although much of the British work on waste disposal remained classified, the AEC had some access to it. Thus Fleming saw British studies from the Atomic Energy Research Establishment, by scientists including J. M. C. Scott, K. D. E. Johnson, and J. Wilkinson, discussing not only dilution but also the absorption of radioactive materials from seawater into the seabed and the nearby shore. These were based on preliminary studies of effluent discharge made into the Irish Sea. The Americans continued to rely on the British, who were very interested in showing the scientific credibility of their practice of discharging wastes into nearby waters. One of the authors working on this problem was Klaus Fuchs, who was arrested as a Soviet spy in 1950, so it was reasonable to expect that little of the secret British work on radiation exposure near British facilities was really secret, except to the lay public.[70]

Still, the British work seemed too incomplete—and tailored to a specific environment—to serve as a sufficient basis for policy. But when the NCRP started asking for research money, AEC scientists recoiled. For many months in 1949 and 1950, faced with the daunting task of judging the behavior of sewage flow from hospitals, the committee proposed that Sergei Feitelberg ask the AEC for project support to study four hospitals with plumbing aged between ten to fifty years. Sewage was just as difficult a problem as ocean disposal because, as in the case of the sea's organisms and sediments, there was a distinct possibility that flushed radioactive materials (typically the radioisotopes iodine-131 or phosphorus-32) might concentrate in sewage. Depending on its characteristics, the slime coating within the sewage plumbing could take up much of the radioactive material, thus keeping it very near to the source instead being carried away. Thus the final act in the mind of a doctor or technician in getting rid of such waste was not final at all. Instead, he was just introducing radioactive substances into the building's pipes and those of nearby sewers, all kept up in the ever-present slime. This perspective, emphasizing *introduction* rather than *disposal*, would stay with Fleming for quite some time.

But a new study was precisely what the AEC did not want. In 1949, Feitelberg failed to convince the AEC that this was necessary, and it took the concerted action of the committee and a letter from its chair, James H. Jensen, to stress the urgency of the need to the AEC. In this case, Jensen's close connection to the AEC helped. In the NCRP's 1951 report of its activities, the waste disposal group congratulated itself on being instrumental in persuading the AEC to support the study.[71] But this also was a pattern to be repeated in subsequent years—the AEC would ask an independent body of scientists to assess existing knowledge, only to be met not with definitive conclusions but with requests for research funds.

Dumping of dangerous waste at sea already was commonplace. In Britain, the Admiralty had designated dumping areas for unstable munitions, such the Hurd Deep, a relatively deep ravine in the English Channel. During the war, the United States Navy (by its own wartime regulations) took explosives and ammunition out to sea and dumped them in water at least nine hundred feet (150 fathoms) deep and ten miles offshore, hoping to avoid any interaction with steamers, fishermen, or submarine cables. Meanwhile the Coast Guard kept records about what was dumped and where. Thus the Navy chose the sites and the Coast Guard did the bookkeeping. But after the war there were no such regulations in effect in the United States. The Coast Guard dumped some of the government's radioactive waste, at least what came from the National Bureau of Standards. The Public Health Service received the waste in Bethesda, Maryland, then put it into "concrete coffins" for storage and shipped it to Norfolk, Virginia. From there the Coast Guard took it out to sea. When Jensen's committee met in late January 1952, Lauriston Taylor mentioned that almost a hundred drums, each fifty-five gallons, thus far had been shipped from the bureau in this way.[72]

But no one seemed to be keeping track, as the Coast Guard had during the war, of when and where all of the material was being dumped. As AEC scientist G. W. Morgan observed, the AEC was not receiving much radioactive material back. That would suggest that hospitals, universities, and other radioisotope users were simply finding their own ways of getting rid of the waste. He suggested it might be wise to provide them, as well as city and state officials, with guidance. California's state legislature, for example, already had acted defensively, by trying to prohibit any ocean dumping of radioactive materials at all. To Morgan, this seemed to be a "poor approach," since it would essentially prohibit working with radioactive isotopes at all. California had major centers of research activity already, thus it seemed unrealistic to ban dumping. Because of these nongovernmental users of radioactive materials, the committee reasoned, perhaps there ought to be some authoritative scientific advice on how to dump it in the wisest manner. What they needed from the NCRP, as AEC biophysics chief Walter Claus suggested, was a "basic philosophy" of waste disposal, because the amount of waste produced was only going to increase and the problem seemed unlikely to go away. In addition, there were private companies that were trying to get into the ocean dumping business; these companies should be furnished, at the very least, with definitions of terms so that they would understand what they were doing. As Oak Ridge scientist William H. Sullivan put it, waste disposal was practiced widely, so "someone should keep up with it."[73]

Jensen's committee started with a basic assumption: deep-sea disposal was safe. Forrest Western said this at the outset, though he acknowledged that he was less sure about disposal at shallow and surface depths. Fleming backed him up, saying that even for shallow disposal the main limitation would not be the sea itself, but rather transporting it to the sea. He stated confidently that as long as the material was dumped far from land, the tremendous quantity of water in the ocean would render the material harmless. To this Western was less sure, and said that the committee would be better off stating an exact figure, rather than telling people that they could dump as much as they wanted.[74] Fleming agreed to take all of their input into consideration and craft a report on ocean disposal to be issued by the NCRP.

In 1952, Fleming accepted an academic appointment at the University of Washington in Seattle; thus the NCRP was going to lose its key oceanographer in the nation's capital. The waste disposal group was sorry to see him go, in view of the importance of sea disposal to long-range radioactive waste plans. As Lauriston Taylor put it to him, "you were beginning to bring the sea disposal into focus." In asking him to continue working with the NCRP, meaning occasional flights to Washington, D.C., Taylor hoped Fleming might make good on this impression and sustain his activities, "at least until this phase of the work is cleaned up."[75]

But in truth neither "bringing into focus" nor "cleaning up" could describe the state of Fleming's findings. His difficulties in arriving at any firm recommendations came from the enormous uncertainty about the behavior of the sea and the quality of any techniques of containing and transporting waste. Especially for bulk disposal, in which pipelines could be used to transport vast quantities of waste to some parts of the sea, Fleming was reticent to make definitive comments. Although he believed that dumping waste far from shore would not pose a problem, he was wary of close-in, continuous bulk disposal because it would mean the concentration of contaminated water in one area. He was unsure if it would be possible to set a threshold value for how much could be dumped into such areas, yet that was precisely what the AEC seemed to want.

By late 1952 Fleming was hoping to be done with it so he could concentrate on his new job at the University of Washington. He resigned himself to the limits of the NCRP's advisory, nonresearch mission. He also accepted that the sea could act as a sewer with a renewable capacity to take in a certain amount of wastes safely. After finishing the first draft of his ocean disposal handbook, he wrote to Jensen that he felt like celebrating. The draft represented "several years of argument, many days discussion and a week's writing."[76] He also told Jensen that he had come around to the belief that he was not the person to make the decision. "The more I thought about the problem of bulk disposal," he wrote, "the more I became convinced that this is primarily an engineering problem." Perhaps they ought to take their cues from people who had knowledge of sewage technology.[77] But to his dismay, his role was far from over. After review by Jensen's group, including AEC scientist Forrest Western, it was clear that changes would need to be made.

In early 1953, the group met again to try to iron out its basic philosophy. Although he was supposed to author the whole report, Fleming's committee responsibilities were limited to the location of dumping sites. Others outlined more specific positions. For example, Western and Taylor made recommendations for packaged waste, saying that the only limitation they could envision would be the amounts that could be conveniently handled. Although the containers ought to be constructed to avoid them falling apart completely, they expected that this would only be important as they were being shipped. Once in the sea, the fate of the containers would be less important—as long as they sank to the sea floor—because the ocean would dilute the contents. Sullivan and Morgan dealt with handling and transportation problems, and Sergei Feitelberg made recommendations for bulk disposal.

It was on the question of bulk disposal that Fleming's initial draft had seemed unsatisfactory to the rest of the committee. In particular, Forrest Western hoped that the report would not imply that bulk disposal (i.e., large quantities, usually through pipelines) was too poorly understood to be recommended. If Fleming's conservative attitudes were published, they would cast a suspicious light on present and projected practices in the United States and Britain. Western proposed including a statement that the NCRP had no reason to object to bulk disposal as long as it could be done safely and economically. They might even suggest tolerance levels (i.e., threshold values) for the amount to dump in any particular area in a certain amount of time. Here, more than packaged waste, they would need to think about the problems of concentration and dilution over time. The committee agreed to have Fleming incorporate some general statements about bulk disposal along with those about packaged wastes. This would give the AEC some breathing room, leaving it "free to make its own interpretations in order to handle their immediate problems."[78]

Other oceanographers knew in early 1953 that Fleming was taking the lead in this subject, and they were eager to have Fleming's report. Allyn Vine had asked one of his colleagues at the Woods Hole Oceanographic Institution to start working on the problem of waste, but no one knew the extent to which they would be repeating ground already covered. "Rumor has it," Vine wrote to Fleming, "that you have been doing considerable thinking and report writing on this in recent months," and he wanted the benefit of these activities for the people at Woods Hole. When Vine revealed his wish to get more involved in work related to radioactive waste at sea, Fleming wrote candidly about the NCRP group's tenuous grasp of the problem. The "innumerable gaps in our knowledge of the oceans" made assessment impossible, for all of the same reasons it had been difficult all along—scientists needed a better understanding of the chemistry and biochemistry of different organisms and isotopes to know where materials might concentrate. Fleming envisioned a future recommendation that would be based on fish consumption of humans. Speaking the language of the NCRP, this meant they would need to set an upper limit for "maximum tolerable quantities" of radioactive materials that could be permitted to accumulate in any body of water.[79] This meant that he had to recommend a threshold despite a sea of variables, most of which were very little understood.

More drafts followed, into the spring of 1953, and Fleming let the others' views dominate over his own sense of uncertainty. The handbook, called "Factors Affecting the Disposal of Radioactive Wastes in the Ocean," strongly reflected the attitudes of the NCRP that regulations should not constrain users of radioactive materials. Fleming started by stating the consensus view that the sea was probably the most appropriate place to put intermediate and large amounts of isotopes with long half-lives and high radioactivity. The two most important factors governing the methods for doing this would be safety and convenience. Often the best course would be storage, flushing into sewage systems, or land burial. Sea disposal would probably be chosen, not just because of suitability of the waste and of the sea, but because of convenience. "It seems reasonable to assume that producers or users of radioactive isotopes located on or near the coasts, or on inland waterways, may find it simpler to dispose of virtually all wastes at sea," whereas those in inland areas might prefer methods closer at hand. Thus the report avoided judging any particular practice and reinforced the idea that the NCRP's recommendations needed not be taken as a mandate to be followed strictly.[80]

On the question of bulk disposal, for which the AEC had been so keen to leave room for interpretation, Fleming wrote very little. He warned about the problem of building up concentrations of radioactivity in certain areas and the possibility of contamination of food fish; but he also included "discharge" as one of the many legitimate means of disposal. The report added obscurely, "Proper safety regulations already in effect govern certain of these procedures but by no means all of them"—a statement that both acknowledged the lack of careful consideration to the consequences of bulk disposal and also allowed for maintaining the status quo.[81]

Even as Fleming wrote of tolerance doses, he shared the growing dissatisfaction with the notion of thresholds and—for the oceans—capacities. He publicly emphasized that whatever course the users of radioactive materials might take, they ought to realize that such disposal would be irreversible. Using language he later would employ with other oceanographers, he focused on the fact that "waste disposal" meant the permanent introduction of radioactive isotopes into the ocean. These could never be taken out. Unlike land burial, which left options for relocation or recovery in the future, there was no conceivable way to take such material out of the sea. "In the case of sea disposal, the act is final." To Fleming, this caveat drew attention to two issues. First, sea disposal should only be used for materials with no foreseeable future use. Second, they had to recognize that if, at some point in the future, radioactive disposal was deemed unsafe, there would be no way to correct past actions.[82]

One of the factors recognized by the committee as one of the many "intangibles" was the importance of public relations. Any recommendations by the NCRP or other body had the potential to cause alarm. Therefore any statements should be designed to minimize or eliminate anxiety by the public. Packages of radioactive material, for example, could end up on bathing beaches or fishermen's nets. Should there be a sharp decline in the fish catch, radiation might be blamed. Rumors needed to be countered by responsible agencies rapidly. There might even be cases in which radioactive contamination became a diplomatic problem, with formal protests lodged about ocean contamination.[83]

These warnings aside, the NCRP handbook acknowledged that there were no regulations for the handling of such material. In addition, there was no established authority for recommending where to dump. The open sea was international space and, whatever the NCRP might recommend, no one was designating specified areas, setting maximum limits, or keeping records. Indeed, the sea seemed to be the ideal choice for dumping—a site for vast amounts of waste, limited only by one's capability of transporting it there, with no accountability whatsoever. That was the state of ocean disposal as Fleming and the other committee members left it in 1953. True to the NCRP's attitude that it should neither judge nor constrain, Fleming's report concluded blandly, "This may ultimately become a matter of international concern."[84]

Chapter 2 Radiation Anxieties

When American biophysicist Detlev Bronk invited British nuclear physicist Sir John Cockcroft to breakfast in April 1956, it became more than a meeting of old colleagues. Bronk was president of the National Academy of Sciences, and in some six weeks the academy would issue its most authoritative study yet, on the biological effects of atomic radiation. Cockcroft, on the other side of the table, was the director of Britain's Atomic Energy Research Establishment (AERE), whose operations hinged upon the recommendations of authoritative scientific bodies. The British were also about six weeks away from issuing a similar study, conducted by the Medical Research Council, and Cockcroft was part of the committee advising it. At breakfast the two men agreed to identify any areas of disagreement, to attempt to minimize them, and to ensure that there were no surprises on either side. This was not the first time Anglo-American discussions about the reports had occurred, and some had been much more specific. Bronk and Cockcroft publicly emphasized that there was absolutely no collusion between the two studies, one American and one British. But the stakes were too high to leave anything to chance.[1]

The radiation studies caused considerable anxiety on both sides of the Atlantic, among academic scientists and atomic energy establishments alike. The hot-button issue for the Americans was radioactive fallout, brought to center stage in 1954 by tests of hydrogen bombs in the Pacific. Although the British were concerned about fallout as well, they had another predicament, too, particularly the fate of their radioactive waste. Plans to store radioactive waste in mines in the Forest of Dean had met with surprisingly resolute resistance, and the population density of Britain suggested that land disposal would become politically difficult. Worse still, newspapers had begun to complain about Britain's use of the sea for waste disposal, and atomic energy officials were being encouraged to say as little about waste disposal policies as possible. So Cockcroft traveled to the United States to collaborate informally with the Americans, believing that any inconsistencies in the two nations' reports would confuse the public and handicap the development of atomic energy.

This chapter traces the developments in Britain and the United States that culminated in these "independent" assessments of the biological effects of radiation. In the British case, the waste disposal problem developed quite early, resulting in decisions to look primarily to the sea to dispose of packaged wastes from several sites, such as the research establishment at Harwell and the weapons establishment at Aldermaston. In addition, they confronted the problem of radioactive effluents into the Thames (from Harwell) and into the Irish Sea (from the chemical processing facility at Windscale), necessitating firm recommendations about permissible levels of chronic (continuous or recurrent over time) exposure to even low levels of radiation. In the United States, the Atomic Energy Commission confronted none of the same pressing political problems about waste disposal in these years. Its chemical processing facilities were not located on the coasts, American discharges into rivers did not worry major metropolitan areas (in contrast to the case of the Thames, which ran through London), and it had plenty of options for temporary land storage. But similar discussions about radiation exposure came about because of nuclear tests, dangers from radioactive fallout, and public disagreements between some leading geneticists and the Atomic Energy Commission.

Although Cockcroft certainly believed that resistance to British waste disposal policies was just as overblown as the general public's fears of all things radioactive, the purpose of informal U.S.-British cooperation was not to provide "whitewash" reports to justify the practices of the U.S. Atomic Energy Commission and the UK Atomic Energy Authority. Instead, the overriding concern—of Cockcroft, Bronk, and many other scientists described here—was the need for continued credibility, which could be ensured by consensus. Bronk, for example, wanted to protect his organization's integrity, as did Harold Himsworth, his counterpart in the Medical Research Council. Any differences between the two nations' reports would undermine the scientific authority of both and certainly would provide ammunition to critics wanting to play up further cleavages of scientific opinion. Still, it was dishonest to claim, as they did, that the reports involved no collusion; if left in isolation, the two reports would have differed markedly, and there would not have been an Anglo-American consensus about the effects of atomic radiation in 1956. That is precisely why Cockcroft had breakfast with Bronk that morning—to ensure that serious differences never materialized.

We Shall Put Our Effluent into the Thames!

As a result of its decision to put its primary atomic research facility near the Thames River, Britain's need to dump some radioactive waste at sea was decided quite early. In choosing the location, one of Britain's primary concerns was close attachment to the universities at either Cambridge or Oxford and convenience to London. The job of constructing facilities at Harwell, near Oxford, fell to the engineers in the Ministry of Works, who assumed that most of the radioactive waste and effluent would be put into the river; the scientists at Harwell, working under the Ministry of Supply, assumed it as well. But other ministries soon played important roles, including the Ministry of Health and the Ministry of Agriculture and Fisheries, both of which contested the legitimacy of contaminating river and ocean waters with radioactive material.

Unfortunately, such questions arose only after construction plans were under way at Harwell. Officials at the Ministry of Health began to wonder, in the summer of 1946, what would happen to the local people when Harwell's radioactive effluent was released into the Thames River. They believed that the levels of effluent envisioned by atomic scientists and engineers would make the Thames unsafe. So officials from the ministries of Works, Supply, and Health convened in London to discuss it. Somewhat outraged by what he considered a last-minute objection, E.J.R. Edwards of the Ministry of Works pointed out that it would be "little less than tragic" if health officials suddenly, after all the work that had been done, found it impossible to accept the disposal of radioactive waste into the Thames. He tried to shift the responsibility by arguing that he was not in a position to solve the effluent problem, if indeed one did exist; it was up to scientists (and thus the Ministry of Supply) and not engineers (the Ministry of Works) to deal with that. The Works officials at the meeting insisted that it was quite out of the question to change the site of the establishment after so much time and money had been spent on it already.[2]

Similarly, the Ministry of Health put the onus of responsibility on the atomic scientists. With the Ministry of Works unwilling to budge, Health officials demanded assurances from their counterparts in the Ministry of Supply that discharges would be made safe. As the minutes of their meetings reveal, Health officials drew a pessimistic conclusion from the elaborate plans for Harwell that "the experts responsible for designing these safety measures realized that they were groping in the dark and dealing with something which they suspected they did not fully understand."[3]

To answer such concerns, British health physicists turned—as the American did—to threshold values established by an ostensibly independent body. In this case, the Ministry of Supply referred to the tolerance levels recommended by the Medical Research Council (MRC). This body had been created in 1914 as an extension of Britain's first national health plan, and it had continued as a separate body after the creation of the Ministry of Health in 1919.[4] When Britain began its own nuclear program after World War II, the MRC established the Committee on Medical and Biological Applications of Nuclear Physics, with various subcommittees, to deal specifically with radiation protection.[5] The MRC's threshold values were based largely on the work of Cambridge radiobiologist Joseph S. Mitchell, who had directed the medical program for the British and Canadian atomic energy project at Chalk River, Canada, during the war. Citing one of Mitchell's papers on radiation exposure, Harwell health physicist W. G. Marley gave his unreserved assurance that the doses set by the Ministry of Supply were designed to protect the consumers of Thames water for as long as fifty years. As long as they enforced Mitchell's figures, they could easily ensure safety. Sir Ernest Rock Carling, representing the MRC, concurred that Mitchell's figures were entirely satisfactory and added that there was far too much apprehension about radioactive elements. He pointed out that the water in Stockholm, Sweden, had higher radioactivity than even some of the health spas who advertised the health benefits of their radioactive waters, and the Swedes did not seem to suffer any ill effects. The meeting adjourned with general agreement that if "Professor Mitchell's tolerances" were respected, no harm could come to anyone.[6]

Harwell scientists realized quite early that it would be impossible to stay within Professor Mitchell's tolerances if they discharged all of the establishment's wastes into the river, so alternative solutions had to be found for some of it. They planned to take about a million gallons of water from the river per day and to return about 80 percent of it. Not all of the water would be dangerously radioactive, but some of it would exceed Mitchell's tolerance levels for drinking water. As in the United States, the British developed different policies based on the levels of radioactivity in the waste. The bulk of it would be discharged into the river, after sitting in delaying tanks to ensure sufficient cooling beforehand. But a smaller portion of highly radioactive waste, from some of the experimental laboratories, would have to be stored in secure tanks or drums and then transported elsewhere for disposal at sea, either through pipelines or from ships.[7]

Scientists at Harwell and at the MRC agreed unanimously that the sea was an obvious choice for its more highly radioactive waste. Their main concern was avoiding contamination of the Thames, as the consequences of dumping at sea were not seriously considered in these early discussions. At the same time, the sea was not perceived as a convenient solution to the waste problem at all; on the contrary, ocean disposal loomed as an expense that atomic energy officials hoped to avoid. Consequently, Harwell scientists worked on how to maximize the amount they could discharge into the Thames while keeping within the MRC's safety standards. From the first meeting, concerns from other ministries were easily dissipated by noting that Harwell planned to stay within these limits.[8] Incidentally, among those who worked on this problem of staying under the MRC's tolerances was Klaus Fuchs, later arrested as a Soviet spy—a point worth emphasizing in light of Russian scientists' later claims that they did not know that wholesale discharge of radioactive waste into rivers might give rise to harmful levels of radiation in the local population. This seems difficult to believe, because in 1946 one of the Soviet Union's principal spies was working directly on the problem for the British.[9]

Despite a few scattered concerns, there never was any serious consideration of changing the site of the AERE facility at Harwell to avoid polluting the Thames. Thus there was no point in belaboring the issue further. As Harwell's Chief Engineer Harold Tongue put it in August 1947, forcibly laying the matter to rest: "The site has been chosen, is being developed, and we shall put our effluent into the Thames!"[10] Tongue established an elaborate system of tanks for radioactive decay and other processes to allow relatively "cooler" wastes to be put into the river through a six-mile pipeline.[11] The higher activity waste from Harwell was sent by road to the northwest, to Cumbria, at a site called Drigg. There they built a pipeline to take the waste out to the sea. Conveniently, Drigg was very close to the site chosen for Britain's plutonium production facility at Windscale, and they used the opportunity as a test case for the future discharge of much larger quantities of waste into the Irish Sea. In 1947, the pipeline went out to sea some 300 to 400 yards, which was alarming enough for one Harwell scientist to complain, "this seems an insufficient length to ensure proper mixing with the sea."[12] For Windscale, the British hoped to use existing toxic drains, but, after consulting with scientists at the National Physical Laboratory in Teddington, they were disappointed to discover that ocean disposal would require more than getting the waste out to the water's edge at low tide.[13]

British scientists felt compelled at that point to make a hydrographic study of the tidal streams and likely diffusions rates near the pipeline. After dumping some dye into the sea, they measured the rate of its spread and dilution and used the results to help determine the length of pipe necessary to prevent concentration of the effluent near the shore. The pipeline extended out to sea about three kilometers. In subsequent years, health physicists at Harwell and the fisheries laboratory at Lowestoft would help to determine the permissible discharges, based on rate of diffusion and the likely concentration of radioactivity in the fish, seaweed, and other organic material in the Irish Sea.[14]

For solid waste, Harwell's director John Cockcroft hoped that dumping it at sea meant putting it, as sludge in steel drums, into the Black Deeps. This was the old name given a deep channel in the Thames Estuary, and it was a favored dumping ground for other industrial wastes. To Cockcroft, these other wastes were far more dangerous than radioactive sludge. Because of the location, he had to consult with the Ministry of Agriculture and Fisheries, which had policy responsibilities in the coastal waters. Now it was in the position of having to give its assent to dump what Cockcroft and his subordinates called "entirely trivial amounts" of radioactive waste in near waters.[15] To Cockcroft's annoyance, the fisheries officials balked. The Black Deeps were a trawling area for fishing boats, they pointed out, and thus the steel drums could easily end up being taken home by unwitting fishermen. They recommended putting radioactive waste into the North Atlantic, beyond the 300 fathom line, the customary depth for dangerous materials such as poison gas and old ammunition.[16]

The disagreement marked the beginning of a protracted struggle for authority in the British government between atomic energy officials and fisheries officials. Although advised that it would be difficult to argue against the Ministry of Agriculture and Fisheries on this point, Cockcroft recoiled at the idea of treating these drums of waste, with only slight amounts of radioactivity, as if they were as dangerous as poison gas. Clearly he thought the fisheries people had an imperfect understanding of the nature of radioactivity. He felt that no one was more qualified than Harwell scientists, whether health physicists or nuclear physicists like Cockcroft, to say whether the practice was risky or not, and yet Harwell was being second-guessed by fisheries people with only a rudimentary knowledge of the atom. Cockcroft wrote to a subordinate, "it would be perfectly safe to sit on top of a drum for about a year" without any apprehension about danger. The expense and logistical problems of taking the waste all the way out to deeper parts of the ocean were, in his view, totally unwarranted.[17]

After pressing the issue again unsuccessfully, Cockcroft called upon the various ministries to convene a "working party" to discuss sea dumping. He probably expected to overwhelm the fisheries ministry by packing the working party with allies: he suggested having it comprise two people from the Admiralty, two atomic energy officials, and one representative from the Ministry of Agriculture and Fisheries. Certainly this reflected Cockcroft's view of the relatively minor importance of fisheries in atomic waste problems. But the Ministry of Agriculture and Fisheries officials insisted on four representatives to represent the myriad levels of expertise required—fisheries, marine practice, hydrography, and marine radioactivity. They seemed to be gearing up for a battle, no doubt believing they ought to have a greater voice in ocean dumping. In the meantime, Cockcroft bemoaned their large numbers in the working party and conceded that he would have to give up the plans to dump in the Black Deeps.[18]

It turned out that the Ministry of Agriculture and Fisheries wanted to be involved, not merely as a matter of asserting influence but also to ensure discretion about the dumping operations. They did not oppose dumping at sea, but they wanted to avoid antagonizing fishing interests, in either the Black Deeps or out in the open sea. That suited Cockcroft perfectly, because he was not interested in calling attention to dumping operations either. Together the members of the working party began to plan Britain's first effort to dump radioactive waste in the deep sea. Initially, they agreed to try to maintain secrecy, to avoid any kind of publicity, and to claim that the purpose of the voyage was to conduct radioactive tracer experiments rather than dump contaminated packages.[19] In March of 1949, as they prepared to dump these solid wastes at sea, about three hundred miles west of Brest, France (in water over two thousand fathoms deep), they considered making a public announcement. But Ministry of Agriculture and Fisheries officials, fearing protests from fishing interests, were "of the opinion that the less said about the dumping the better."[20]

Secrecy, in practice, proved difficult. With so many parts of government involved in decision making, Cockcroft recognized that it would be nearly impossible to keep its dumping operations secret. Furthermore, secrecy conflicted with his own view that openness and public education was the best way to curb bad publicity. In the end, despite widespread desire for secrecy, the establishment at Harwell released an announcement for the press, believing it was highly likely that there would have been an information leak from one of the ministries involved. Cockcroft wanted to shape the public's views rather than leave the possibility of sensational stories. Harwell sent a note to the Foreign Office, getting its advice in case the French objected to the site (it was closer to France than Britain). To the relief of Harwell scientists, just two months earlier the United States Atomic Energy Commission had issued an announcement, reported in the newspapers, that it was disposing of some of its waste at sea. Thus the British could not expect any criticism from the United States, and they could claim that they were just following a precedent.[21]

One secret that was briefly kept in this case was the full disclosure of future discharge operations, and the ones in the dark were the officials at the Ministry of Agriculture and Fisheries. They were not privy to the scale of ocean dumping that the whole atomic energy establishment had in mind. When bureaucrats met in the "working party" to set rules for where these packages should be dumped, the fisheries people believed real negotiations were taking place about dumping policies. They favored isolation over dilution; in other words, they hoped to put the waste in stagnant water rather than in water that had high circulation, which was the opposite of most atomic energy officials' views. The fisheries people agreed to a site that had a high rate of mixing, but only grudgingly and only for an experimental period of six months. They made clear that they had been "very perturbed" by Harwell's initial plans to dump in the Black Deeps. Consequently, scientists at Harwell went out of their way to accommodate the ministry's wishes. But the fisheries ministry seemed oblivious to the fact that, from pipelines at the plutonium works at Windscale, a much larger amount of radioactive waste would be entering the sea much closer to shore using the principles of mixing favored by health physicists. Expecting fierce opposition, Harwell was in no hurry to clarify this to the Ministry of Agriculture and Fisheries. Their representatives in the working party did not appear to know about it, W. G. Marley admitted to a colleague connected to Windscale, "and we have been careful to avoid reference to this project." Naturally they would find out eventually, but it would be necessary to mount some careful preparation to avoid strong resistance.[22] Thus began another pattern in the history of radioactive waste disposal: a disproportionately large amount of attention devoted to a few dumping operations at sea, despite larger-scale, continuous operations from pipelines.

As a result of these discussions, on 5 April 1949 the UK began its first ocean-dumping operation with twenty drums carrying a total of about three tons of waste. Another cruise, with about twice the cargo, would follow in June. For the first dump, after an uneventful journey from Harwell, the trucks carrying the drums of waste arrived at the Devonport Dockyard at Plymouth and proceeded immediately to begin loading onto the frigate HMS *Tremadoc Bay*. It was not a particularly smooth operation. Someone had forgotten to order a crane, so the ship's crew manually loaded the drums. Initially the Admiralty hoped to load the drums into the depth-charge racks to dispose of the drums using the same equipment used to drop antisubmarine explosives. But the Harwell drums did not quite fit, and they were simply lashed to the deck. A forceful gale blew, prompting the captain to request a postponement. It was denied, so the ship pulled out slowly and made its way to the site with winds gusting over ninety miles per hour. Eventually the ship hands rolled each of the twenty drums out on a plank extending from the ship's stern, allowing them to drop into the sea. When all twenty of the drums were gone, the crew washed down the deck and their clothes and took a few more water samples before returning home.[23] Harwell released information about the dump to the press, but only after the ship had left port, stressing that the total quantity to be dumped was "much too small to have any harmful effects on fish or human life."[24]

Despite the intense negotiations beforehand, not all went according to plan. After all, ocean dumping required the cooperation of the capricious seas. Although the Ministry of Agriculture and Fisheries had insisted on sampling the water before and after the dump, the crew abandoned much of its preliminary sampling because of rough weather.[25] In addition, the carefully chosen site itself was abandoned at the last minute; after the *Tremadoc Bay* arrived, it was soon directed to steam to a different spot about two hundred miles away. The reason was that the general manager of Western Union Telegraph Company, L. C. Smyth, noticed the announcement of the operation in the newspapers and wrote immediately to the Ministry of Supply, asking for assurances that it had considered the placement of telegraph cables in the sea. Harwell's eventual reply gave the impression that they had considered it, noting that the dump site was far from undersea cables. But in reality they had not thought it through at all; Smyth's letter precipitated an abrupt last-minute shift in the site, thus changing the plans of an operation months in the making.[26]

No one working in the Ministry of Supply regarded these operations as harmful to the sea or to humans. But they disliked dumping at sea because it was a logistical nuisance. Unfortunately, they expected more, and larger, consignments to be ready in the coming months. Meanwhile, the Royal Navy was trying to extract itself from participation in ocean disposal, not wanting to bear the expense and any political repercussions. It offered the Ministry of Supply a couple of ships, but the ministry could not imagine paying the costs of equipping and maintaining them for just a few consignments per year.[27] Thus it considered turning to a private firm. Still, the firm in question realized that the profits would be pretty low and that its main business of carrying mail across the Atlantic would always take priority, which meant that the firm might end up taking British radioactive waste all the way to New York Harbor and then dumping it on the return trip.[28] This was not exactly what the British had in mind, and they continued to debate their options. When the *New York Times* published an article in August 1950 on the problems with radioactive waste in the United States, Harwell engineer Harold Tongue commented to colleagues, "It would appear that our American friends are also in trouble!!"[29]

Looking at their experiences in trying to negotiate with other government bodies, first with the Thames Conservancy Board, then with the Ministry of Agriculture and Fisheries and the Admiralty, Harwell officials held bleak views of the future of waste disposal. Advising colleagues at the still-secret military nuclear production facility at Aldermaston, Harold Tongue warned that dumping radioactive waste at sea in the future would be very costly. William Penney, the director of the Aldermaston facility, desperately needed a solution to his waste problem, which he expected to reach a crisis point within the year. According to meeting minutes between scientists at Harwell and Aldermaston in 1950, Tongue felt that "the only real solution to the general problem was a relaxation in the maximum amount of radiation permitted to be discharged into the Thames." Indeed, he and his colleagues currently were lobbying both the Thames Conservancy and the Medical Research Council to get the threshold value raised by a factor of 100, to avoid the cost of transporting waste over land to the sea pipeline at Windscale. The waste disposed into the Thames was above the threshold already, but Harwell's policy presumed that the river itself would dilute it enough to fall below the threshold. If they could justify revising the threshold upward, they could decrease the amount they would need to dump at sea by increasing what they put directly into the river.[30]

In fact Harwell had numerous headaches with the Thames Conservancy because it occasionally came very close to exceeding the agreed-upon thresholds and had to inform the water board about these instances.[31] But in 1951 Harwell convinced it and the Medical Research Council to relax its restrictions. In the end, it did not require an explicit change in the MRC's threshold levels. Instead, it meant measuring the effluent differently; for example, instead of limiting daily discharge, Harwell calculated average flows over several days or even longer. This method, according to Harwell's assessment, "takes account of the fact that the effect of radioactivity is a long term one depending upon intake over a very long period (in this case a life time) and provided a given total is not exceeded it is immaterial if, over a short period, the concentration is greater or less than average." Although the change implied no major alteration in the recommended threshold of radiation exposure, in practice it enabled Harwell to increase its disposal into the Thames threefold.[32]

The AEA leadership decided that its press notice during the first dumping operation in 1949 sufficed for all subsequent operations. Consequently, a dump off the west coast of Ireland in 1951 went unheralded. The public now had been notified that such operations occurred, and (the argument went) there was no reason to take special note during any particular operation. All of the interested government offices concurred with this strategy, not desiring to call attention to their associations with radioactive waste. They did keep public statements ready to hand, however, in case questions were asked, because the operations were not, strictly speaking, secret ones.[33]

The dump off Ireland in 1951 was ambitious, disposing of more highly radioactive wastes, including drums of polonium and plutonium. They experienced some disturbing operational difficulties before getting to sea; the area used to fill the drums became seriously contaminated, and the crane used to move the drums stirred up so much radioactive dust that measurements of the open air in the vicinity was more than twice the radiation tolerance level. An outraged official pointed out that everyone involved in the operation should be checked by doctors, have blood and urine tests, and "the next operation should be more adequately prepared for and supervised."[34] Indeed, once this particular voyage ended, and the 100 drums were put into 1,500 fathoms of water off Ireland, the ship called at Londonderry, Ireland, to secure medical attention for two of its crew. Although the actual operation went smoothly, the official report of the journey was not enthusiastic about doing it again, chalking up the success to exceptionally fine weather:

> A similar operation attempted during the unsettled weather that is the rule rather than the exception in the North Atlantic would be hazardous. There is no shelter for stormswept ships on the wild and rocky Coast of North Western Ireland. Tide races and overfalls make any approach to the land impossible until Lough Foyle is reached. This is a full 18 hours steaming from 11° W., during which time severe weather may develop. Even accepting all this as a reasonable risk inherent in the seaman's calling, another operation in less favorable conditions may take weeks to complete.[35]

With such unpredictable seas, ocean dumping had few enthusiasts; it was expensive, uncertain, and dangerous. Alternatives, however, had not materialized.

If the operation off Ireland could not be a good precedent for the future, what was to be done with these plutonium and polonium wastes? The Chief Engineer, Harold Tongue, struggled for an answer. Perhaps they could be dumped in the Hurd Deep along with lower-activity packages. The Hurd Deep was a deep part of the English Channel, near the islands of Guernsey and Jersey and close to the coast of France. But health physicists had already been disturbed by the level of contamination around the drums at the start of the recent operation. As the minutes of a meeting with Tongue revealed: "This type of material could not, therefore, be classed as non hazardous and the Hurd Deep was ruled out." Perhaps, then, they could ask the naval authorities to supply a larger ship, to make the operation less dangerous.[36] However, these authorities were not willing to do so; they would only help with routine operations like those in the Hurd Deep. Harwell already had decided that it could not contemplate the cost of operating even small ships themselves, so getting an oceangoing one seemed out of the question. Thus Tongue and his colleagues found a way to include even these higher activity wastes in the consignments for the Hurd Deep. The attractions of the Hurd Deep were simple: the naval authorities would not complain about taking part in operations that were in traditional dumping areas; the diplomatic effects would be minimal;

the costs would be lower because no large oceangoing ship would be required; and the likelihood of routine successful completion would be much higher. Also, if the various government ministries could accept the Hurd Deep as a dumping ground for a wider range of solid wastes, the short-term waste disposal problem would be solved. The highest-level wastes still existed in only small quantities, and they would not need to be dumped routinely; although the deep sea was the likely solution for these wastes, the problem was not as immediate and could be addressed through periodic special arrangements.[37]

Another reason for wanting to dump in a longstanding dumping area like the Hurd Deep was that the ship crews would see their jobs as routine, making news reports less likely. Such publicity was a rising problem. During the first two dumps in the Atlantic, the crew had been enthusiastic, with one deck hand insisting on working through the whole operation even after his shift had finished. The novelty of being part of this special operation appealed to the men. However, this was not something that Harwell officials wanted to encourage. Although they were happy about the enthusiasm, it made for a memorable experience bound to be retold at home and in the pubs. By incorporating radioactive waste disposal into existing procedures for other kinds of waste, British atomic energy officials hoped to routinize dumping so that no one would think it novel at all.[38]

By cooperating with naval authorities and trying to act with the limits set by the Ministry of Agriculture and Fisheries, Britain dumped about 2,027 tons of radioactive waste into the ocean between 1949 and 1953. Most of this (1,931 tons) was put into the Hurd Deep, about once per month, as opposed to once or twice per year for Atlantic disposals. The constraints under which the Authority operated had disadvantages, to be sure, but respecting them helped Harwell and Aldermaston to avoid having to maintain their own oceangoing vessel. Buying a ship would have necessitated the creation of a maritime culture at Harwell. The thought repulsed Harwell scientist G. W. Clare: "We at Harwell are not fitted to run a marine enterprise. It would mean employing suitably qualified staff apart from the crew, and so on (another little empire is born!)."[39]

From this point on, one of the principal concerns would be packaging: if the containers could be encased in concrete and made strong enough to prevent them breaking open during dumping operations, they could be considered nonhazardous and dumped in the Hurd Deep.[40] This would avoid antagonizing the Ministry of Agriculture and Fisheries. At Harwell, officials turned from disposal sites and logistics to experiments with sea water-resistant cements, plastic linings, and drum designs. Although other options remained possible, such as incineration and dumping into mine shafts, the focus was ocean disposal, specifically into the Hurd Deep. As a matter of policy, the Authority settled into their disposal decision, particularly because it had resulted from long negotiation and approval by other government bodies. As R. H. Burns put it to a colleague who had suggested a change: "It must be remembered that we have only one avenue of disposal available at the present time and it is essential that we do nothing to endanger our good relations with the Admiralty and the Ministry of Agriculture and Fisheries."[41] Indeed, this relationship was consolidated and codified in Britain's Atomic Energy Authority Act of 1954. Not only did the act create the Atomic Energy Authority (with Harwell, Aldermaston, and other facilities under its jurisdiction), it required that the AEA attain permission from the Ministry of Housing and Local Government and the Ministry of Agriculture and Fisheries (in 1955, this became the Ministry of Agriculture, Fisheries and Food) for every single disposal operation. Summarizing its practices, the AEA gave assurances that "the precautions taken are such that, on the basis of our present extensive knowledge of the effects of radioactivity, there can be no conceivable hazard to human beings nor to farm animals or fisheries."[42]

Denying Dangers

By the early 1950s, both the United States and United Kingdom had developed the key facilities of their nuclear weapons programs and identified sources of dangerous waste. Both disposed of large quantities of contaminated material into rivers, usually below thresholds of safety identified by health physicists. Beyond that, they envisioned storage on land or disposal to sea, through pipelines and (for solid wastes) from ships. Scientists in both countries saw the sea as a potential problem, not because of contamination of the sea, but because of the logistical, financial, and potential diplomatic problems related to putting radioactive waste there. But thus far, no insurmountable crises had developed, despite hints of differing perspectives from those whose professions focused on the oceans. The catalysts for change, raising the stakes of radioactive waste disposal at sea, were the prospects of the global development of nuclear power and the controversies about radiation effects associated with nuclear fallout. Both of these problems originated in the United States because of nuclear weapons testing and a major initiative to promote the peaceful uses of atomic energy worldwide.

The waste disposal problem intensified in the mid-1950s, not simply because of demands by the military, by scientists, or by business, but also because of President Eisenhower and a foreign policy initiative he created after his first year in office. Addressing the United Nations, Eisenhower announced his plan to promote the peaceful uses of atomic energy as a way to turn the nightmare of atomic war into the dream of cheap nuclear energy. He promised to provide fission materials to countries around the world who wanted to develop nuclear power, and he supported the creation of an international atomic energy body to coordinate the expansion of the industry and control the proliferation of the materials needed to split the atom. Atoms for Peace, as the initiative soon was called, emerged from the need to control public anxiety about atomic energy, to discourage other countries from developing atomic bombs, and to take steps toward disarmament. The initiative tied the presidency to the success of technologies based on atomic energy, such as commercial reactors, nuclear-powered ships, and even nuclear-powered airplanes. The Eisenhower administration supported all of these in subsequent years.[43]

The Eisenhower administration reveled in the good feeling surrounding Atoms for Peace, despite the reservations expressed behind the scenes that it was a "cruel deception," in the words of advisor Robert Cutler, to hold out the false promise of cheap power to most of the countries of the world. Some in Eisenhower's cabinet doubted the wisdom of leaning so heavily on Atoms for Peace, linking the administration's reputation to a specific technology. Secretary of Defense Charles Wilson worried about the costs of atomic power even for developed countries, and "he wondered if we were not overextending the promotional aspects of the process."[44] In March 1955, just over a year after the initiative's announcement, members of the National Security Council confronted Eisenhower about the dubious economic viability of civilian atomic power. The president defended his initiative, insisting that every one of his speeches on Atoms for Peace had been checked thoroughly beforehand by the "technicians."[45]

Eisenhower tried to answer such concerns, especially by those on his National Security Council, by saying that even if atomic energy was not economical now, it might well be so in the future, given a little faith in the engineers. For the present it was crucial to emphasize positive aspects of the atom. Besides, as Secretary of State John Foster Dulles argued to Wilson, it would be disastrous for foreign policy if the United States gave up on the Atoms for Peace promises, even on technical or economic grounds. This was something that the United States was supposed to be giving as a benefit to humanity, and it was crucial that it not be shown to be "a dud and a bluff."[46] In an internal 1955 statement of policy, the National Security Council admitted that atomic power was not likely to revolutionize the world economy, but that the United States had to promote it to preserve psychological advantages.[47]

One peculiar aspect of Atoms for Peace was the partisanship it encouraged on the part of the AEC and its chairman, Lewis Strauss. Initially, Strauss became a special assistant to the president on atomic energy to advise on the appointment of a chairman. Strauss's status gave him easy access to the president and he also attended high-level meetings of the National Security Council. Ultimately, it was Strauss himself who took the job as chairman. He initially enjoyed bipartisan support and was confirmed as chairman without any significant objections. But Strauss also held onto his previous post, making him privy to an array of Eisenhower's political aims and to some state secrets, which he kept from the other commissioners. Thus Strauss transformed the chairmanship into something more like a directorship, in which he retained considerable authority for himself at the expense of the other commissioners while keeping a strong party bias in favor of the president. These subtle steps led Strauss's AEC to downplay any negative effects from radiation that might endanger the Atoms for Peace initiative, such as the biological effects of exposure to radioactive debris from fallout, reactors, or wastes.[48] Conversely, the AEC under Strauss vigorously promoted the peaceful development of atomic energy, including the industrial-scale production and dissemination of radioisotopes.[49]

In March 1954, Strauss went to the Bikini atoll to witness the Castle series of explosions to test a deliverable hydrogen bomb. The success was marred because a Japanese fishing boat, the *Fukuryu Maru* (Lucky Dragon) was blanketed with radioactive debris (fallout) from one of the tests in the Castle series. The blast, code-named Bravo, was a thermonuclear weapon and turned out to be much larger than expected. The fishermen had radiation sickness when they returned home (one of them eventually died). The incident made international headlines and heightened tensions with Japan. Fallout also fell onto some inhabited atolls in the Marshall Islands, and over two hundred islanders developed radiation sickness, including loss of hair and skin lesions. At a press conference, the president admitted that the results were unexpected.[50]

When Strauss returned to the United States, he gave a press conference of his own. He played down the danger, noting that the Marshallese—as well as some exposed Americans—were receiving excellent care. He acknowledged an unexpected wind shift during the test but insisted that nothing went out of control. As for the Japanese, he remarked that they had ignored American warnings and had accidentally trespassed into a danger area. During the question period, Strauss made a tactical error by describing the hydrogen bomb itself, which he said could "take out a city," even one as large as New York.[51]

Although Strauss's remarks about destroying a whole city gave rise to considerable sensational coverage by the press, this was not the statement that appalled scientists and began a major period of antagonism between the AEC and some in the scientific community. He also stated that although American and Soviet bomb tests had raised the background level of radiation, it still was "far below the levels which could be harmful in any way to human beings." The statement was unambiguous, implied a definite threshold of safety, and struck many scientists—especially geneticists—as dishonest.[52]

Because of these political problems with military uses of the atom, gaining political support for the peaceful uses of atomic energy became a high priority for Strauss and the president. At home, Strauss already was guiding the construction of the first commercial nuclear reactor, which was a collaborative effort between Westinghouse Electric Company, the U.S. Navy, and Duquesne Light Company. Built in Shippingport, Pennsylvania, the reactor was financed largely by the AEC, which was required by law under the 1946 McMahon Act. One of the principal aims of Atoms for Peace was to change that law to enable more active participation of private companies in the peaceful development of atomic energy and to encourage the proliferation of reactors to generate electricity in the United States and elsewhere.

The administration's commitment to the expansion of atomic energy emboldened some of the critics who worried about the biological effects of atomic radiation. At the International Congress on Nuclear Engineering in 1954, Walton A. Rodger spoke publicly of the dangers of storing radioactive waste near the plants; instead, he argued, they should be stored in desert regions, far from humans, or in abandoned mines. For now, the sea might suffice, but when such plants were used commercially, they might generate tons of material on a daily basis, and even the ocean would lack the capacity to dilute it effectively. Such warnings fueled feelings of uncertainty about constructing new plants and irritated companies wanting to build them. Westinghouse Electric Company, for example, rebutted Rodger's warning, claiming that the problem was not insurmountable and that there was no danger of contamination in the power plant planned for Shippingport.[53]

Such concerns about civilian nuclear power blended into a rising critique of nuclear weapons testing. One such critic was California Institute of Technology geneticist Alfred H. Sturtevant. In 1954, Sturtevant was president of the American Association for the Advancement of Science, a prestigious office at the head of the body that promoted all disciplines of natural science and published the country's premier scientific journal, *Science*. In his presidential address at the annual meeting on June 22, 1954, Sturtevant warned about the harm from fallout. The head of the AEC's Division of Biology and Medicine, John C. Bugher, publicized a strongly worded correction, including the words "absurd and scientifically indefensible" to characterize Sturtevant's views. One of Sturtevant's colleagues, R. A. Brink, wrote to offer encouragement after visiting AEC colleagues at Oak Ridge. Brink observed that "there was a hint in some of the things said by the biology group at Oak Ridge with whom I visited briefly last week that as a matter of policy AEC officially denies all public statements that its experimental work involves any hazards, genetic or otherwise, to the population at large." That policy, Brink believed, applied not only to scientists but to anything that "the man on the street may see seemingly run counter to the Commission's official line." Brink believed that this was the reason for the vehement categorical denial of damage.[54]

The critique of the AEC from such a high-profile scientist helped to turn radiation effects into a genuine public controversy. Sturtevant received many personal responses to the publication of his presidential address and the coverage of it in national newspapers. Some were requests to reprint, as in the case of the *Bulletin of the Atomic Scientists*. Others were inquiries from major publishing houses hoping for a book contract, as in the case of Princeton University Press and G. P. Putnam's Sons. Some were suggestions from a variety of individuals about possible experiments he could conduct—most of these found their way into a special file that Sturtevant called "crank letters." Many letters, however, were from private citizens who, for the first time, read about a major scientific figure criticizing official United States policies regarding radiation. Some congratulated him on taking a stand, while others just wanted advice from someone who seemed less enthusiastic about the wonders of radiation. Mrs. L. Rowan of Chicago, for example, wrote that her nine-month-old daughter had two birthmarks that doctors had recommended removing with radium. "Do you feel, personally," she asked, "that this could be dangerous, genetically?" Sturtevant wrote back that she should consult with radiologists at the University of Chicago Medical School rather than relying on her doctors' word alone.[55]

The most effusive response came from fellow geneticist Hermann J. Muller, who had conducted some of the work upon which Sturtevant's conclusions were based. Muller agreed with all of Sturtevant's points but warned about the future. Their worst course of action, he believed, would be to alienate the AEC so much that they could be identified with the segments of American society whose hostility to nuclear testing led them to make ludicrous statements. Such people may be well-meaning, but they "try to picture the genetic effects of radiation on man as meaning that we shall have an obvious increase of freaks and monsters." It was best to avoid associating themselves with such people, he felt, because it would be easy for the AEC to refute such outrageous ideas.[56] Muller recognized that defenses were being constructed on both sides of the question. He hoped to avoid a polarized, politicized debate. The important goal was not to tear the AEC down but to convince it to recognize damage rather than deny it.

American physicist Lauriston Taylor, the chairman of the National Committee on Radiation Protection, took a more benign view of the risks of radiation from fallout and wastes. He became a principal advocate for the view that radioactive waste, despite becoming a growing problem, ought to be equated with the by-products of other technological advances. In 1954, a headline in the *Los Angeles Times* characterized his attitude as belittling the peril of the wastes, though this exaggerated his view. Taylor said that he thought the best course of action would be to concentrate radioactive waste into tiny pellets by incinerating wastes and then condensing the smoke; these pellets could be put in containers and dropped into the sea at least a mile offshore. Most of the waste "will be half dead in 20 years," and thus human beings would not have to worry about it. "As a matter of fact," Taylor pointed out, "I suspect we shall have to increase our physical toleration to radioactivity when this becomes a real world of atomic production. It may be something like getting used to the fumes of automobiles."[57]

Taylor's role as the leader of the NCRP gave these statements considerable authority, for this body ostensibly served as the basis for most radiation protection policies in the United States. But the NCRP had no power to enforce its recommendations, and Taylor did not attempt to empower his committee to set enforceable standards, which he felt might needlessly retard the development of atomic energy. Taylor's viewpoint reflected his understanding of radioactive waste as another technological challenge of industrial society. It was ultimately a factory problem. In the 1920s, he argued, radiation protection was predominantly a medical issue, and in the 1930s its industrial use became important. Today, he said, radiation was potentially usable by anyone, and thus it would affect all walks of life. Radioactive materials and radiation-producing machines could be expected to find their way into every industry, making occupational exposure the primary concern. In addition, radiation could escape the atomic plants to affect the rest of the population, and some uniform system of regulation ought to be recommended. Taylor did not insist that such recommendations be enforced by the federal government, however. The *Los Angeles Times* concluded, "Dr. Taylor doesn't seem to feel the same forebodings about damage to the human race as do some biologists."[58]

In fact, the NCRP had done all it deemed necessary on the subject of radioactive waste, and neither it nor the AEC had much desire to dwell on the problem. Taylor's goal had been to suggest broad guidelines, not to establish exacting standards. Through the subcommittee on waste disposal and decontamination, the NCRP had thought through some of these problems and issued a few handbooks. As for the oceans, Richard Fleming had made his mark by stating that the oceans should probably house a considerable amount of waste, though more study was needed on the problem. With this permissive attitude in writing, Taylor decided in 1954 that the subcommittee did not need to exist any longer. "All good things have to come to an end," he wrote to Fleming. It had been a harmonious group, and the "meetings were always fun to attend," so it was with regret that the tight-knit subcommittee would no longer need to meet.

Taylor's specific reasons for dissolving the NCRP's subcommittee on waste disposal were vague. Yet the action was consistent with his view that more specific standards went against the fundamental outlook of the NCRP. "The nature of our protection program, at the present time, is such that it does not seem wise or profitable to continue further the lines along which your subcommittee activities have been directed," he wrote to Fleming. Taylor assured him and the other members that the work would still continue if need be, but that the NCRP did not need a standing subcommittee on the problem.[59]

Attention to the waste disposal problem could not be so easily set aside, however, due to the prominence of its sister subject, the radiation effects of nuclear fallout. The complacency of government bodies such as the AEC and the NCRP simply strengthened the resolve of critics who emphasized the dangers of radiation. In January 1955, Sturtevant gave another critical review of radiation effects before a group at Caltech, and he was buoyed by letters of encouragement and inquiry. Robert Rugh, a radiologist at Columbia University, saw the warnings about the promiscuous use of X-rays in the newspapers. "I have been particularly incensed by papers such as those of Dr. Ira Kaplan who advocates ovarian x-radiation to overcome sterility and (not knowing any genetics) he presents photographs of first generation and second generation offspring to prove that there are no ill effects."[60] Mrs. James W. Miller of San Gabriel, California, read about Sturtevant's address in the *Los Angeles Times*. She immediately thought about her teenage years, during which her acne problems had been severe enough that two different doctors each had administered over two dozen X-ray treatments to clear up her complexion. She also had a mole removed through irradiation, as well as several other X-rays. She wrote to Sturtevant wondering if this might affect her offspring. "I had a mongoloid baby who lived only three months," she wrote. Were there any studies showing a direct relationship between X-rays and children with mental retardation? She offered herself and her husband to Sturtevant to help his work by their experience. Floyd Mulkey, from the Woodlawn Methodist Church in Chicago, was inspired to commend Sturtevant for his courage in saying it was inexcusable for the AEC to deny there were hazards. "I am sure the government—and the professional patriots—have been trying hard to silence you on this issue." Mulkey found himself disillusioned with his government, because even when the AEC claimed to be studying the biological effects, there was never any doubt that it would exonerate itself. "It is tragic that so many of us have completely lost faith in the honesty of our government today."[61]

Other scientists also spoke out against what appeared to be the official line of the AEC. A number of them were dismayed by an article that appeared on 25 March 1955 in *U.S. News and World Report*. The article, entitled "The Facts about A-Bomb 'Fallout,'" stated that geneticists were divided about whether radiation exposure to a population was harmful or helpful. It described an experiment on a strain of fruit flies that passed 128 generations in a highly radioactive container, resulting in a "much improved race of fruit flies with more vigor, hardiness, resistance to disease, better reproductive capacity." From this the writer concluded that the fear of radiation was not well founded, and that some experts believed mutations usually worked out in the end to improve species. Upon reading the article, Muller sent a scathing letter to the AEC criticizing this "fallacious idea," claiming instead that the situation lacked the slightest ambiguity among geneticists. There was clear evidence, he said, "that the vast majority of induced mutations have eventually to be eliminated and are therefore damaging," and that the AEC ought to step in and correct such misstatements by widely read newspapers.[62]

The AEC did not feel obligated to correct such statements. After all, this was a newspaper article, not an official report. Although Muller felt that people might reasonably expect that an article such as this reflected the official AEC line, that was not the case. As AEC scientist Earl Green wrote to Muller, there appeared to be no fundamental misunderstanding on the part of scientists inside and outside of the AEC. Green hoped that an open public discussion of the issues would lead to a more balanced presentation of ideas, since it seemed clear that magazines and newspapers tended to exaggerate the views of both sides. Green quoted from some internal AEC memoranda to the effect that the fruit fly experiment in question did not yield any interpretation like the one made by the newspaper article. The same memorandum objected to the implied conclusion that there was no genetic damage to human beings. Yet the AEC was not willing to concede to the view, which was just as extreme in its opinion, that there was certainty about damage. This was a matter of risk, and these risks ought to be balanced with other risks to which the population routinely is subjected, as well as the risk of not testing nuclear devices.[63]

Green was trying to appear reasonable and conciliatory, pointing out that the positions of both sides were put into an extreme light by the press. Sturtevant, after reading a copy of the letter, wrote Green that he could not agree even to this. The statement that fallout was certain to produce defective offspring in a future generations, which Green would have preferred to phrase as a matter of risk, was not extreme at all from Sturtevant's point of view. "The critical word is 'certain,'" he pointed out, and "surely any scientist understands this to mean that the probability is so great that we may safely disregard the extremely remote opposite possibility." He went on to make a conservative calculation. Assume that one mutation is induced per 10,000 human gametes by 1 r of radiation. Also suppose that the average inhabitant of the United States has received 1/10 r from fallout. If these inhabitants will produce 100 million offspring, the likelihood that no mutation has been induced was "distinctly less than 1 in 10^{50}."

The argument against this view, to which many in the AEC held dear, was that there was a threshold value of radiation below which no mutations were induced. As long as exposures were kept below this level of safety, one could be exposed routinely to low amounts of radiation. Alternatively, if there was no threshold, then all radiation exposure had to be considered cumulative, adding to the number of induced mutations regardless of how low the exposure was at any given time. Sturtevant assumed there was no threshold, though he admitted that experimental evidence did not directly prove this was the case. However, there was no reason to believe, other than wishful thinking, that there was a threshold, because "theory gives no basis for a threshold, and experiment does not suggest it." To Sturtevant, the only way to soften the word "certain" would have been to say "practically certain"—but there was no justification for playing down the certainty of harmful mutations by implying any meaningful degree of probability or risk.[64]

The latter interpretation did not lead all geneticists to oppose nuclear testing or nuclear power. In fact, Muller told Green that he agreed that the present rate of testing was acceptable in view of the potential benefit from the tests and the existence of greater hazards routinely accepted.[65] Sturtevant took a harder line. It was not fair, he said, to equate genetic damage from radiation with other risks. Even though the risk to the individual was less than that of driving a car, the point had no meaning because radiation was not a personal choice and the individual could do nothing to avoid the risk or minimize it. "Also," Sturtevant pointed out, "most of the other risks apply to the exposed individual, not to his descendants."[66] But both Muller and Sturtevant agreed that genetic harm from fallout must be balanced with the possible gains of testing nuclear weapons. It was the spirit of the AEC's interpretation, rather than ultimate conclusion, that geneticists found troubling. Muller suspected that the supposed disagreements between scientists came from broad understanding on the part of the general public about who the scientists were. The chief opposition to accepting the fact that any amount of radiation damaged genetic materials appeared to come from physicians who used radiation for diagnostic purposes. He told a colleague that it seemed to the AEC "as if 'doctors disagree' since it is not commonly realized that few medical doctors have sufficient understanding of genetics to be judges of the matter." Somehow geneticists would need to penetrate through the biases of nongeneticists who felt that they already had a thorough understanding of the dangers of radiation.[67]

Courting Public Opinion

Amidst these disagreements about nuclear weapons testing, the consequences of peaceful atoms became a matter of international discussion. To help publicize and gain international support for Atoms for Peace, Eisenhower proposed a worldwide conference to discuss the range of technical possibilities. Chairman Strauss announced this proposal in April 1954, with the conference to take place the following year in Geneva. The delegates would cover an array of topics about civilian atomic energy, such as reactor technology, radiation protection, and the uses of radioisotopes in medicine and industry.[68] Although many of the attendees were scientists and engineers intimately acquainted with the uncertainties surrounding radiation exposure, the AEC ensured that potential critics were barred from the American delegation.[69]

Most of the spectators of the International Conference on the Peaceful Uses of Atomic Energy emphasized the great technological potential of the peaceful atom. Their attitudes were historical; they viewed atomic energy as part of a long process of industrial change. Thus they saw the conference as an unprecedented opportunity to shape an industry in its infancy, to study and prevent environmental and health problems rather than let them go unchecked. Atomic energy could, they hoped, stand in stark contrast to the laissez-faire approach to industrialization of the early nineteenth century. Setting the tone for this cautious but hopeful attitude was the director-general of the World Health Organization, Brazilian physician Marcolino Candau. He argued that "time is short, and public health must fulfill its obligation by intelligent control so that general exposure to radiation background will not soon reach levels from which there is no return." This warning, despite its concerns about danger, contained no hints of disparagement over the industry as a whole. It implicitly sanctioned the large-scale, albeit adequately controlled, development of atomic energy as a natural aspect of technological progress. The World Health Organization acknowledged the practical dilemma: public health officials must control the problem intelligently, probably through genetics research, yet they also "must not be accused of hindering the development of nuclear power and thus depriving the world of its benefits."[70]

Far from taking a cynical view of Atoms for Peace as propaganda, most of the scientists and engineers present took the conference as evidence that the United States was serious about helping the rest of the world with technological progress. Studies of waste disposal were not particularly controversial. The British reported on dye experiments they had used in planning the proper length of their pipeline at Windscale. They also explained how the circulation of the sea and the probable high concentration of radioactivity in organic matter, particularly harvested edible seaweed, had helped them formulate permissible levels of discharge.[71] On the general subject of peaceful nuclear power, most newspapers hailed the conference as a success. As Strauss told the National Security Council, the conference was a "victory for our fundamental national policy," because everyone saw how committed the United States was to atomic technology and that Eisenhower was giving the peaceful atom more than lip service.[72]

Atomic energy establishments in the United States and Britain used the success of the Geneva conference to promote an expanding atomic energy industry. Willard F. Libby, a politically conservative commissioner on the AEC, and John Cockcroft, director of Britain's Atomic Energy Research Establishment at Harwell, both utilized *The Scientific Monthly* as a venue for doing so. Libby observed that the peaceful uses of atomic energy would require a major effort to incorporate radiochemistry into high school science classrooms. With the expansion of atomic energy into civilian use, in addition to the medical and industrial use of radioisotopes, "the student would have a continuing use for his knowledge of radiochemistry just as he has a continuing use for his slide rule." Certainly such knowledge would also come in handy, he added, in understanding the risks from fallout or from an atomic war. Libby emphasized that all people, regardless of whether it convinced people to enter scientific fields, needed to develop a better understanding of the role of radioactivity.[73]

On the question of radioactive wastes, John Cockcroft argued that "it seems likely that the radiation from these waste products will find important uses." He called the long-lived (and dangerous) radioactive isotopes of cesium and strontium useful as sources of radioactivity. The rest of the waste would be stored until its activity was low enough to be jettisoned into the sea or otherwise discarded. The cost of storage would be low, he said, perhaps less than 2 percent of the total cost of nuclear power, and in any event this cost would likely be offset by revenues from sales of the strontium and cesium to hospitals and industry, where radioactive sources were needed. Cockcroft looked forward to the creation of Eisenhower's proposed international agency for atomic energy, which would ensure that any dangers inherent in peaceful applications of atomic energy would be avoided. "We must all hope," he wrote, "that its birth will not be too painful or too long delayed."[74]

Cockcroft's optimism about the short-term was not universally shared, even among those who expected a bright future in atomic power. As it stood in 1955, atomic power cost considerably more than efficient coal-burning power sources. Glenn T. Seaborg, the codiscoverer of plutonium (and later AEC chairman), told a conference on atomic energy in April 1955 that the expansion of atomic energy, and its future viability as a power source, might be hampered by the problem of waste disposal. Later that month, an audience at a meeting of the American Society of Civil Engineers listened as scientists from the AEC's Reactor Development Division provided similar warnings. Although the problems of disposal thus far had been adequately met, sanitary engineer Joseph A. Lieberman said, looking at the future "it becomes obvious that safe, more efficient and economical ultimate disposal of radioactive wastes is one of the major challenges of the industry as it progresses."[75]

At the Nuclear Engineering and Science Congress in Cleveland in December 1955, essentially the same conclusions were reached with some two thousand participants in attendance. With the development of nuclear power, the production of wastes would intensify. In a report sponsored by the American Chemical Society, E. I . Goodman and R. A. Brightsen predicted that the United States would produce a ton of fission products per day by the year 2000. World production probably would triple that. If dumped at sea, the long-lived isotope strontium-90 would contaminate about 5 percent of the ocean's total volume. Dilution at sea, they concluded, could not be considered a permanent solution.[76] This was one of the first times that ocean disposal was publicly cast as a source of pollution rather than as a solution to the land storage problem.

As 1955 drew to a close, a *New York Times* article observed that nuclear technology was "coming of age" and that some hard realities about safety needed to be faced. This remained one of science's toughest problems, journalist Robert K. Plumb observed, and the recent Cleveland conference was proving it. The report by Goodman and Brightsen had indicated that the problem of disposal should be addressed at the reactor design stage and should be considered at every stage, including reactor operation and reprocessing methods. Industry professionals, Plumb charged, were making no effort to do this, focusing instead on maximizing output. The conference highlighted the apparent widespread inattention to the after-effects of atomic power. Plumb quoted one participant, University of Rochester physician George Hoyt Whipple, saying that radiation safety was a problem of conservation—"the conservation of the human race, if you will." Whipple, who was a Nobel Prize–winner in 1934, focused not on waste disposal but the larger problem of unnecessary exposure to radiation, and he urged that permissible levels accepted by the AEC be drastically reduced. All areas of exposure needed to be addressed: wastes, occupational exposure, and fallout from nuclear testing. He posed the question, "Is this risk acceptable to the average individual?"[77]

Amidst these public statements in newspapers by American atomic scientists, critics of waste disposal singled out Britain as the primary offender. Although the United States also dumped at sea, the British bore the brunt of most criticism. For example, the *New York Times* in 1955 observed that "most physicists object to dumping the radioactive wastes into the sea" because ultimately marine life would be damaged. "Yet that is exactly what Britain has been doing," it noted, calling to mind that in January 1955, 1,500 tons of waste were dropped off Land's End. Although the newspaper exaggerated the opposition by physicists, this was, by the Atomic Energy Authority's admission, probably the largest amount that had ever been dumped at once. The containers would hold for about thirty years, the AEA maintained, by which time most of the radioactivity inside them would have become harmless.[78]

The negative tone in the report of British activities in the *New York Times* actually was the result of a public relations misstep by the UK Atomic Energy Authority. Whereas the radiation issue in the United States turned primarily on the question of radioactive fallout, there was a growing political consciousness in the United Kingdom about radioactive waste. In late 1954, atomic energy officials were surprised to find stiff resistance in local communities to the plans to store radioactive waste in abandoned mines in the Forest of Dean, on the Welsh border. Bowing to political pressure, and not wishing to provide a negative story for the newspapers, they soon abandoned the idea. But the incident sparked some soul-searching among the AEA's leadership about what would happen if ocean disposal met with the same resistance. Unlike the Forest of Dean, which simply seemed like a convenient solution, the ocean was an indispensable part of the nuclear cycle, particularly if land disposal were to prove so politically difficult. In the aftermath of the Forest of Dean controversy, a letter appeared in the *Times* of London, asking questions about the magnitude of the waste disposal problem. After a couple of other newspapers began to initiate their own inquiries, AEA chairman Edwin Plowden decided that he did not want the oceans to turn out like the Forest of Dean, and he tried to take active steps to educate the public about the issue. He and other officials, including John Cockcroft at Harwell, agreed that the best strategy was education, not secrecy; the British public would, they believed, act rationally if they were adequately prepared to understand the need to dump radioactive waste at sea. Plowden turned to the editor of the *Times* to explore ways to address public concern through a carefully worded article about ocean waste disposal.[79]

The result of this strategy of openness was a minor disaster within the Atomic Energy Authority, and it led to a setback for those who wished to focus on education rather than secrecy. In late 1954, two health physicists working for the AEA, W. G. Marley and R. H. Burns, met with some reporters to give an overview of their practices and to answer some questions. In the course of those discussions, Burns revealed that there was currently a dumping operation taking place, though it had not been announced beforehand. The treatment of this meeting, in the *Times* and other newspapers in and outside Britain, embarrassed the Atomic Energy Authority because the reporters focused primarily on the unheralded dumping operation rather than the reasons behind Britain's policies. In January 1955, they reported—truthfully but uncomfortably for the AEA—that the current operation was the largest one ever undertaken, and that the sea was the final destination for the hotly contested radioactive waste originally intended for the Forest of Dean.[80]

Immediately officials at the AEA started to point fingers about "the public hubbub which has resulted from newspaper 'stunting'" and sensationalizing of these activities. Although looking for someone to blame, many within the AEA realized that the incident highlighted a disagreement about how to handle public relations. Which was more effective: openness or secrecy? Should Burns and Marley have denied that an operation was actually under way? British atomic energy officials considered their 1949 announcement of a radioactive waste disposal operation enough evidence that they dumped at sea and that periodic reminders were unnecessary. That did not mean, however, that the operations were national secrets. Defending Burns and Marley, AEA Public Relations director Eric Underwood argued that the AEA ought to be providing *more* information, not less, about the dumping, to show the public how harmless it was. Like others, he believed that the anxiety surrounding the newspaper coverage could have far-reaching implications; "I would recommend strongly," he pointed out, "that we do not let it have the effect of making people within the Authority feel that this is a secret subject." Still, it would be important to develop a briefing system for officials before they interacted with newspaper reporters, "so that they can be warned away from dangerous ground."[81]

Although regretting the news coverage, Plowden agreed with Underwood that it would have been improper to expect Burns or Marley to lie to reporters by denying that the Authority continued to dispose of radioactive waste at sea as it had in 1949. On the other hand, he now had the headache of a number of official questions raised in Parliament about it.[82] He had to explain exactly, as one official put it, "what we do with our radioactive waste and why it's all right and how we know it's all right."[83] Plowden tried to assert firm control over information, insisting that he should personally approve any further statements from employees about waste disposal. Some, like John Cockcroft, thought this was an overreaction. After all, Burns and Marley were trying to educate the public, and the newspapers simply reported on what they heard. Cockcroft agreed to submit any further statements, adding, "but I do think we must pursue our policy of education. Otherwise we are lost!" He asked Plowden, "Why do we worry so much?"[84]

Plowden's worries were caused by the new public visibility of the waste disposal problem and the top-down pressure from politicians. The Atomic Energy Authority achieved high status when it was created in 1954, but it also courted new political pressures. Instead of being a fiefdom within the Ministry of Supply, statutory changes now had the AEA answer to the lord president of the council, a high cabinet position dating back to the sixteenth century. Lord Salisbury (Robert Gascoyne-Cecil), the lord president at the time, complained about the publicity. He pointed out that the problem of the newspaper articles had been the AEA's own doing, when it confidently had tried to shape public opinion. Inviting reporters to interview atomic energy officials had been intended to reduce anxiety, but the opposite had occurred. "This only shows how delicate the subject is," he surmised. "We need, I feel, have no regrets about the 'Times' article; but we shall do well, so far as we can, to let sleeping dogs lie with regard to this matter in the future."[85] Lord Salisbury was not declaring a policy of secrecy, but neither was he impressed by the benefits of trying to provide more information to the general public. His message was clear: stop talking about it with reporters.

Plowden's idea to educate through public media was thus somewhat stifled by the delicacy of bringing the subject up at all. He wrote back to Cockcroft, who had asked why they worried so much: "The answer is that our political masters worry a great deal when they are confronted with a spate of Parliamentary Questions. This particular interview provoked no less than five." Plowden agreed that education should be pursued, but they clearly had not thought it through properly. They would have to develop a much more thoughtful public relations program for the AEA than just an occasional lecture to reporters. For the time being, all statements needed approval by him; the alternative would be that Lord Salisbury would have to approve them. Given Salisbury's "let sleeping dogs lie" attitude, approvals of any public information seemed unlikely.[86] One result of the affair was the recognition of the important role within the AEA of the public relations director, Eric Underwood. Afterward, Underwood planned frequent visits to Harwell and other facilities likely to attract the attention of headlines or parliamentary questions to minimize harmful (and maximize beneficial) publicity.[87]

Independent Assessments?

Given the public controversies, the most pressing needs for the AEC and the AEA were scientifically ironclad statements upon which officials could base policies. Neither believed secrecy was the best solution, particularly given the increased political and scientific attention to their policies, and given the expectation of a civilian nuclear power industry. Some publicly available yardstick was needed to demonstrate consistently that their practices fell in line with scientific findings. That meant that such statements would have to be reliable scientifically but also impeccably nonpartisan. Given the controversies over fallout and the recalcitrance over waste disposal, a new consensus was needed.

These unexpected political tensions prompted both the AEC and the AEA to encourage authoritative bodies to conduct independent assessments about the biological effects of radiation. The American version grew from recommendations by the Rockefeller Foundation to finance an independent evaluation of radiation effects, one that could not be accused of being an AEC whitewash.[88] Conducted under the National Academy of Sciences, it was expansive, covering six subjects—not only genetic and pathological effects of radiation, but also the effects on oceans and fisheries, meteorology, food irradiation, and waste disposal. By contrast, the British study under the Medical Research Council confined itself to general assessments of the genetic and pathological effects of radiation, leaving aside the other, more policy-oriented, subjects. Ultimately the academy and the MRC issued their authoritative reports simultaneously in June 1956, and they claimed there was no collusion between the two groups. The opposite was in fact true; there was a great deal of collusion, and the leaders of both studies were keenly aware of the political stakes. Had they issued contradictory reports, they knew, both would have been open to criticism, and both projects would have been undermined.

The American study was based on the National Academy of Sciences' committees on the biological effects of atomic radiation (the BEAR committees). When the American geneticists met in February 1956, they were faced with a conundrum. They agreed that the essential feature of their views was that all radiation caused genetic effects. This set them apart from those who believed there was a threshold of safety. But what, then, could they usefully do for the BEAR study? Warren Weaver, mathematician and longtime Rockefeller Foundation administrator, chaired the genetics panel. He urged his panelists to look beyond their general principles and establish some guidance for safeguarding the population. It was not enough to say "as little as possible," he pointed out. They must acknowledge that their goal was to establish proper safeguards for a growing industry. Gioacchino Failla, who had worked on this problem for many years in the NCRP, agreed—they had to set some figure, despite their recognition that it might lead people to believe that the figure reflected a threshold value of absolute safety. Because of its firm grasp of the problem, the genetics committee needed to be the one to set that figure. "If it doesn't," he said, "some other group will." Some of the exposure levels being discussed for the entire population were ludicrous, he believed, such as 10 percent of what was permissible for occupational exposure. No geneticist would agree to that, he said. It was just too high. But the AEC was beginning to set down codes of practice, and they might very well take the higher figure unless the geneticists set a different, but still definite, figure.[89]

The American geneticists were afraid of the implications of any figure they might set. They appreciated the argument that it needed to come from them, rather than some other body that did not understand the harmful effects to human descendents. But they also recognized that setting a figure would itself contradict their own position that all radiation was damaging and that there was no threshold of safety at all. Caltech geneticist George Beadle wanted to abolish the term "permissible dose," because they could trace almost all of their problems to the use of that word. It could be manipulated easily by those who wished to play down the negative effects of radiation, such as AEC commissioner Willard Libby. "A beautiful example of a true, but misleading, statement comes right out of Libby's remarks," Beadle complained. "He says the amount of radioactive fallout is only a small fraction of the permissible dose. This is accurate, but completely misleading, because he doesn't say that 'permissible dose' doesn't mean a thing at all."[90]

They argued about the effects their decisions might have on the standing of the United States in its struggle against the Soviet Union. Failla hoped they would avoid setting extremely low figures that might constrain hospitals and universities, where a great deal of research was being done. "What with the international situation as it is, and the attitude of the Russians, who are now trying to beat us in the scientific as well as the military field, it would be rather unfortunate to place an added handicap on us at this time." Another panelist, Alexander Hollaender, pointed out that low figures might hamper the nuclear power industry, too. The no-threshold view might require them to cut the level of exposure to a point at which the industry would suffer. The development of reactors might be delayed. "If you could do it all over the world, it would be all right," he noted. But if the geneticists in the United States started curbing

American industry, he pointed out, other countries were unlikely to be so enlightened, and then Americans would fall behind. Hollaender also warned that if the limit was too low, no one would pay any attention to it.[91]

These concerns persuaded the geneticists to abandon their basic argument, namely the no-threshold concept. Instead they offered an arbitrary figure to guide policy, 10 roentgens for population exposure. It was Oak Ridge National Laboratory geneticist William Russell's figure, and he noted, "but I won't be pinned down on the reason for picking it." The others consented, not because of any scientific reason to set that particular figure, but because they thought it seemed reasonable. Weaver felt that if any group should pick an arbitrary figure, it ought to be this one. But they realized that they were making political judgments. James V. Neel, who spent years studying the genetic effects of Japanese atomic bomb victims in the academy's Atomic Bomb Casualty Commission, said that he would accept Russell's figure if they all were willing to say that such a level of exposure was necessary for the moment, but not ideal in the long term. It would not hurt the human race very much, but the figure ought to be presented to the public in the proper way. "If this is a statement for practical reasons, which the national interests demand, that is one thing, and I am all for it." But if it were to be issued as a statement which they, as scientists, certified to the public as safe, on the basis of detailed calculations, that was another matter. To this Herman Muller pointed out that they were part of the committee as scientists, and based on that reasoning they should not give out a figure if it had no scientific basis. But the others argued that something had to be decided in the meantime—something practical.[92]

With such arbitrary figures coming from the geneticists, the likelihood of matching precisely with other independent studies seemed small. In fact, some of the AEC commissioners were nervous that the American and British bodies might issue contradictory reports. If that occurred, the job of assuring the general public of the harmlessness of radiation—from fallout, for example—would be much more difficult. One of the commissioners who strongly supported nuclear testing and often played down the genetic risks of radiation was Willard Libby. In 1955, Libby telephoned National Academy of Sciences president Detlev W. Bronk and suggested that he contact Sir Harold Himsworth, Secretary of Britain's Medical Research Council.[93] Soon Himsworth and Bronk began a fruitful correspondence that lasted throughout the next year as the committees prepared their reports. It was a relationship that Bronk later described to Himsworth as "international friendship and negotiation at its very best."[94] The two chiefs shared information and collaborated informally, avoiding any appearance of official connections between the two studies. They were, after all, supposed to be independent. "What I have particularly in mind," Himsworth wrote to Bronk at the start of their correspondence, "is a desire that the individual scientists on our respective Committees should feel perfectly free, as individuals, to discuss with each other any problems in the field with which we are engaged and that you and I, as the respective Chairmen, might feel free to enter into private correspondence on any points which it might seem good use to do so." Himsworth observed that his suggestion amounted to little more than keeping the normal scientific channels open.[95]

Although Himsworth did not envision any major points of conflict, he did not want to leave it to chance. To ensure the harmony between the two reports, Himsworth made a trip to the United States in early April 1956. He wrote Bronk beforehand that "my main concern is to compare notes so as to make sure that we are as much in accord as I have been assuming," and he put himself at Bronk's disposal for the three days of his visit.[96] Genetics panel chairman Warren Weaver met personally with Himsworth during this visit. By that time, each group had preliminary reports that they discussed with each other. Weaver was comforted to see general agreement between the two groups' findings. Some minor differences resulted in different measurement assumptions. For example, figures for medical and background radiation exposure during one's reproductive lifetime were not the same, because of different estimates of what constituted one's reproductive lifetime.[97]

The Americans' decision to recommend specific figures surprised the British. In Weaver's words, Himsworth found "our recommended dose figures a little, shall I say, startling." The reason was not so much the figures themselves, but rather the inclusion of any specific figures at all. Why blend scientific evaluation of the biological effects with policy instructions? The Americans were planning to recommend 10 roentgens for population exposure over one's reproductive life, and 50 roentgens for occupational exposure. Himsworth found the whole idea of specific recommendations unpalatable, because they inevitably would be misinterpreted by someone and could be manipulated in the press. They also had the potential of undermining the objectivity of the study. Also, recommending different levels of population exposure and occupational exposure would be difficult to justify if a specific figure was named; it would open a host of social and legal problems.[98]

Weaver later wrote to Himsworth defending his committee's decision to offer a specific figure. It was true, he said, that their knowledge about effects was incomplete. "And yet decisions must be made," he said. "The geneticists do not escape their social duty by standing mute—for that decision leads to consequences, just as clearly as does a decision to face the difficulties and give as much help as possible." [99]

Both groups began to recognize that they were on the verge of issuing two reports with different findings. This made the atomic energy leadership nervous. So in late April, Sir John Cockcroft paid Bronk a personal visit. Over breakfast, the two men agreed that they needed to find a way to present a common front. Cockcroft worried about public relations and felt that any disagreements between the American and British groups would be confusing to the general public. Bronk then suggested that the two groups make a list of significant questions and issues and state the various positions of each committee. Then the information could be exchanged between the two committees, so that there would be no surprises and each group would be well prepared to address any inquiries from the press about differences between the reports, making it easier to play down the differences. Bronk wrote to Himsworth, "I recall that last summer you and I agreed that it would be well not to have collaboration; but now I wonder whether you would not think it well to have a mutual understanding before release. Cockcroft, I think, liked this idea."[100]

As the release dates of the reports approached, the informal exchanges grew more specific. The boundaries between the two groups blurred, and autonomous evaluation turned to careful negotiation. In May, Himsworth wrote to Bronk about "a particular bone-seeking isotope," which their data seemed to show was accumulating at a rate that would sooner or later "encroach significantly upon the available margin of safety." They had not yet decided how to handle the issue, and Himsworth wrote that "I should be very glad to know what your people are thinking on this subject."[101] Himsworth also sent in confidence the drafts of the British committee's findings regarding hazards, and the specific recommendations the MRC planned to make to the British government. As a friendly reminder, Himsworth emphasized the confidential nature of the report: "Our reports to Parliament are regarded as in the confidential-secret category until they are presented so, unless you want to visit me in the Tower, would you confine the circulation of these papers to those concerned with your report with whom you are in confidential relationship."[102] Just as Himsworth sent copies of these preliminary findings, Bronk sent copies of the American one to the MRC prior to its official release in June. Himsworth apologized that the Americans did not receive a copy of the official, final version of the British report until it was released—bound, as it was, "by the rules of Parliamentary privilege."[103]

The two reports were released simultaneously on 12 June 1956. The benefits of making consistent points at the same moment outweighed the need to avoid the appearance of collusion. Besides, both the MRC and the NAS had enough influence to ensure that selected media outlets emphasized that the two reports had not been authored together and that despite minor differences, the results were essentially the same. Himsworth happily reported: "At this end, therefore, the timing with regard to publicity has gone just as we wished. I hope it was equally satisfactory over on your side of the Atlantic." He was very pleased that all of the newspaper articles he saw made due notice of the similarity between the two reports, while "each, fortunately, reproduces my assurance that the Reports were prepared without collusion." The biggest difference that most people observed was not on scientific matters, but rather that the American report was more of a popular account that a layperson could understand, whereas the British report was more technical. To Bronk, Himsworth beamed, "Need I say how much your frank and candid cooperation in this matter has meant to me. It is a good basis on which to continue our trans-Atlantic relationships."[104]

The National Academy of Sciences and the Medical Research Council had worked together in order to depoliticize radiation effects by establishing a consensus of scientific opinion. But behind the scenes, their conclusions were negotiated; in other words, these reports were not a reflection of an existing consensus in the scientific community, but were themselves new creations based on science, politics, and a little scientific diplomacy. Aside from the pressures they both were receiving from atomic energy establishments, as in the case of Libby and Cockcroft, they were both motivated to ensure institutional integrity. Both bodies had been chosen because they carried the weight of authority and credibility, and both were keen to preserve it. Thus the reports by the BEAR committees and the MRC were not designed to toe a particular line; they were, however, designed to agree with each other and to avoid any taint of political motivation. In addition, despite repeated references to the harm from any amount of radiation, they included dosage recommendations; thus they represented the epitome of the threshold view that health physicists used in order to advise on atomic energy policies, including waste disposal.[105]

Meanwhile, as most scientists debated the wisdom of nuclear testing and analyzed the genetic and pathological effects of radiation, other scientists used the new studies to come to the forefront of science policy. In particular, oceanographers seized the moment in the BEAR committees to attempt to assert a role in atomic energy policy—in effect, to become atomic scientists in their own right. The role of the oceans in waste disposal policies had encouraged them, particularly in the United States. When the BEAR committee panel chairmen gave a press conference in June 1956, a reporter asked whether there were any political recommendations. Bronk said that there were not any—after all, this was a scientific report, not a political one. But oceanographer Roger

Revelle spoke up and referred the reporter to the specific threshold figures the academy had recommended. Bronk interjected that he would prefer not to refer to them as political. But Revelle knew quite well that these numbers would be used as policy guidelines by the AEC, even if Bronk did not wish to emphasize it. He also knew that because of the BEAR committees and the problem of radioactive waste at sea, oceanographers had begun to carve out a place for themselves in the world of atomic science policy.

Chapter 3 — The Other Atomic Scientists

In his 1957 charter address at the University of California, Riverside, oceanographer Roger Revelle mused about the new roles of scientists and politicians in the postwar era. Between them, he said, they held the future of the human race in their hands, and each should pay close attention to the other. Politicians should become more science-minded, while scientists needed to consider the political dimensions of their actions. More specifically, he outlined some steps that scientists needed to take in the realm of politics. First, he said, they ought to emphasize the uncertainties that surround political action rather than adhere to a set of inflexible political principles. And, second, scientists should not shy away from promoting their own interests. For Revelle, this meant protecting research projects from interference, lowering the barriers of secrecy, and planning international cooperative projects. But, fundamentally, it meant that the scientist "must use all the political wiles of which he is capable to gain financial and material support for his basic research."[1]

These lessons touched on Revelle's past experiences finding patrons for oceanographic research in his years at the Navy's Bureau of Ships, the Office of Naval Research, and as director of the Scripps Institution of Oceanography.[2] But they also reflected his recent experience as the chairman of the panel on oceanography and fisheries within the National Academy of Sciences study of the biological effects of atomic radiation (the BEAR committees). For oceanographers charged with the responsibility of gauging the effects of radioactive contamination on the world's seas, the BEAR experience highlighted both of Revelle's first two principles of action: emphasize the uncertainties and promote oceanographers' interests. The BEAR committees catapulted oceanographers, and especially Revelle, to national stature, giving them an unprecedented power to influence policy on major national issues such as nuclear fallout and civilian nuclear power. The BEAR committees were an opportunity to make a difference in policy and to establish a need for continued reliance on oceanographers in the years ahead. Particularly if Eisenhower's plan to develop nuclear power on a global scale were to succeed, knowledge about the oceanic dumping grounds would be crucial.

The BEAR committees served as a vehicle for scientists to assert authority over scientific questions that pertained to national policy. Part of the frustration some scientists felt thus far was that conclusions reached within disciplines outside of physics appeared to be neglected by policymakers. Until now, atomic scientists were just the physicists interested in the workings of the atomic nucleus. But atomic science touched other disciplines as well, as revealed in the sharp criticism of the Atomic Energy Commission (AEC) by geneticists who felt that the government ignored a basic truth from their discipline, namely that genetic harm was probably the most common way in which atomic radiation damaged humans. Oceanographers also felt like outsiders, though few took a strong line against either nuclear tests or dumping radioactive waste at sea. Rather, their strategy was to embrace governmental policy and to benefit from it. They hoped that the prospect of ocean disposal would lead to more patronage for oceanic research, or at least necessitate greater involvement of oceanographers in policymaking. Thus the BEAR committees not only saw scientists confronting the issue of radioactive contamination at sea, they also became a forum in which oceanographers could gain new roles of authority and influence.

Because the political ramifications of the BEAR committees were abundantly clear, they provoked oceanographers to think deeply about the consequences for themselves, for the nation, and for their science. Some followed Revelle in seeing an opportunity to solicit patronage for future research, viewing radioactivity as an experimental tool rather than as a danger. Others focused on the potential risks: for example, Woods Hole Oceanographic Institution scientists such as Allyn Vine agreed with Revelle that oceanographers should play a larger role in the atomic energy establishment, but he emphasized the need for research as a precaution against the uncertainties surrounding waste disposal. Others such as Richard Fleming wondered if waste disposal—like the genetic threshold—was a conceptual illusion, a distraction from what really was happening to the sea. In their desire to set acceptable rules for dumping, he feared, oceanographers might disingenuously imply that there was an identifiable amount of waste that would keep the oceans normal. Yet no one quite knew what the "normal" ocean was.

Still others groused that Atoms for Peace would make American oceanographers the custodians of the world's oceans, responsible for the environmental consequences of civilian nuclear power all over the world. As Milner Schaefer put it, "We take the garbage contract for everybody."[3] Although his complaint referred to Americans having to dispose of all of the world's waste, which never happened, he was correct in a larger sense—the BEAR oceanographers would be responsible for how marine scientists all over the world saw and used the oceans. After all, these oceanographers stood at the forefront of a major policy initiative from the United States designed to convince the world that the future lay in nuclear power. They knew that the world would look to the sea to dispose of the vast stock of waste products, and they were in a position to make the crucial decisions to inform global action in the years to come.

Promising Possibilities Timidly Explored

Marine scientists' involvement in atomic matters did not begin with the BEAR committees, and they already had experience looking for new research opportunities in government projects. A number of them were involved directly in atomic bomb work. During the war, for example, University of Washington scientist Lauren Donaldson began a study of the effects of radiation upon salmon. Working for the Manhattan Project through the university's Applied Fisheries Laboratory, Donaldson irradiated salmon with X-rays from 1943 and continued this work long after the war, managing to court funding from the AEC throughout most of the commission's life.[4] The laboratory worked with the AEC and General Electric to assess the environmental effects of the Hanford site, and the laboratory participated in the 1946 atomic bomb test at Bikini atoll (Operation Crossroads) and the Bikini scientific resurvey the following year. The benefits for scientists involved in these tests seemed enormous. Assessing Operation Crossroads a couple of years later, Donaldson observed that there was still a vast number of samples leftover from the blast, such as fish and animal tissues, that scientists had not yet begun to study.[5] Donaldson called the resurvey "one of the most important biological expeditions of all time," not only because of all of the scientific problems that might be revealed by studying the effects of radiation on the life zones of Bikini, but also because it was "the forecast of things to come if men can't learn to live together in peace and harmony."[6]

In California, Scripps scientists were well aware of the research opportunities that came with a strong relationship between science and policy. Not only was Revelle instrumental in turning the Navy's attention toward basic research in oceanography, but Scripps was crucial in helping the state of California assess the causes, and possible remedies, for a catastrophic decline in the sardine catch in the late 1940s. In the case of sardines, despite cautions in the 1930s about the need to respect a maximum sustainable yield for the yearly sardine catch, fishermen and government officials refused to enforce limits.[7] The first lesson, for the oceanographers involved in this analysis, was that policy should balance the short-term needs of industry with its long-term interests, perhaps by setting—and respecting—a threshold value for sardine fishing. But this conclusion was never definitively reached, and the research funds to identify the cause of the decline continued to flow from the state for many subsequent years (the scientific consensus that overfishing was the primary problem did not emerge until the mid-1960s). Thus a second lesson was that active involvement in major policy issues could lead to an extraordinary amount of interest in, and financial support for, scientific research.

Oceanographers saw such projects as opportunities. Although scientists had been involved in planning Operation Crossroads, Revelle made certain the subsequent resurvey was more closely attuned to the needs of scientists. He also fostered a strong cooperative relationship between oceanographers and the Navy and seemed willing to find ways to incorporate scientific projects into military contexts.[8] Over the years, he and other leading American oceanographers adopted strategies to define the marine sciences broadly enough to appeal to a wide spectrum of patrons.[9] In the mid-1950s, the Atomic Energy Commission was ripe for plucking: it had a major policy mandate (Atoms for Peace), it had a major problem to solve (radioactive waste), and it needed authoritative voices outside of itself to satisfy critics.

Although oceanographer Richard Fleming had been frustrated by the AEC's inattention to the problem of ocean disposal, the political winds had become more favorable to oceanographers by the time of the BEAR committees. Oceanographers and sanitary engineers worked together to assess the possibility of doing as the National Committee for Radiation Protection (through Fleming's work) had suggested in 1953, putting high-level packaged radioactive waste into the ocean. The Johns Hopkins University and Woods Hole Oceanographic Institution organized a two-day conference in June 1955 to assess the problem. Fleming did not bother to come, citing his past frustrations and expressing his doubt that the AEC had experienced a change of heart.[10] But some of the most prominent American oceanographers were slated to take part, parsing out the disparate components of the problem: Henry Stetson and W. Maurice Ewing on marine geology; Edward Goldberg and Dayton C. Carritt on sediments; Henry Stommel, Walter Munk, and Donald C. Pritchard on large water masses; Columbus Iselin, Roger Revelle, W. Maurice Ewing, and William von Arx on overturn processes; Allyn Vine and Arnold Arons on sampling and monitoring; Vaughan Bowen on biology; and Bostwick Ketchum and Albert Parr on ecology. In addition, scientists at Johns Hopkins had created a special project group on sanitary engineering and radioactive waste that would lead discussions on structural concerns for containers, problems in methods, and the logistical problems of warehousing, transportation, and handling the wastes. Some staff of the AEC and the Oak Ridge National Laboratory also participated.[11]

It remained to be seen, however, whether oceanographers' major successes in working with the Navy could be repeated with the AEC. Acquiring funds for oceanographic studies, particularly if they were not connected directly to weapons tests, proved difficult. Woods Hole oceanographers such as Bostwick Ketchum and Allyn Vine tried a number of times to interest the AEC in such studies. Should the AEC plan to dispose of large amounts of wastes at sea, Ketchum argued in one proposal, it would need to know more about ocean circulation and about marine ecology. Oceanic processes, as yet poorly understood, would determine the ultimate fate of any wastes being dumped or discharged into the sea. Such discharges already were being conducted by the AEC. Ketchum wrote: "It is clear to oceanographers, however, that no completely isolated body of water exists in the sea, and that wherever wastes are discharged they may ultimately spread to all parts of the ocean."[12]

Oceanographers were equipped with tools to make quantitative evaluations, and Ketchum presented a detailed proposal of oceanographic work to assess potential effects of waste disposal in coastal areas; they were asking for $144,391. He and other Woods Hole scientists met in November 1955 with Walter Claus, the chief of the Biophysics Branch in the AEC's Division of Biology and Medicine. Ketchum reported that although Claus politely admitted that such a study was probably a good idea, there were "no funds currently available" for such a large-scale program. He promised to look into the matter, perhaps to get money for some aspect of the work, but not for the whole thing. To the oceanographers, however, the vast sums spent on the production of atomic bombs made the proposed expenditure to study disposal seem very small.[13]

Oceanographers were becoming increasingly frustrated with the AEC's attitude that ocean disposal studies were not considered essential to the development of atomic energy. Preparing for the BEAR meeting of oceanographers, Vine wrote a memorandum in February 1956 outlining what he considered to be the basic conundrum of waste disposal. He captured the uncertainty by describing the problem as "somewhere between nonexistent and insolvable." Partly this was due to a lack of information, but more important, it was not clear what would control the outcome more—scientific, economic, or political aspects. It was likely that the long-term problem of disposal would prove more difficult in all three of these than the production of atomic energy itself. By now, he wrote, billions of dollars had been spent on production. "Now research on the basic problem of disposal has barely been started and engineering effort is much less," he wrote. "Hence, the DISPOSAL problem is many years behind the PRODUCTION problem."[14]

If disposal was going to be easy to do, Vine argued, there would be no problem at all. But if it was going to be difficult, the government had to recognize the need for a major program of research and development. It may very well be, he suggested, that the three aspects already mentioned, along with the psychological factor, may diminish the feasibility of radioactive waste disposal. If so, the governments of the world that had begun to base their economies on nuclear power would be in an awkward position. If the problem were to remain unsolved, dumping would occur anyway, on land and at sea, which would cause innumerable conflicts. "It is doubtful if any country, state or county will want to receive their neighbor's radioactive waste," Vine observed, and the indiscriminate dumping in international waters or territories, or in rivers shared by several countries, would become a source of discord between nations. If one could expect such problems, why was there so little money being spent on solving them? "There is too large a chance," Vine wrote, "that we are all drifting along hoping that science and luck will take care of the waste products which a world-wide effort is producing."[15]

While Fleming and some of the scientists at Woods Hole such as Vine were attempting to gain the AEC's ear, scientists at Scripps took a lead in the field and ultimately Revelle became chairman of the panel on oceanography and fisheries within the BEAR committees. One of the reasons for his success was a difference in tactics: although all of these oceanographers stressed the need for research, Vine's approach emphasized the potential dangers of radioactive wastes, whereas Revelle's pointed out the potential benefits of radioactivity. This might have been the result of political savvy or just plain optimism on Revelle's part. But it would have been savvy, indeed, given the goals of the Eisenhower administration and the AEC. Why should the AEC want to support scientific work whose rationale, at best, would be to validate existing dumping policy and, at worst, would be to criticize it? It would be preferable to support scientists who were trying to point out new ways that radioactivity at sea could benefit science. The whole rationale for the president's Atoms for Peace initiative was to offer something positive to the world and to use science and technology to turn the atom toward constructive purposes. Scientists and the AEC already had claimed that radioisotopes were as revolutionary as the microscope, enabling useful tracer experiments in a number of disciplines.[16]

Roger Revelle saw radioactivity as an important tool in science, and he had argued as much to the AEC and the Navy in recent years.[17]

It is no accident that Revelle played a prominent role in the first international conference, in Geneva in 1955, on the peaceful uses of atomic energy, which AEC chairman Strauss devised. Scientists with critical attitudes toward American policies were not welcome in the American delegation at the conference, a point revealed later when it became clear that Strauss had barred geneticist Hermann Muller from participating. The conference's purpose was to highlight positive, peaceful uses of atomic energy, not to introduce potential negative consequences. When Scripps oceanographers had taken part in some of the Pacific tests, they did so to use the tests as giant experiments, not primarily to monitor for dangers. Scripps's contribution to the conference (authored by Revelle, Theodore Folsom, Edward Goldberg, and John Isaacs) reflected Revelle's desire to see his institution play a major role in nuclear affairs. Entitled "Nuclear Science and Oceanography," the paper emphasized the "promising possibilities" of using radioactive materials as tracers and the fact that these techniques had only been "timidly explored."[18]

Revelle's emphasis on the uses of radioactive materials in science turned out to be an effective strategy for oceanographers wanting to play a greater role in nuclear affairs. In contrast to the geneticists, it was a relatively nonthreatening approach, not designed to judge the AEC's past policies or question the future development of nuclear power. Instead, it began by accepting the assumptions of the AEC and President Eisenhower that weapons tests should occur and that civilian nuclear power was the wave of the future. Revelle and others did not challenge these assumptions, but instead built on the notion of wondrous progress by claiming that science stood to gain a great deal from the introduction of radioactive wastes into the ocean. "A more vigorous and imaginative application of nuclear tools in the marine sciences," they argued in the Geneva paper, "would certainly result in important breakthroughs." Although they did not ignore the potential problems of radioactive waste disposal, they stressed the importance of research, which would lead to a better understanding of waste disposal in due course.[19]

By embracing Atoms for Peace, Revelle and other oceanographers carved out a niche for the marine sciences in the world of atomic energy. Experience with the sardine problem and with naval interests had given oceanographers an entrée into the world of "Big Science," as historians have dubbed it, with large-scale projects supported with generous public funding. Atomic energy, and the prospect of radioactive waste disposal at sea, was another such opportunity. Atoms for Peace was a boon for oceanographers wanting financial support and a voice in policy.

The Nuclear Future

The question of waste disposal at sea had been reopened because of the political force of the Eisenhower administration and its Atoms for Peace plans. American scientists played their part by attending the Geneva conference and praising the technological possibilities. Revelle did the same, but he also saw an opportunity for sustained research should this political climate continue. At the Geneva conference, Revelle emphasized to reporters how little was known about the oceans and the need for greater understanding by scientists. "Geneticists in general are chary about dropping things in the sea," he told them. "I am personally of the opinion that it might—I repeat, it might—be done safely providing we carry out some detailed observations right now."[20] The challenge to Eisenhower's nuclear policies posed by the nuclear fallout controversy did not dampen oceanographers' enthusiasm. Now that the National Academy was putting together scientific panels to help inform the government and lay public about the biological effects of atomic radiation, there was an even greater opportunity for public visibility and possible patronage.

The AEC had, in fact, begun to look more seriously at oceanographers and the roles they could play in assessing the seas for dumping purposes. At the Geneva conference, Revelle and other American oceanographers revealed that the AEC had asked them to look into the problem of waste disposal at sea. Revelle acknowledged that at the present time "we know almost nothing about the long-term genetic effects of radiation on future generations." Closer to his own realm of expertise, he also acknowledged that oceanographers did not have much idea "about what goes on in the depths of the oceans." He listed the known options: ocean dumping; desert or arctic dumping; pumping wastes into emptied oil wells; encasing waste in new, stronger materials that would only become stronger over time; and rocketing the wastes off the planet. British scientist Eugen Glueckauf (AERE) joined Revelle in hoping that the oceans could be used. He believed that by the year 2000, Britain would produce about twenty tons of waste annually, with world production at about two hundred tons.[21]

At the time of the Geneva conference, waste disposal had not yet achieved the status of an extraordinary problem, largely because of a widespread belief that technological solutions might be found. Glueckauf, for example, served on a British committee to design ways to separate cesium and strontium on a large scale, partly to address their dangers as radioactive waste but also to develop major sources of radioactivity for irradiation purposes in future industry. As D. H. Everett later recalled of Glueckauf's work, radioactive waste was at the time "a Cinderella subject compared with more glamorous topics such as reactor physics and fuel design," and engineers for nuclear facilities had not yet begun to think broadly about radioactive wastes at the design stage. For the most part, scientists and engineers expected that the wastes would have uses or that scientists would develop ways of minimizing harm.[22] The precise scientific problems that would become crucial were not well understood. In fact, the conference itself was far more noteworthy for showcasing the promise of atomic technology than for its scientific merit, and it helped to stimulate international interest in the Atoms for Peace initiative.[23]

In the month following the Geneva conference, a group of scientists at the University of Hawaii reported less encouraging news about the possibility of dumping waste at sea. Their work, led by Robert W. Hiatt, focused on the concentration of radioisotopes in marine organisms after weapons tests. He and his colleagues tested a number of fish species, with a particular interest in those that might be utilized as food sources for humans, including large species such as the black skipjack, dolphin, and yellowfish tuna, and smaller ones such as the papio, aholehole, and the *Tilapia mozambique*. They targeted a particular isotope, strontium-89, in the hope that it would give an adequate picture of its much more dangerous sibling, strontium-90, a major component of nuclear fallout and waste (they chose this isotope to minimize the hazards to scientists in the laboratory and to cut down on the dangers of the waste they produced). Although they made little effort to measure radiation damage, they did find that some parts of the fish retained their radioactivity for extended periods. Some visceral tissues such as kidney, spleen, liver, and heart, showed a continuous decrease in radioactivity in as little as an hour after being administered the dose. But structural tissues such as the head, gill arches, and especially the skeleton, concentrated strontium rapidly and retained it at an apparently constant level for a long period of time.[24] The problem of biological concentration would become a crucial area of research for scientists trying to gauge the effects of ocean contamination. For the moment it appeared that dangerous materials could reach humans by concentration in fish tissue.

The specific question of waste disposal was included in the BEAR committees as a panel chaired by Abel Wolman of Johns Hopkins University. When the study had concluded, he was asked whether ocean disposal should be regarded as unsafe; he pointed out that, to his knowledge, the disposal of high level waste was not being practiced by anyone, either in the United States or elsewhere. He replied that a "certain amount" of low-level waste was being dumped, but "under control and with no, I would judge, discernable objection." Also some of it was "released to nature" in a variety of ways under strict control and with "no hazard to surrounding life." Speaking for the entire BEAR committee, Wolman stated, "it is certainly our feeling" that the high-level waste should be held back for the time being because of the lack of information and adequate research on the possible effects.[25] Wolman was expressing the general feeling not only of his own group but also of Roger Revelle's committee on oceanography and fisheries as well that thus far nothing dangerous was occurring and that there ought to be more studies to see if even more waste could be dumped in the sea. Otherwise, the atomic energy industry would have to keep piling up waste in temporary (one hoped) storage until a permanent solution to the waste problem could be reached.

Although the Wolman committee was concerned with waste disposal generally, most of its members were connected primarily either to radiation safety (i.e., health physics) or to geology. The specific question of ocean contamination was handled by the committee under Revelle. This was not a new subject for the oceanographers, as they had been involved in atomic energy since the first Pacific atomic bomb tests in 1946. Over the years, their work was incorporated into planning the investigations to follow the fission, and later fusion, blasts. Oceanographic work was integrated closely into the Castle series, during which the controversial 1954 Bravo shot took place.

Richard Fleming had passed on the opportunity to participate in the 1955 conference, but he did become involved in the BEAR project when invited. The National Academy had considerable clout, and there might have been a chance to make a real difference. But Revelle, the chairman, was not primarily interested in setting limits to radioactive waste disposal. After forming the committee he wrote to Fleming that their principal job was twofold: their first task would be to show how radioactive materials could be utilized in the marine sciences, and the other was to outline what needed to be done to get the right information to protect the environment and its users. This order of priority reflected Revelle's plan to use the BEAR work to promote scientific research and expand support for oceanography by linking it with radioactivity. At the same time, he recognized that the BEAR reports would be used by countries all over the world and would form the basis of American contributions to any scientific effort under the UN or other bodies. Thus they needed to prepare a "fair, honest and lucid report."[26]

The controversy over fallout prompted the BEAR work, and the genetics and pathology panels proved the most controversial; still, the importance of ocean studies also had become abundantly clear during the *Fukuryu Maru* incident. Upon the ship's return to Japan, the crew was hospitalized, but the tons of shark and tuna in its hold went up for sale. When the Japanese press realized and published this fact, fish prices collapsed. The AEC sought to find out where other fishing vessels had been at the time while publicly trying to reassure American consumers that they should continue to buy fish caught in the Pacific. The AEC also tried to placate the Japanese by providing them with money so they could conduct independent scientific studies. Meanwhile the AEC chairman, Lewis Strauss, made matters worse by insisting that it was the *Fukuryu Maru* crew's fault for being off course, and by suggesting that radioactivity itself had not caused its medical problems. Compounding this was the suspicion by Japanese victims that the American AEC doctors who had come to help were more interested in studying them than treating them, as was the case in the Atomic Bomb Casualty Commission.[27]

Concern about fisheries was the clearest reason that nuclear tests at sea—and certainly waste disposal at sea—would necessitate a much fuller understanding of oceanography. American and Japanese scientists alike knew that the difficulty, despite the AEC's assurances, was that the extent of danger to humans or fish was just as unknown as the circulation and mixing patterns of the ocean, and oceanographers were needed to assess such problems. Japan was more responsive to this need than was the United States. After the Bravo shot and subsequent fishing boat scandal, it planned an expedition and sent a ship, the *Sinkatsu Maru*, into and around the blast area to take water samples. These samples, four to six weeks after the event, were still highly radioactive, with the activity running deep into the water. The Japanese study, soon described in a paper by chemist Yasuo Miyake, observed that the radioactive area of the ocean drifted about a thousand miles in two months, contaminating a large area. Miyake's study was one of the major ones on hand when the BEAR committees were formed.[28]

After the *Fukuryu Maru* incident and the *Sinkatsu Maru* scientific cruise, Miyake became a leading figure on the Japanese side on the controversy over nuclear fallout. He worked for the Japanese government at the Meteorological Research Institute, leading one of its units, the Geochemical Laboratory. The institute published his analysis of the fallout from the American test in its journal, *Papers in Meteorology and Geophysics*.[29] Although Miyake had no reason to be skeptical of radioactive waste, the politically charged issue of fallout gave his work on ocean contamination a wide audience. The distribution of radioactivity in the ocean was important for both issues. After the Bravo shot, the laboratory put up atmospheric monitoring stations in several locations in Japan, and the laboratory reoriented much of its work toward problems of environmental radioactivity.[30] Unlike the Americans, Miyake had little incentive to look on the bright side of nuclear testing and play down the dangers, as the vast majority of harm from all things nuclear had fallen on the Japanese.

American oceanographers did not dismiss Japanese work out of hand. Doing so would have undermined years of work cultivating goodwill between American and Japanese marine scientists. Revelle already had been grooming the Japanese oceanographic community as a partner in international collaboration and it would have been senseless to demean Miyake's work. Revelle had fostered professional relationships with colleagues at the Japanese Hydrographic Office, Japanese scientists had welcomed the 1953 American TRANSPAC expedition to its ports, and a large-scale synoptic data survey of the North Pacific (NORPAC) was undertaken by Japan, the United States, and Canada in 1955. In fact, part of NORPAC's scientific plan was to study the radioactivity in the water. American oceanographers from Scripps even had met with Emperor Hirohito, who had a passion for sea slugs and had conducted marine biological research of his own.[31]

In the summer of that year Revelle and Gordon Lill, the head of the Office of Naval Research's (ONR) geophysics branch, went to Japan to visit institutions and coordinate further studies. Though not everyone was glad to see them—Revelle described the president of Tokyo University as "a very cold fish"—most of the visit was marked by signs of appreciation and possible cooperation. They came bearing gifts and one evening distributed fountain pens, bowls, cloth, and other presents to their Japanese guests, including Miyake and his wife.[32] In Japan, Revelle's taste for connecting science to larger problems came out in full. During a talk on science and natural resources, he did not bother to begin with technical details about minerals and fish. Instead he asked, "What is the future of the human race?" The prospects of peace, he said, opened up the possibility of thinking in long-range terms, and it was within this larger question that he wanted to situate the ocean sciences. He would have found it difficult, and there is no evidence that he bothered to try, to ignore Japan's fixation with the global problem of radioactive contamination. Further, Revelle and other oceanographers at Scripps were committed to reinforcing the reputations of Japanese colleagues and potential collaborators, not to dismissing them.[33]

Nevertheless, Revelle did attempt to persuade the Japanese that nuclear power was the technology of the future. In his speech in Japan, Revelle argued that the material needs of the world could not be met with the resources currently available. If everyone in the world lived at the standards of the people of the United States, it would take twenty-five to fifty years to deplete the world's resources, a situation exacerbated by the population explosion. All of the area of the world, he warned the Japanese, would one day be one large city like Tokyo. Without the development of new resources, he pointed out, "the future of mankind is very black," with death by malnutrition. Revelle reminded his listeners of the lessons of Thomas Malthus, who described populations outpacing the food supply—man would be essentially a wild animal, and in the long run his fate would be similar to the beasts that competed for resources. "Will our children's children look forward only to a slow decline into misery and fear?" There was some hope, Revelle pointed out. There were many substitutes for natural resources available, from advances in metallurgy, chemistry, and "above all in nuclear physics."[34]

With nuclear physics, humanity could avoid its Malthusian fate by bombarding abundant materials with neutrons, thus creating new substances for old purposes. Revelle dreamt of an age of using vast amounts of power to gather dispersed elements and concentrate them for future use. He believed that it was possible to recycle most natural resources. Natural resources do not disappear, he told the Japanese; they simply become dispersed. Every rock contained trace amounts of all the necessary resources—aluminum, iron, copper, manganese, zinc, lead, and other metals. But they needed energy to concentrate these resources to make them useful. "The problem of resources," Revelle wrote in his speech notes, "thus is essentially a power problem."[35]

Revelle had high expectations for modern technology. His visions of the future were not just for Japanese ears. He would repeat similar arguments in his 1957 charter address in California, noting that by the year 2000, "it should be possible to step into a rocket plane, after you have had a comfortable breakfast at home, and arrive anywhere in the world before lunch." As for natural resources, the perennial quest for mineral deposits no longer would be necessary. "The air and the sea and the ordinary rock that you see in your back yard will furnish all the necessary primary substances for industrial civilization." He predicted regular excursion trips to the moon, scientific plans for exploring Mars, and the disappearance of dishwashing, laundry, and housecleaning—"the housewife and the manual laborer will have become electronic technicians." All of this would be possible because of the vast amount of power from the atom. Crucial substances like oil and coal would become simply organic chemicals of interest to scientists, while "power from the fission or fusion of atoms will be ten times more abundant than all the electrical, mechanical and heating power we use today."[36]

Thus the chairman of the panel on oceanography and fisheries for the BEAR committees was a man whose vision of the future was nuclear. There was no hint of skepticism about its role in the world, and indeed he seemed to believe that without nuclear power, humanity would face a serious crisis of natural resources. Moreover, he believed that oceanographers could play an important role in facilitating that process—a role that also would benefit oceanographers by opening avenues for patronage, influence, and political power.

In Search of the Normal Ocean

Miyake and his colleagues had made significant claims about the change in radioactivity from normal levels, due to natural sources such as minerals and cosmic rays, to elevated levels, due to the spread of radioactivity from American nuclear tests. The work of Japanese scientists on the dangers of ocean contamination had to be treated with care. In part, this meant ensuring that the techniques used by the Americans and the Japanese were consistent with each other, to allow easy comparison. Revelle had discussed the measuring techniques with Miyake while in Japan. When he returned home, he mobilized scientists at Scripps to start planning a follow-up to Miyake's study. He also had one of his colleagues phone Richard Fleming at the University of Washington, giving him the precise details of container sizes and sampling techniques that Miyake had used, so that Fleming could attempt similar work on the NORPAC expedition, subjecting the Japanese work to scrutiny.[37] In the subsequent EQUAPAC expedition in 1956, Scripps oceanographers would coordinate closely with Miyake. They agreed to take stations two degrees in latitude apart, collecting five-liter water samples at the surface and two-liter samples at five hundred feet; they also collected plankton at the surface.[38] Both the Japanese and Americans recognized the importance of ensuring that their work remained credible, because political forces in both countries would pay careful attention to the results.[39]

Before these closely calibrated expeditions materialized, American oceanographers used the initial Japanese study to plan Operation Troll, a joint project between the AEC, the Scripps Institution of Oceanography, and the University of Washington. Designed as a follow-up to the *Sinkatsu Maru* cruise, the American expedition started in March 1955 at the Marshall Islands and then worked its way westward, to judge what had happened to the radioactivity in the water over the past year. Using a vessel ill-equipped for the task of fishing, which they needed to do for adequate sampling, they took water samples, plankton hauls, and any fish they could acquire, from the Marshall Islands to the Philippines, then north to the Japanese home islands. They found radioactivity everywhere they sampled. Unfortunately, the expedition used crude methods that did not allow as precise an idea of radioactivity distribution as Miyake's work. One of the important findings, however, was that there appeared to be uniform distribution of radioactivity to a depth of about six hundred meters. Why it was mixed, they did not know: it could have been physical mixing or it could have been biological (i.e., due to plankton and other fauna). This appeared to be a good sign that the radioactivity would be dispersed, but the mixing dropped off considerably to the west and thus few firm conclusions could be drawn.

The frank discussions between American oceanographers during the BEAR meetings revealed that the techniques used in Operation Troll were crude to the point of embarrassment. To give some idea of the radioactivity measured in Operation Troll: the oceanographers determined that their samples of water emitted about twenty disintegrations per minute per liter of water on average, compared to eight thousand disintegrations per minute per liter a year before in the Japanese study. Thus the radioactivity appeared to have cooled considerably. But these estimates were not very reliable—the data was only accurate to plus or minus twenty-five disintegrations due to an ineffective counting technique. Warren Wooster, who took part in the expedition, described the method, putting it lightly, as "not very good." Nevertheless, they determined that there was still some high activity, off Minanao, near the Philippines, at about sixty disintegrations. During the BEAR meetings, when Wooster was asked what a normal amount of radioactivity might be, he could not give a definitive response: "We didn't look at any normal sea water because all the water we looked at was not normal any more."[40]

Oceanographers lamented the lack of what scientists call baseline data, which records the figures for all of the variables prior to an experiment. Baseline data is needed to allow comparison to judge the effects of an experiment. In the case of ocean contamination, without baseline data there was no way to compare the real level of natural "background" radioactivity to the levels created from weapons tests. Despite this, those analyzing the data, such as Allyn Vine, were impressed by the mixing, determining that "it looked like over a tremendous area things were getting down pretty close to background. The mixing process had been pretty effective." When pressed by Milner Schaefer about what could be considered normal background radioactivity, in order to understand what the figures actually meant, Wooster admitted: "I really don't know. The division of effort was that the AEC did all the counting. I just handed them samples. They have been revising their figures."[41] Thus, in addition to crude methods, the oceanographers relied on AEC data, which itself appeared subject to a certain degree of fluidity.

American oceanographers had another opportunity to study the effects of a nuclear blast on the sea a few months after Operation Troll. On 14 March 1955, five hundred miles southwest of San Diego, California, the United States detonated its first deepwater nuclear device. Compared to the Bravo test, it was relatively small, only thirty kilotons. Operation Wigwam, as it was called, was a long-anticipated replacement of a canceled test during Operation Crossroads in 1946. The purpose of the test was primarily a military one, pushed by the Joint Chiefs of Staff, to see the effects of a deep blast (some 2,000 feet in depth) on the operations of naval forces. The dominance of the Navy in the project provided oceanographers with a more integrated role. Scripps scientists spearheaded the part of the experiment that was intended to study the effects of the blast on marine biology. Wigwam was being implemented by the Navy and oceanographers, and it was not the AEC's brain child. In fact, the AEC tried to stop it. It was afraid of the potential public relations problems, because the blast would have an unknown effect on marine ecosystems and food fish. When there were other tests that the AEC considered more important, AEC scientist John Bugher reasoned, why court trouble needlessly by blasting underwater? Eventually the operation was approved, but only after the AEC, Department of Defense, and the Food and Drug Administration worked together to devise a public relations plan and a radioactive monitoring program, just in case there was a repeat of the international censure following the *Fukuryu Maru* incident.[42]

Although Wigwam's purpose was to ascertain the effects on ships, particularly submarines, oceanographers used it to analyze the spread of radioactivity. One of the principal findings was that the area of contamination appeared to double every twenty-four hours. The speed, however, was largely due to the momentum of the blast, so there was not much way to tell from this sort of test what the spreading rate would be from either radioactive waste or fallout. The spread was in three dimensions, and on successive days it could move in different directions. Revelle interpreted this as a kind of sloshing effect, much like using one's mouth to blow water against the side of a bathtub; when one stops blowing, the water sloshes back. The Wigwam tests gave some indication about mixing at the surface and at intermediate depth, but almost nothing about mixing at the bottom of the ocean. The latter, particularly the rate of exchange between the deep water and surface water, was one of the principal pieces of knowledge needed to understand the problem of disposal. Thus by the time of the BEAR study, oceanographers knew few facts with certainty; they were, however, coming to grips with what they would like to know.[43]

Even as oceanographers agreed that more needed to be known, they were beginning to disagree about the meaning of ocean disposal. When the BEAR oceanographers met at Princeton in the early days of March 1956, Richard Fleming brought to the BEAR meetings a profound distaste for the word "disposal." This was likely due to his experience with the NCRP, when scientists had to recognize that disposal often meant simply moving the radioactive material into pipes, sewer systems, and other sites meant to transport the waste. The term "disposal" seemed illusory, because really they were just introducing the material somewhere else. In this case, the waste was being introduced into the ocean. Fleming recommended that oceanographers abandon the word "disposal" altogether. To Fleming, this was not just a matter of semantics. "If you talk to a radiologist in a hospital," he said, "when he flushes radioiodine down the toilet and he has disposed of it, this is a final act, but here he is just introducing it into the environment." For Fleming, the whole idea of getting rid of wastes was false. From the environmental point of view, he argued, "there is no such thing as disposal."[44] Another BEAR participant, Douglas Whitaker, countered that there were some plans to put radioactive waste in salt beds at great depths, to keep it away from people for years to come, long enough to let it decay and "take care of itself." Was that not disposal? Fleming said it was not: "This is isolation."[45]

The minutes of the Princeton meeting reveal an apparently frustrated Fleming trying, with limited success, to change oceanographers' understanding of their role in waste disposal policy. Fleming's argument was that they were starting in the wrong place: they were identifying a problem (dangerous waste), and trying to get rid of it (by dumping it into the sea). Instead, they should start with the environment. One must understand the ocean first to comprehend the effects of introducing new substances into it. This way, "it doesn't matter where this material comes from—whether it comes from military action, bomb tests, or reactor waste or factory smoke or anything else." Bostwick Ketchum backed him up, noting that once they recognized that the real need was in understanding general oceanic processes, it would be easier to determine the effects of any quantity from whatever source.[46]

Roger Revelle did not appear to understand Fleming's point, saying that such ideas seemed completely vague to him. After all, they were there to talk about radioactive waste disposal, and they had to make recommendations. An argument ensued in which Revelle thought it was important to study the effects of specific radioactive substances, Fleming objected to the notion of even including radioactivity in their discussions at all, and Milner Schaefer was frustrated that they seemed to be arguing about nothing. Of course they should consider oceanic processes, Schaefer said, but the committee was there specifically to address the effects of radiation. But Fleming pressed his point, insisting that there was a substantive difference between his and Revelle's outlook. It did not matter who put radioactive material into the ocean, or where or how it was done. They could debate all day and arrive at certain limits to recommend to the government, but the introduction of such materials into the environment was essentially beyond their control. There was radioactive material in the ocean already, and that could not be helped. There was no way to establish a "normal" condition of the ocean from which to base recommendations, because it was already filled with waste and natural radioactivity. Any attempt to assess acceptable annual limits on waste material based on change from the norm seemed meaningless.[47]

But Fleming's outlook flew in the face of Revelle's. For Revelle, the oceanographers were being asked to make a recommendation—a policy recommendation that reflected their special knowledge, based on their understanding of what was normal for the ocean and what it could absorb without harm. It was the political aspect of their role that appealed to Revelle; they were being asked to set a limit, and they were poised to act as an authoritative voice on the subject. Why abandon this power by acknowledging their impotence, as Fleming proposed, and avoid making specific recommendations about this specific political problem?

Revelle, despite periodic objections by Fleming, moved on to assess the specific effects of bomb tests and nuclear power. Within a hundred years, he said, the full-scale development of atomic energy would present the same risk as exploding a 100,000 megaton nuclear bomb each year. Whatever environmental approach they took, this was going to be a problem. Their task was to talk about numbers—specifics, not generalities.[48] To aid in this, Ketchum outlined three problems relevant to disposal upon which oceanographers disagreed. One was the overturn of the ocean; another was the rate of plankton productivity; the last was the effect of radiation on ecological cycles. The overturn of the ocean was fundamental because it would help to determine how fast radioactive material would be mixed, and where—if at all—mixing was prevented for long periods. Revelle did not see the relevance of plankton production rates, but Ketchum reminded him that plankton studies would be the key to understanding the concentration of radioactivity in biological material; this concentration would be in a growing, not static, population. As for the ecologic cycle, it was not clear what the effects of killing individual plankton would have on the fish population that feeds on plankton. David Carritt also added a fourth problem to address, that of transient stages. Even should one accept that radioactive waste would be dispersed by the ocean, the process would not occur overnight. As Revelle truncated the problem, even if you expect to "spread this stuff all over the ocean, in the intermediate stage you may very well concentrate in the Chesapeake Bay and ruin the crabs."[49]

With these specific problems identified, oceanographers consciously moved beyond their assigned roles as objective experts and paused to consider the political ramifications. The meeting turned to international issues. Revelle recalled his experience in Geneva at the international scientific conference on the peaceful aspects of atomic energy. The impression Revelle received there was that most of those involved were concerned primarily with cost. "They say you cannot spend more than 8 cents a gallon on the disposal of radioactive waste. In my opinion this is absolutely nonsense. You can pay what it costs if you have enough of it." In Revelle's view, there were all kinds of ways to get rid of the wastes, and the same solution did not have to apply to all situations. The solution would simply depend on magnitude: "You can put them in the ground. You can put them in pellets and disperse them in the state of Nevada. When you are dealing with geochemically significant amounts, then you have to do something else."[50]

Revelle was convinced that oceanographers could help government and industry establish rules for disposal. He felt that basing them on magnitude was crucial, especially given the talks at Geneva. In small amounts, future conventions probably would require countries simply to ensure sufficient dispersal; but if the amounts rose, more stringent regulations would be needed. It seemed very clear that there were going to be international conventions governing the disposal of wastes, particularly if the world bought into the idea of cheap nuclear energy solving their economic problems. "We just can't afford to have Indonesians or the Malayans and all sorts of other people all over the world indiscriminately dumping things into the ocean if in the long run this is going to end up with the ocean being like Lake Geneva is now—a mess." Again Fleming interjected, claiming that the cumulative effects of pollution, rather than catastrophic effects, were the hardest to correct. "In the case of these fission products," he lamented, "it is hopeless."[51]

The others came to less dire conclusions and insisted that it was their task to define a safe level. Vine reminded the group that they needed to figure out if a real future problem existed or not, and whether or not they were going to do something about it. For the foreseeable future, lots of installations—military and peaceful—were going to be producing radioactive materials, and they would face the problem of waste disposal. Again Fleming objected: terminology was important in seeing where the genuine problems lay.

> DR. FLEMING: I would prefer the term "introduction" rather than disposal.
> DR. VINE: I think you have a very good point there.
> DR. [HARMON] CRAIG: It is essentially a garbage problem.
> DR. FLEMING: We are on the receiving end of this.
> DR. VINE: I think that is a very fine terminology to carry throughout.[52]

Fleming hoped to persuade the others to move away from the perception that they were helping the atomic industry take out its garbage, and instead to evaluate the effects upon the ocean of introducing into it large amounts of radioactive materials. They were oceanographers, after all, and they ought to see themselves as receiving, rather than disposing. For Fleming, calling the process "disposal" or the materials "wastes" or "garbage" simply reinforced the wrongheaded view.

Although Fleming seemed to have an effect on the group, most of them realized that they had been called together for a specific purpose, and that was to assess the biological threat posed by atomic radiation. Like the BEAR genetics panel and its dilemma about thresholds, they knew that if they did not set a figure, someone else would do so. Even if they could not set a truly definitive and scientifically justifiable capacity level, not setting one might have undermined their role altogether. Because of the high level of interest in the BEAR committees, from the press, from the academy, and from the Atomic Energy Commission, their work was likely to have a profound effect on policy decisions. So the group continued to meet over the next couple of days, attempting to outline a plan of action for assessing the oceanographic aspects of the BEAR problem, with a view toward making specific policy recommendations.

The crucial point for arriving at any conclusions about waste disposal was the action of deep water. Did it stay in place for relatively long periods, undisturbed, or was it periodically flushed out and replaced? There were a number of ways to approach the problem: carbon dioxide exchange between ocean and atmosphere; heat flow measurements; analyzing carbon-14 in deep water; and measurement of deep current velocities. All of this work had indicated that the deep ocean had been there for less than a thousand years—the problem was to know precisely how much less. Also, recent work on the deep Atlantic by Valentine Worthington at Woods Hole analyzing decreases in oxygen content over the past decades suggested that the deep water either was moving rather fast or that it was formed as little as 150 years previously. Unfortunately, Revelle observed, "if Worthington is right, and the deep water actually is flushed out occasionally, this may happen at any time or it may not happen for a very long time." This could happen hundreds or thousands of years in the future, or it could happen next year. Worthington's data, however, was just on the Atlantic. "We have no dope in the Pacific," Revelle conceded. What little "dope" there was indicated that, in the Pacific, oxygen levels were increasing, which might mean that the water in one ocean was being replaced steadily by the water from another.[53]

The suggestion that there might be a long-term, oscillatory phenomenon—in other words, "there must be some flushing around," as Schaefer put it—made the prospect of long-term, isolated disposal at the bottom of the sea seem questionable. Alternatively, it might be a good thing from the point of view of mixing, because the radioactivity would be diluted before it reached the surface layers of the ocean. It might, according to Vine, be wise to put this material in the areas where there were deep currents, to maximize the amount of water flowing past them each day. The only areas that seemed feasible for long-term isolation, with no mixing, were the deep ocean trenches, where water might well be trapped for centuries. But was there any evidence that the water in the trenches was really that old? Age—based on carbon-14, heat flow, oxygen content, or whatever method—was still only vaguely known, with no certainty. And in addition, there were no Pacific measurements at all.[54]

The problem of isolation, Fleming pointed out, was exacerbated by the fact that the mere act of dumping would disturb the stability of the environment. When introducing drums of radioactive waste to deep areas of the ocean, "you are creating an unstable situation initially," which will increase with the spread of radioactivity—"in other words, you are adding energy all the time." Heavier water would be added because of the radiation, and it would sink, pushing other molecules out of the way.[55] All of these ideas threatened a basic hope within the atomic energy establishment—that there were isolated places in the deep ocean where high-level wastes could lie undisturbed for centuries.

The Happy Exploitation of the Oceans

The BEAR oceanographers mused about the wider problem throughout the world. Douglas Whitaker observed that the president's plan, based on his Atoms for Peace initiative, was to offer uranium to the world for use in power reactors. It also meant that atomic waste might become an especially American problem. The reason was that the United States was planning, Whitaker said, to take back the spent fuel to be reprocessed, "to bring these things back home and do the dirty work here." Particularly in the "backward" countries, he said, where they simply do not have the technology or know-how to reprocess the materials themselves, "we may get quite a slug of their problem, out of proportion even to our own consumption." It was here that Schaefer added, "We take the garbage contract for everybody."[56]

There were certainly good political reasons for taking such a contract. This was American technology, and it was in Americans' interest to ensure that it was done properly on firm scientific grounds. Thus it might have been wise to find ways of asserting control over waste disposal, to avoid having other countries, perhaps not as attuned to the biological effects, ruin the practice for all. This was Revelle's view. He put in his own reservations about island nations such as Japan and Britain, both destined to dump a lot of waste at sea. Japan had no depleted oil wells or waste land at all, yet they had a great need for atomic energy. "They are completely surrounded by the ocean. They are kind of irresponsible about these things." He and the others appeared to agree that the rest of the world, whatever the United States decided, was probably going to see the ocean as an inviting place to dump radioactive waste. And if the

Americans were promising to help get rid of the wastes, then other countries were unlikely to see waste disposal as their problem. Americans, Vine said, appeared to be the only ones who were taking an interest in the "introduction problem"—using Fleming's preferred terminology. Most countries were devoting their energies primarily to production, without much thought about the by-products. Echoing Whitaker's concerns about the president's plan, he said: "The 'Atoms for Peace' had better include a good resting place for the atoms when we get through with them."[57]

The more they talked about it, the more indignant some of the oceanographers became about the whole concept of Atoms for Peace, because of its inattention to the possible global consequences. In perusing a recent report of the Panel on the Impact of Peaceful Uses of Atomic Energy, Vine observed that nowhere in the table of contents was there a section dealing with disposal. "I can't believe this is right," he said. It should have been included in considerations at all levels—research, ecology, energy development, and in the overall national economy. As it stood, the thinking about nuclear energy production was easily fifteen years ahead of the thinking about disposal. Revelle joked, "My only desire to comment on this is that I hate to work on high priority problems. I am glad it is a low priority problem."[58]

Certainly Revelle's comment reflected his preference for highlighting opportunities rather than problems. But it also masked his deep interest in drawing attention to the importance of such work. Although the committee had a very specific purpose, to assess biological effects at sea, the oceanographers discussed an array of other matters, not the least of which was the need to promote themselves as an integral part of the atomic energy establishment. They discussed at length the role of oceanographers in past nuclear tests, and the need to include more science-oriented thinking in future tests. Such tests were scientific tools, especially useful in oceanography—there might come a time, they envisioned, when there would be no military justification for a nuclear test at all. In that event they would still continue tests for purely scientific purposes. Harmon Craig suggested they draft a proposal asking that the AEC and the military cede authority over planning nuclear tests to a body like the National Academy of Sciences. This would ensure not only efficient scientific planning, getting the most possible knowledge out of each test, but it would probably be good for international relations as well, giving other nations confidence that the scientists were in charge. Craig said, "I feel that bomb tests are actually too important a thing to entrust to the military and AEC." Revelle joked, "It is kind of like war." Not that any of them expected the proposal to achieve anything, or that scientists would soon take over the tests; rather, they hoped to plant the suggestion that science should have a higher priority. "This is the way you get military people and government commissions to do things," Craig reminded the group. "You propose that somebody else take over their job, and they start doing the job a lot better."[59]

These discussions reveal that there was no hint of the kind of political opposition to nuclear testing that frames historical understanding of some scientists' roles in other disciplines. Oceanographers were more concerned about asserting their role in nuclear tests than critiquing the tests. The prevention of radioactive materials being deposited at sea was the furthest from their minds. What they wanted was more knowledge, more planning, and more consideration of the biological and geophysical effects at all levels. Clearly American oceanographers had spent a great deal of time preparing for Pacific tests, ensuring enough knowledge of the environment beforehand to make postdetonation assessments useful. The same, they argued, ought to be done about waste disposal.[60]

They also bantered about the undue dominance of physicists in such matters. If the physicists could be pushed aside, they would know a lot more about not only the biological effects, but also geophysical ones and other environmental effects of enormous importance for earth science. If the military wanted to set off a bomb, why must they only turn to nuclear physicists? "Electricians don't control the lighting authority in the United States," Craig said, "because they put together the wires and make the lights work." Revelle added that physicists were connected with bomb work because they were interested in learning about the nucleus, and "it is very difficult to get them to do anything which didn't apply to that objective." Vine chimed in: "They owe it both to the country as a whole, and also to the taxpayers, that these [tests] be regarded as the sources for one kind of controlled experiment." Rather than think of detonations as weapons tests, they ought to be regarded as scientific experiments on the environment.[61]

After meeting separately on more focused subjects over the next weeks, Revelle's group drafted a report for the BEAR committees that was cautiously optimistic about the effects of radiation on the sea. In general, the report insisted that any introduction (Fleming's terminology was retained) of radioactive waste at sea should be based on and followed up by intensive, well-funded, and independent research efforts. No blanket assurances or warnings could be made, as there simply was not enough knowledge about the sea to be sure. Of the three ways to introduce radioactive material—bomb tests, radioactive waste, and radioactive tracer studies—the report was unsurprisingly most enthusiastic about the third, devoting considerable space to the benefits of controlled, scientific study. But the report also was strongly cautious, emphasizing the need to avoid the past mistakes of industrial pollution, because the effects of radioactive pollution would be irreversible. One must not repeat the past, when "short-range solutions, based on inadequate knowledge, special interest, and what we now know was a fatuous confidence in the capacity of the atmosphere and the waters to absorb noxious substances, were employed." But following this caution against fatuous confidence was the following statement about radioactive material: "There is no question of trying to keep all of this material out of the sea. It is certain that some of it can be safely added."[62]

The BEAR oceanographers' report attempted to respond to the most contentious questions. Yes, there was natural radioactivity in the sea, at much lower amounts than was found on land—so in general, organisms on land (such as humans) received a greater dose of radiation from nature than their counterparts at sea. Yes, weapons tests had altered the level of radioactivity considerably. Taken as an average of the total ocean, they felt this was negligible, but in local areas it could be immense, especially in the short-term. Yes, both the United States and Britain already were adding waste to the oceans, through pipelines, in containers, and even river discharge. Had marine life been seriously damaged by the United States atomic energy program? "Probably no." Again, there had been an enormous amount of local damage, but on the population scale, there was little evidence of serious harm. Figuring out what could be considered high levels and low levels was a principal reason to fund more research. On genetics, the oceanographers toed the AEC line. The report stated that genetic damage by even low levels was not measurable at present, and it made a bizarre assertion about genetic effects: "but there is no reason to believe that this influence will be an undesirable one." Yes, living things took up radioactive materials into their bodies, so physical and chemical dilution of radioactive wastes was only part of the solution. And no, not all of these materials were harmful in the same way, because different isotopes were taken up by different parts of the body. Radioactive strontium, for example, was dangerous to humans because it concentrated in the bones.[63]

The report also tried to predict the future, make specific recommendations, and set the record straight for the public. Nuclear power would expand, to be sure, and within fifty years the radioactivity in waste products would reach eight times the natural radioactivity of the oceans. The sensible place to disperse wastes, rather than try to isolate them, would be the seas, and there might be places in the ocean suitable for isolation as well. The Black Sea, for instance, seemed to be a place where the bottom water had been sitting undisturbed for two and a half millennia. Deep-water isolation, however, needed to be studied further before any conclusions could be reached. No, it would not be safe to introduce wastes indiscriminately into the ocean, especially in coastal areas or even the upper layers of the ocean far out at sea. "There is no place in the sea where large amounts of radioactive materials can be introduced into the surface waters without the probability of their eventually appearing in another region where human activities might be endangered."[64] This did not mean that the ocean could not be used, but rather that intensive studies needed to be made to ensure that the right materials were introduced in the right places, at the right depths.

Revelle's report also echoed the typical treatment of the lay public as irrational actors in making policy decisions related to nuclear matters: "Ignorance and emotionalism characterize much of the discussion of the effects of large amounts of radioactivity on the oceans and the fisheries. Our present knowledge should be sufficient to dispel much of the over-confidence on the one hand and much of the fear on the other that have characterized discussion both within the Government and among the general public." It might help matters, the report suggested, if the secrecy around all atomic matters was relaxed and the public better educated about the issues.[65]

Revelle's panel made an important distinction about audiences. Although the overall BEAR reports tried to avoid the semblance of policy making, insisting instead that the study was essentially an objective scientific assessment, the oceanographers under Revelle took an entirely different approach. The summary report, "Oceanography, Fisheries, and Atomic Radiation," appeared in the journal *Science* in July 1956, explicitly addressing research administrators, statesmen, scientists, and the public—listed in that order. Certain scientific questions needed answering, the report said, and thus policy-minded patrons of science ought to listen up. The BEAR report acted as a roadmap for future patronage. There were also recommendations about the expected national and international agreements needed for "the happy exploitation of the oceans in the new atomic age." For scientists, there was a technical summery of present and potential effects. But for the public he made a further distinction. The report summarized the levels of calculated risk, which had to be balanced against the "wonderful promise" of atomic energy for the general welfare of humanity.[66]

For the skeptics of atomic energy, and for those who questioned the wisdom of dumping waste at sea, the BEAR oceanographers appeared to have little sympathy. They seemed willing to characterize critics as ignorant and emotional and looked forward to the exploitation of the seas for the benefit of mankind. There were uncertainties, to be sure, but unproven dangers should not close down the whole peaceful atomic energy initiative. Revelle's attitude was encapsulated in the same 1957 speech in which he described the relationship between science and politics. He closed his speech by recommending how conscientious scientists should act. Politicians set the courses of action, and the scientists' responsibility was to follow their lead, or to persuade the politicians to follow a different course. The commands were set for each person to follow. Quoting Socrates, Revelle said, "we do not rudely impose them, but give him the alternative of obeying or convincing us." This outlook fits the American oceanographic community in the mid-1950s. Rather than oppose dumping of radioactive wastes, they were ready to serve the political ambitions of Atoms for Peace. On one hand, the dangers were hypothetical, giving no reason to oppose the use of the sea for radioactive waste. On the other hand, the uncertainties worked to their advantage, providing them with a voice of authority and justification for patronage.

Chapter 4 Forging an International Consensus

ALTHOUGH THERE WERE SOME tough questions at the press conference about the BEAR committees on exposure from fallout and waste disposal, academy president Bronk and the six committee chairmen conveyed the unified message that no harm had been done yet. But there were some bumpy spots, as when one reporter asked why there were no political recommendations. Bronk had explained that the National Academy of Sciences was not charged with that responsibility. This was, after all, simply a scientific evaluation. As noted in previous chapters, both Roger Revelle and Warren Weaver understood quite well the implication. What were the policy recommendations? Did they challenge the AEC's? What was the government going to do about them? They referred the reporter to the parts of the document with specific figures, but Bronk firmly interrupted to point out, "I prefer not to refer to them as political."[1]

The subject in question was of course nuclear fallout, and the chairmen jointly played down any harm that might have been done. Pathology panel chairman Shields Warren pointed out that the tests themselves were done safely: "There is absolutely no effect on the individuals at the tests over and above those that are discussed here [on the population as a whole]." There were some exceptions to that, including accidental exposure in 1954 to the inhabitants of the Marshall Islands and some Americans near the test site, and another instance when some Americans were handling radioactive materials improperly and received burns on their hands. "Those are the only two exceptions. There is absolutely no risk in the witnessing of a test or taking part in the general preparations." Later in the press conference Warren interrupted to correct himself: of course the Japanese fishing boat incident also had to be considered an exception to the rule of absolute safety. But the Japanese had disregarded American instructions, he pointed out.[2]

Despite his detached stance, Bronk was well aware of the political, and even diplomatic, significance of his academy's study. Not only had he acted to ensure that the American and British reports did not explicitly contradict each other, but he had also worked with the State Department to help it integrate the report into foreign policy. George C. Spiegel, of the Department of State's Office of the Special Assistant for Atomic Energy, had kept in touch with the academy over the past year. Spiegel did his best to put pressure on the Academy to arrive at conclusions sooner rather than later, because already international discussions were taking place and resolutions were being made about radiation, under the auspices of various organizations. If the United States hoped to influence such discussions, he believed an authoritative statement would be needed soon. In 1955, he told Bronk that he did not want the academy's inaction to put the American delegation to the United Nations General Assembly in an embarrassing position during the debates about radiation. But Bronk was not to be rushed: he told Spiegel that a hastily prepared report, specifically put out in time for the general assembly, would look improper to everyone. According to Spiegel's memorandum of conversation, Bronk observed that such reports could be justly accused of being based on inadequate evaluation of all available information, and further "for appearing to be a 'rubber stamping' of statements that had been made by others."[3]

Bronk's aversion to being seen as rubber-stamping the AEC's views reflected his belief that the academy ought to produce disinterested, nonpolitical scientific advice. Bronk was not naive; he knew that the academy was steeped deeply in government influence, yet he was keen to protect the independent character of the academy's findings. In the BEAR study he had ensured the incorporation of the widest possible variety of opinions to protect it from the criticism of being an AEC whitewash. Bronk warned Spiegel that, especially in the area of genetics, "the final NAS appraisal may well include minority findings." And as for the timing, he said, the State Department would have to wait just like everyone else. Some of the genetics results may not be available for generations, and certainly not in time for the American delegation's use at the general assembly.[4]

Political and diplomatic realities could not be put off forever, however, and the State Department was right to point out the urgency of the matter. American diplomats craved a definitive statement representing the consensus of the country's best scientists. Between 1955 and 1957, international delegates from both sides of the cold war met to discuss the implications of the peaceful uses of atomic energy, and the United Nations conducted its own study of the biological effects of radiation. A new organization was born, the International Atomic Energy Agency, which took on the responsibilities of speaking authoritatively about radiation. After the publications of the BEAR reports, the Americans were not the only ones wanting to achieve consensus. Particularly on the subject of waste disposal at sea, the British and Americans, who thus far had worked together to align their national reports, attempted even closer collaboration to avoid making inconsistent statements in these international forums. American oceanographers had succeeded in gaining a foothold in the American atomic world, emphasizing the opportunities and uncertainties of waste disposal and arguing for increased patronage. Few were more disturbed by this tactic than the scientists at the British Atomic Energy Authority, who relied on the sea as a receptacle for radioactive discharges from the chemical reprocessing facility at Windscale, and who dumped packaged wastes into the English Channel. They were not content to argue such issues out in international forums, and they met directly with the American oceanographers to achieve an Anglo-American consensus before ever bringing issues to the international community. The crucial negotiations occurred between scientists inside and outside atomic energy establishments. American and British scientists and officials met informally to hash out the broad outlines of what they, collectively, would argue in any international forum. They agreed that waste disposal at sea was permissible despite the lack of knowledge, and when international bodies requested scientific assessments from them, they responded instead with policy recommendations.

The "Somewhat Out of Channels" Approach

When the BEAR report was released, the United States government finally could rely on an authoritative scientific document that carried the blessing of many of the AEC's harshest domestic critics. Likewise, Britain's Medical Research Council (MRC) reinforced assumptions being made in the Atomic Energy Authority. In the international arena, diplomats were eager finally to integrate these findings into foreign policy. A month after the report's release, State Department official Gerard Smith wrote to Bronk wondering what exactly the calls for international action in the report should entail. After some informal discussions between the academy and the State Department, Bronk formally responded several months later. The international problems, he felt, were largely related to any activities by the United States or other nations that could affect the biosphere (i.e., the parts of the earth that support life), which was shared by all. Specifically, he meant nuclear tests, experiments, and the disposal of radioactive waste.[5]

Bronk identified one of the most immediate of these problems as waste disposal. International cooperation was urgently needed to study, and then to utilize, the oceans of the world "as a reservoir for radioactive wastes." The reason for making this an international endeavor was not to develop means of limiting disposal, but rather (as BEAR oceanographers pointed out) because scientists did not wish to spoil the ocean as an experimental laboratory. In other words, once nations began to dump waste indiscriminately at sea, it would be useless to conduct experiments using radioactive tracers. Anglo-American discussions, Bronk argued, must precede diplomacy. Such discussions on an international basis "in our opinion must underlie and precede conventions on the disposal and dispersal of radioactivity in the sea and the air." For the moment, sufficient discussion had not occurred, and there was no need to rush into diplomatic negotiations.[6]

The director of Britain's Atomic Energy Research Establishment (AERE), Sir John Cockcroft, felt differently. Despite general agreement between the United States' and the United Kingdom's reports in 1956, Cockcroft disliked how American oceanographers characterized British waste disposal practices, and he wanted Anglo-American scientific discussions to occur as early as possible, prior to any United Nations meetings on the subject. While most rejoiced that the two reports meshed so easily on the contentious issue of fallout, Cockcroft bristled at what he viewed as an implication by the Americans that the British were harming the environment. The Americans had noted that low-level radioactive waste was discharged routinely into the sea, particularly by Britain. Oceanographer Roger Revelle had said so during the press conference after the American report was issued, and it deeply perturbed British atomic energy officials. Cockcroft wrote to Revelle to complain. It was not correct, he wrote, to say that the United Kingdom dumped most of its fission products into the Irish Sea. That discharge was limited to one hundred curies a day or less. They had arrived at that figure through "careful oceanographic work, followed by detailed biological monitoring." They, like the Americans, stored most of their waste in tanks, and "only the weak effluent, such as plant washings, are disposed of into the sea."[7] Perhaps Cockcroft was right to correct Revelle on this point. After all, the British did not put all of their waste in the sea. But the suggestion that the policy was based on careful oceanographic work overstated the case, and it obscured the history of pipeline disposal, in which oceanographers had played little part.

Cockcroft might have been content to bicker with Revelle on this point, but he was quite aware that the United Nations was also studying the effects of atomic radiation, including waste disposal at sea. If any future international agreements were to regulate waste disposal at sea, they undoubtedly would be based on the UN studies; consequently, Cockcroft wanted to protect the British point of view against contradictory statements from American oceanographers. Cockcroft immediately wrote to the British embassy in Washington, D.C., ostensibly following up on American calls for international action. In light of the UN discussions, Cockcroft reasoned, it made sense to have the American and British scientists discuss the matter together beforehand. He proposed sending one or two of his staff to talk the issue over with the American BEAR oceanographers.[8]

American and British scientists met together near the Woods Hole Oceanographic Institution, in North Falmouth, Massachusetts, in September 1956, to discuss their positions. They hoped to balance the need to use the ocean as a sewer and the need to use it as a laboratory. This was just the beginning, Bronk told Gerard Smith of the State Department, of the close collaboration between national delegations to take place in the future. There was a role for the State Department, he said, in encouraging and facilitating such meetings to discuss the problem. This would be especially pertinent during the upcoming International Geophysical Year (IGY), planned for 1957–58, which presented an unprecedented opportunity for sampling the oceans and the atmosphere on a collaborative basis. In fact, both Revelle and Harry Wexler, who headed the BEAR panel on meteorology, were to be deeply involved in IGY work. Presumably, all this work would lead to some international conventions on the release of radioactive materials into the air or into the sea.[9]

The meeting at Falmouth was the first clear indication to the British of how successful American oceanographers had been at gaining influence in atomic energy affairs. The opposite was true of the British, whose interests in waste disposal were entirely operations based. Only one of the four British scientists who turned up in Falmouth was an oceanographer—John Swallow of the National Institute of Oceanography. This undoubtedly was a concession on Cockcroft's part, since almost all of the Americans were oceanographers. The other British representatives were certainly more relevant to Britain's specific operations; they came from the Atomic Energy Research Establishment, the reprocessing facility at Windscale, which made most of the British discharges to sea, and the Ministry of Agriculture, Fisheries and Food, which conducted biological studies of the Irish Sea and also was partly responsible for authorizing discharges. Most of the fourteen American scientists present were members of the BEAR panel on oceanography, including Revelle, who acted as the meeting's chair. Cockcroft, who was not present, had objected to Roger Revelle's characterization of British disposal practices, claiming that they discharged far less than Revelle supposed. Whether this was true or not, the British wanted to dump a lot more. The British came to Falmouth to discuss and evaluate the research necessary to "estimate the feasibility of very large scale atomic waste disposal at sea and the means whereby international collaboration in this research could be accomplished." The purpose of the meeting, in addition to this, was to arrive at some general scientific recommendations upon which they could all agree, based on information currently on hand.[10]

The Americans and British agreed that the maximum permissible amounts should be set by two considerations, namely the transfer of radioactive substances back to man and the effects of the waste upon the marine environment. There already were some extant recommendations concerning the direct hazards to man from radiation, issued by the International Commission on Radiological Protection (ICRP). These the oceanographers accepted; the problem was that there were no standards at all, from the ICRP or elsewhere, regarding the effects of radioactive substances on marine resources. The (American) National Committee on Radiation Protection had begun to assess this in the late 1940s, but the work had been discontinued. Marine resources, aside from the harm done to them, also represented a pathway for harmful substances to reach humans. There were limits, then, to the usefulness of ICRP recommendations, which gave little attention to these ecological pathways through which harmful radiation could reach man. This was where the oceanographers, fresh from their experience on the BEAR committees, felt they could—and should—play a leading role. Ocean dynamics, chemistry, and marine biology were the primary disciplines to study in order to arrive at definitive conclusions, they believed, about how the ICRP recommendations could apply to the sea.[11]

This meeting between American and British scientists was not called to ask if disposal should occur, but rather to determine how, where, and how much. They divided the problem into two parts, coastal sea disposal and deep ocean disposal. Although more research was recommended on all aspects of the problem, including monitoring after the fact, the consensus of the meeting was that there was already sufficient knowledge to recommend coastal dumping. Actually, this was simply providing scientific justification for preexisting procedures. This was already practiced on a large scale for three types of waste: bulk liquids, packaged liquids or sludge, and packaged solids. Packaged wastes were designed either to rupture under the pressure of the sea at a certain depth, thus releasing waste in liquid or sludge form to be mixed with the sea, or to sink to the bottom without rupturing, to ensure that the solid waste never reached the surface. The basic requirement here was to ensure proper density of material so that it would indeed sink. In both cases the major consideration would be to avoid having these containers picked up by fishermen or salvagers.[12]

"Bulk liquids" simply referred to the vast amount of radioactive liquid discharged through pipelines into the sea, whereas packaged waste was put aboard ships and dumped farther away from the coast. What levels were safe for discharging through pipelines? The British shied away from general statements, insisting that the problem must be broken down to all of the factors in each particular situation. One had to understand precisely what isotopes were involved, because each isotope presented different risks for different periods of time. In addition, each oceanic region had its own peculiarities—physically, biologically, chemically, and in terms of how much seafood was eaten by the local population. If local people drew upon the sea for most of their food, this would be crucial to understanding the limits of dumping. These were precisely the kinds of studies that Britain, through its Ministry of Agriculture, Fisheries and Food, had been conducting for several years in the Irish Sea, where the bulk liquid waste from the Windscale facility were discharged on a routine basis. Although Cockcroft had stated to Revelle that they kept the discharges below a hundred curies per day, this was likely a question of semantics. Even if policy indicated keeping it under that amount, the AERE could dispose of more and simply call it an experimental operation. Such experiments gave them the necessary data set to justify increased discharges. At the Falmouth meeting, the British scientists confidently stated that their investigations of discharges into the Irish Sea indicated that "fission products can be safely released in that area at an average rate of several hundred curies a day," and it was likely that similar quantities could be "safely liberated" in other areas as well.[13]

Deep ocean disposal posed the greatest uncertainty and the greatest promise. Although confident about the safety of disposal in coastal waters, the oceanographers pointed out that they simply did not understand the basic properties of the deep ocean well enough yet to make definitive recommendations. Still, making what they considered conservative assumptions, they reasoned that "it is possible to arrive at a lower figure which appears completely safe." The reason for differentiating between coastal and deep ocean areas was that only the deep ocean seemed to be a potential graveyard for the most highly radioactive waste because of its immense capacity to dilute it. Pipeline discharge and coastal-water dumping entailed relatively low-level wastes, whereas the deep sea held promise for the higher levels of radioactive waste then kept in storage tanks on land.

One of the isotopes that most concerned them was strontium-90, the same long-lived isotope that concerned people about fallout from nuclear testing. Based on conservative estimates about the dispersal of these substances in the ocean, they judged that it would be safe—as an experiment—to release about ten megacuries (ten million curies) of strontium-90 into the deep sea on an annual basis, which would be equivalent to the amount of strontium-90 waste from the fission of about four tons of uranium-235 per year, the amount needed to generate about four thousand megawatts of electricity. Given the prospect of the peaceful development of atomic energy, they believed, such an experiment seemed necessary to judge the feasibility of ocean disposal. However, this would be just the beginning. The calculations were, after all, based on conservative estimates. The oceanographers speculated that if the waste was more perfectly mixed in the ocean than they (conservatively) allowed, then one might expect the ocean to absorb the products not of four tons of U-235, but rather a thousand tons of U-235, every year. But in order to make such claims, according to the meeting report, "an extensive program of research will have to be undertaken."[14]

All of this was based on the assumption that both the United States and Britain—not to mention the Soviet Union—would invest heavily in atomic energy for military and civilian uses. Although such large-scale disposal was premature given their knowledge, they confidently observed that present knowledge about biology and oceanography was sufficient to assume the safety of certain amounts of radioactive material to be dispersed in shallow coastal waters and much larger quantities in the deep sea. The probable extent of the latter was not yet known.

The Falmouth meeting set the stage for a consolidated Anglo-American position about wastes. Although American and British scientists, they reasoned, were the best equipped to deal with the problem because they had the expertise—and, after all, they were nuclear powers—other countries should be involved later, to facilitate wider understanding of the issues. The Americans were certainly gratified to have the meeting, although it took place at the initiative of the British Embassy. Negotiating directly with British atomic energy officials reflected a new reality: American oceanographers had carved out a niche in atomic energy affairs, and in the United States they were becoming the designated speakers for the sea. The British were satisfied with the meeting, too; they wanted to dampen the kind of implicit criticisms made by Revelle when describing how much waste Britain put into the sea. What they discovered was not a group of scientists opposed to dumping, but a group wanting acknowledgment of the need for research. They hoped to discuss the problem together, to arrive at a consensus, before debating the issues in the open at a meeting of some international organization such as the United Nations. The British approach, as Philip J. Farley of the U.S. State Department put it, was "somewhat out of channels" but had the benefit of allowing the free exchange of ideas and information. Like the collusion between the two countries prior to the BEAR and MRC reports, the meeting was designed in part to iron out some of their differences prior to airing them in an embarrassing international forum.[15]

Aside from the solidarity of purpose between American and British scientists, the Falmouth meeting also gave British atomic energy officials a taste of what to expect from the American oceanographers. In the course of the meeting, the latter had made repeated calls for extensive research programs in oceanography. The meeting itself, though ending in general consensus, marked the beginning of a decidedly negative attitude toward oceanographers by atomic energy scientists and officials in the United Kingdom. Cockcroft, though he did not come himself to Falmouth, in particular would later complain about oceanographers' opportunism. Health physicist John Dunster came to Falmouth as Harwell's representative, and in subsequent years he shared Cockcroft's pessimism and warned his colleagues that oceanographers' desire to "solve" the waste disposal problem amounted to little more than a justification for more research funds. It may be that this impression was gained at Falmouth, where oceanographers dominated the American group and made it clear that the future use of the oceans as repositories of waste would require major scientific studies concurrently and/or in advance. Moreover, these studies were to be directed at the effects of radioactivity on the marine environment—a never-ending research agenda—rather than the specific ways in which dangers reached man. Because the oceanographers strongly believed that their expertise should govern these decisions, rather than relying purely on radiation safety levels set by health physicists, perhaps Cockcroft and Dunster perceived a grasp for influence and patronage. If oceanographers were implying that they could, should they choose to do so, deem waste disposal as unsafe, the future relations between oceanographers and health physicists had the potential to be rather strained.

International Studies of Radiation

The prospect of authoritative reports on radiation effects being published by the United States and Britain could not be expected to satisfy all skeptics outside these countries. The United Nations was undertaking its own assessment through its Scientific Committee on the Effects of Atomic Radiation (UNSCEAR), led by Gunnar Randers, the director of Norway's Institute for Atomic Energy. It had its first meeting in New York in the middle of March 1956, a few months before the American and British reports were published. Like the BEAR study, the committee broke its work down into categories such as genetic effects. Some of these categories were markedly different, however, and were either more specific or more general than the American group. For example, rather than simply having a panel on pathology, UNSCEAR narrowed the problem further by differentiating the effects of external radiation from the effects of being irradiated by internally absorbed isotopes. The lack of attention to the latter subject would be, in fact, one of the criticisms leveled at the BEAR report in subsequent years.[16] Other areas were more general, such as environmental contamination. On this score, the scientists at the United Nations were almost entirely preoccupied with the problem of radioactive fallout. The committee did suggest, however, that in light of the possible long-term use of atomic energy, and the obvious importance of fish to large segments of the whole population, a worldwide study of ocean contamination should be undertaken.[17]

Though already conducting its own assessment, the United States was instrumental in creating UNSCEAR. The State Department reasoned that other countries were more likely to trust a scientific study done by the United Nations than one done by Americans, and there was every reason to believe that the United Nations would take American views very seriously. As Brookhaven National Laboratory scientist W. A. Higinbotham put it, such a move would "prove that we have nothing to hide" and that Americans supported international cooperation. An international study could assuage the global discord about weapons tests, reassure the world about peaceful uses of atomic energy, and "calm many of the present irrational fears on this score." Higinbotham emphasized the point: "No pronouncement by the US AEC or even by the Academy can carry such weight of authority outside our shores."[18]

Asking scientists from other countries to study the problem was, however, fraught with uncertainty. Although American and British scientists could be expected to be sensitive to the need to conduct nuclear tests, the same could not be said of scientists elsewhere. The prospect of an important role to be played by the United Nations distressed the AEC, and its scientists looked for potential troublemakers on the national committees of other nations. This was particularly true of Japan, where scientists were rather vocal about nuclear testing and the effects of radiation on the sea's resources. One of these was Yasushi Nishiwaki, a Japanese biophysicist who had measured high levels of radioactivity in tuna in Osaka after the 1954 *Lucky Dragon* incident, leading to widespread fear of contaminated fish. In 1956 Caltech geneticist George Beadle received a letter from Nishiwaki saying that the conventional methods of measuring radiation exposure, using gummed films, tended to underestimate the real levels. It was his belief that the bones of the people of northern Japan were 10–100 times more radioactive than they had been two years previously. Beadle asked his colleagues at the academy and the AEC if they had heard anything about it. He was stunned to find that key individuals in the AEC—particularly Charles L. Dunham and Merril Eisenbud—considered Nishiwaki to be a communist subversive. They believed his comments were now, as in the case of the *Lucky Dragon*, intentionally sensationalist and were not to be trusted. He was, after all, married to an American girl with alleged communist leanings and was at the center of anti-American agitation in Japan. Thus his views about radiation effects on fish, they believed, could be dismissed.[19]

Although Beadle acknowledged that Nishiwaki was "an emotional fellow inclined to some extravagance," with a strong bias against the AEC's activities, it seemed extreme to label him a communist sympathizer because of it. Beadle had met him a couple of times and spent a day with him in Osaka when he visited there. He found him charming, intelligent, and "amazingly well informed," albeit opposed to nuclear weapons testing. In addition, he had quite a high reputation in Japan. It seemed unwise to label him a communist and try to alienate him. This was precisely the kind of person they ought to cultivate, if they really believed in what they were doing. "Here is an able, intelligent, influential fellow who can, if handled right, be on our side. On the other hand he can do us a lot of harm if we handle him in the wrong way." Rather than push him aside, he ought to be brought in, even invited to the United States for study or training. That, Beadle believed, would help relations with the Japanese rather than harm them.

Beadle tried to generalize the point, emphasizing that the AEC ought to cooperate with other scientific establishments rather than browbeat them. His BEAR genetics panel, for example, planned to do this by keeping in close contact with geneticists in Japan. He was convinced that the difficult problems that lay ahead would be solved better "if we're all pulling in the same direction," rather than simply assuring everyone that the American way was the right one and protecting AEC policies against all critics.[20] Beadle's views accorded well with Revelle's; the oceanographer's handling of Miyake's work stood in stark contrast to the AEC's handling of Nishiwaki's.

Nishiwaki was not the only example of the AEC's fear of foreign scientists' exaggerations of radiation dangers. American concerns about foreign criticism were particularly sharp when it came to deciding what body should conduct the international radiation study. Some urged a study based on the expertise of the International Council of Scientific Unions (ICSU), which was not made up of political appointees. But scientists working under ICSU showed no signs of the kind of sympathy toward nuclear weapons testing that the American and British scientists had shown. At an April 1955 general assembly of the International Union of Biological Sciences, in Rome, geneticists led by Norwegian Knut Faegri drafted a resolution that called attention to the "already demonstrated" hereditary damages incurred by the human race from weapons testing. This made the Americans at the meeting rather uncomfortable. They sensed considerable hostility in that scientific community about weapons tests and a profound suspicion that the dangers of peaceful atomic technology were being ignored by the United States for political reasons.[21] At a meeting in Oslo in August 1955, the ICSU resolved to take matters into its own hands by asking its members to conduct studies of radiation effects independently of the United States.[22]

The U.S. State Department and the AEC did not favor an ICSU study; instead, they hoped to create a higher-level committee in which governments could nominate the scientists. AEC chairman Strauss firmly objected to any existing international scientific organization second-guessing what the Americans were doing. An independent scientific body would undoubtedly act like a "packed jury" against them, he told Secretary of State Dulles and others.[23] Dulles agreed, and noted that if any body should take over this subject, it should be a high-level one formed directly under the United Nations, which he believed would be easier to control than an ICSU study. The UN group should be new, not laden with existing institutional prejudices. The president, too, was also much more comfortable trying to influence the United Nations. He deftly preempted ICSU's action by making a positive proposal of his own by sending Dulles again to the United Nations General Assembly to recommend that all nations pool their knowledge under the United Nations. This led to the creation of the United Nations Scientific Committee on the Effects of Atomic Radiation (UNSCEAR).[24]

ICSU had already appointed a special committee for this purpose, but the flurry of activity at the United Nations convinced the committee that their own work would be superfluous. There were already plenty of bodies examining the problem—not just the BEAR scientists and UNSCEAR, but also the International Conferences of Radiobiology, established by the Faraday Society in Cambridge in 1952, and of course the International Commission of Radiological Protection, which had been active since 1928. Granted, such bodies were concerned primarily with medical uses and occupational safety; these, the greatest dangers, might soon be overtaken by the general public health issues in developing atomic energy and in nuclear weapons tests. But at a meeting in May 1956, the committee stated that there was no point in duplicating UNSCEAR's efforts. Although it abandoned its idea of making an independent evaluation, it urged that specific international rules be laid down for the future use of atomic energy, with strict international control.[25]

If they were worried that the UN study might make competing statements against their national studies in 1956, the governments of the United States and Britain had very little to fear. It would be a long time before the United Nations and its specialized agencies got their acts together. Whereas the informal Anglo-American gathering in Massachusetts resulted in specific recommendations about type and location of waste, making definitive statements about what could be considered as safe, the UNSCEAR meeting the following month (October 1956) did little more than identify subjects worthy of assessment. After all, the United Nations had different sorts of problems. For example, the delegates spent a lot of time deciding how they were going to circulate an ionization chamber to all the interested countries so that they could use it as a model to build their own. The model would be one constructed by the National Bureau of Standards in the United States. Brazil, for example, was keen to get a hold of one so that they might contribute to the radiation studies taking place in 1957 and 1958 during that International Geophysical Year. They also debated issues such as standard measurements, which might differ across international lines.

Although oceanographers had been calling the waste disposal issue the most important one facing the entire realm of atomic energy, most still viewed it as a background issue. Nuclear weapons testing and fallout made most of the atomic-related headlines, and these became serious campaign issues in Democratic presidential hopeful Adlai Stevenson's effort to unseat Dwight Eisenhower in 1956. At the UNSCEAR meetings, the United States put ocean disposal on the agenda for debate, but then backed off, not wanting to be the initiator of even more contentious discussions. Its delegates asked their colleagues at the British embassy if they would prefer to raise the question. The British were eager to comply: "it is all to the good," wrote Foreign Office official G. G. Brown, " that the Committee should at this time devote some of its meetings to this comparatively non-controversial topic." Sir Pierson Dixon, the United Kingdom's representative, opened discussion of ocean disposal and got precisely what he desired—no controversy, no discord, and general agreement that the British practices were safe. All seemed to agree that ocean disposal was going to become the preferred method, especially for small countries, and the British were setting the standards.[26]

Studying waste disposal at sea in UNSCEAR entailed reining in an enormous network of bureaucracies. They had to assemble the expertise of all the specialized agencies involved—such as the World Health Organization (WHO), Food and Agriculture Organization (FAO), the United Nations Educational, Scientific and Cultural Organization (UNESCO), and the International Council of Scientific Unions (ICSU). In short, they were a long way from making definitive statements. The British and American "out of channels" approach allowed the two powers to arrive at a consensus well in advance of genuine debate.[27]

Scientists and governments were in fact rather busy on the international level making preparations for the International Geophysical Year, set to begin in the middle of 1957. Both the United States and the Soviet Union invested a great deal of political capital in the IGY, calling it an international scientific undertaking to ease the tensions of the cold war. For many scientists, the IGY would be the ultimate test for scientific internationalism because all of the scientific data was to be shared. Typically, access to geophysical data, especially involving the atmosphere and oceans, was restricted for security reasons. It was the first major collaborative scientific project between the United States and the Soviet Union, a prospect that had been impossible before Soviet leader Joseph Stalin died in 1953. This international undertaking did not diminish UNSCEAR's importance, but it did influence its emphasis toward one of experiment and coordination with IGY work. American oceanographers' vision of turning major policy problems into research opportunities fit rather well with the need to advise UNSCEAR and plan the IGY at the same time.

UNSCEAR asked the specialized agencies of the United Nations for some pretty specific information to help draw conclusions about waste disposal. For example, it wanted oceanographers to identify the rate and variability of diffusion and dispersion of deep water, the rate of exchange between the deep and surface waters, the paths and velocities of important water masses in the oceans, and the duration of the geochemical cycle of different elements between the water and the sediment. It also wanted some biological details, particularly on the concentration of radioactive isotopes by different marine organisms, the possible rate of biological transfer of these isotopes, and the distribution, abundance, and rate of growth of the ocean's organisms. If marine scientists working with UNESCO, FAO, and WHO could arrive at a consensus about all this information, then UNSCEAR could develop sound recommendations about the disposal of radioactive waste at sea, based on the best possible information.[28]

These specific questions proved difficult for marine scientists to answer, given the present state of knowledge about the oceans. When UNSCEAR convened a meeting of oceanographers in Göteborg, Sweden, in January 1957, most of the participants were gearing up for the IGY—in fact, the same oceanographers were meeting in the same city that week to plan the IGY. They realized that the problem of radioactive waste disposal could provide a powerful justification for coordinated scientific research. They lamented the fact that they were being asked to advise on a subject upon which very little research had been done. The physical oceanographers pointed out that almost nothing was known about deep sea currents. And after listening to a hastily prepared report on marine ecology by Woods Hole Oceanographic Institution scientist Bostwick Ketchum, they recognized that the concentration of isotopes in marine life needed greater study as well. According to one report of the meeting, all of the participants "recognized that the problem of radioactive waste disposal presented an opportunity for oceanographic research." They hoped that knowledge about coastal disposal would be exchanged openly between nations or published in accessible journals, in the spirit of the IGY. If nations hoped to dump high-level waste into the deep sea, they pointed out, intensive studies would be needed at national and international levels. But even if studies of the effects of radioactivity at sea were not completed, it was essential to assess the levels of natural radioactivity in the sea, before nuclear weapons tests and the expansion of nuclear power rendered it impossible.[29]

In the meantime, a few leading oceanographers (Roger Revelle, British oceanographer Henry Charnock, and German oceanographers Günther Dietrich, J. Joseph, and Georg Wüst) were to provide critical information for a final report to UNSCEAR, based on consultation with their own compatriots. It became Charnock's responsibility to compile all of their comments and draft a final report to represent the views of UNESCO, FAO, and WHO. This report was supposed to answer UNSCEAR's questions.[30]

The oceanographers did not make their contribution to UNSCEAR in a timely fashion, and its original due date of July 1957 passed without any results. By that time, the IGY had begun and Revelle, whose failure to write his contribution to the report was one of the reasons for the delay, was very busy with it. The eventual report was about half as long as the specialized agencies had expected—so there were further delays. By mid-October, excitement about the IGY had turned into anxiety, because the Soviet Union had used the international project to launch the first artificial satellite, *Sputnik*. But finally the report on ocean disposal was ready. Unfortunately, the results were inconclusive. When Charnock sent his report to UNESCO, FAO, and WHO, he pointed out that most of the scientific questions related to sea disposal, in the present state of knowledge, were impossible to answer. "The problems involved," he wrote, "are among the most difficult in the physical and biological sciences." Thus he was only able to give a brief and general account of the problem, with few definitive statements about the behavior of radioactivity in the oceans.[31]

Unsurprisingly, the basic conclusions of Charnock's report were identical to those produced in Massachusetts over a year earlier by Anglo-American consensus, though now expanded somewhat. Charnock described the problem in generalities, pointing out the differences between high-level and low-level waste, and observing the importance of ensuring dissipation and dilution when making pipeline discharges. Charnock used Windscale as an example, noting that it had conducted experiments with fluorescent dyes to judge the likely diffusion of waste in the vicinity of its pipeline. He made a vague mention of the possibility of biological concentration of radioactive material but included no details. He remarked only that the "Windscale experience has confirmed that a coastal discharge of some hundreds of curies per day was safe," but that each site should be evaluated independently.[32]

Charnock wrote vaguely about the open sea, far from the coast. He acknowledged that it was difficult to make reliable estimates or measurements there, even in the upper layers, "but it seems clear that permissible quantities can be greater than in coastal waters." That was for disposing of low- or medium-level wastes at the surface. As for putting high-level waste in drums meant to sink to the deep sea, the oceanographers' estimates were "crude" and "rough" at best, based on a range of scenarios put forth by Americans Roger Revelle and Harmon Craig about how quickly the material would rise to the surface and mix with surrounding waters. These estimates were totally inconclusive, of course; some said that a continuous introduction of 1,000 curies per day would ultimately be harmless, while others warned of that much smaller amounts would greatly increase radioactive strontium in human bones by the year 2000. All of the scenarios were based solely on circulation and dilution, not on the chemical processes in the sediment or on any biological factors. "Clearly such estimates must be refined," Charnock observed, "before any large-scale dumping is done."[33]

Charnock said even less about the relevant biological processes, though he acknowledged that these would be important considerations. Man could be endangered by radioactivity indirectly by eating sea food that had concentrated radioactive isotopes in its tissues. Sea organisms were known to concentrate elements "by factors as large as several thousand." Scientists' understanding of such processes was not sufficient to assess the direct dangers to the organisms themselves, Charnock stated, "but the risk is by no means negligible."[34]

The overriding conclusion was simple: they did not know how much could be dumped safely, but preliminary studies showed that at least several tons per year would be safe. At the same time, more research was needed. These mirrored the negotiations that had already taken place, between oceanographers seeking atomic energy establishments' support for more research and atomic energy establishments seeking oceanographers' blessing for putting certain amounts into the sea safely.[35]

International organizations were not blind to the fact that the scientific assessment they had requested turned out instead to be an expression of discussions between oceanographers and atomic energy establishments in the United States and Britain. Charnock's report was so obviously an Anglo-American position paper that scientists in international bodies tried to distance themselves from it. The mention of Windscale as evidence of safety was itself suspect, because UNSCEAR's explicit goal was to evaluate such discharges, not to see an existing British practice as evidence that they were safe. Although his report was supposed to be a product of several different international organizations, some were hesitant to sign on to what seemed a mere reflection of existing American and British policies. G. L. Kesteven, the chief of the Food and Agriculture Organization's (FAO) Fisheries Biology Branch, had advised Charnock that he was pleased with his work, to a certain extent, but he would like to see some edits: "the small changes proposed in the report will concern the 'policy matters.'"[36] When he saw that the final report did not reflect his suggestions, he wrote a little more harshly. Perhaps, he intimated, Charnock had believed that FAO's previous comments "either were harshly critical or indicated a naïve ignorance about the subject matters on which the UNSCEAR asked for information." FAO's principal objection to the oceanographers' report was that the oceanographers seemed to think that they were part of the UNSCEAR group itself—that is to say, Charnock's report tried to recommend policies rather than to provide information that UNSCEAR could use to make its own assessment. This was true enough, for Charnock's report clearly stated that ocean disposal at some levels could be done safely. From FAO's point of view, that was for UNSCEAR to decide, not a few oceanographers.[37]

World Health Organization official I. S. Eve reacted in much the same way. He liked the report but saw too many policy recommendations. The important data from the international agencies had been relegated to appendices. Rather than calling it a joint report, he suggested changing the title to "Sea and Ocean Disposal of Radioactive Wastes—some conclusions which may be drawn from data submitted by UNESCO, FAO and WHO (Appendices A, B, and C)," by Dr. H. Charnock, UNESCO consultant.[38] He wrote apologetically to Charnock that although it was an excellent paper, the organization "has to be very careful as to the statement to which it commits itself, especially when, as in this case, there are so regrettably few experimentally verified facts to go on." Eve refused to let WHO's name be associated with the report's conclusions. He wrote to a colleague at UNESCO that the oceanographers had indulged in too much interpretation in a controversial subject. Ultimately, the report emphasized that it was an assemblage of information edited and correlated by an individual, not the official scientific consensus of any of the bodies.[39]

Despite such reservations, these bodies did not conduct separate studies to compete with Charnock's, so his report became UNSCEAR's de facto main source representing the views not only of some of the world's leading oceanographers, but also of the United Nations' specialized agencies devoted to world health, fisheries issues, and science. UNSCEAR had little choice but to take the Anglo-American findings at face value; this they did not only in oceanography but also across the board. When preliminary reports were circulated to selected scientists in mid-1958, the whole report, not just the ocean disposal aspects, did not differ markedly from the findings already announced by the American BEAR committees and British Medical Research Council two years earlier. Harold Himsworth, who headed the Medical Research Council, joked to academy president Detlev Bronk, "I think that you and I will both be in danger, although completely innocent, of being accused once again of collusion—this time with the United Nations." Himsworth was "astonished" yet reassured that the UNSCEAR report seemed to follow the American and British lines closely, despite two more years of work. This Bronk readily admitted in August 1958, after the UN report finally was issued, noting in a press conference the "remarkable degree" of agreement with the "quite independent" reports of 1956.[40]

The Challenge and Promise of 1957

The broad range of agreement at the international level with the pronouncements of the American and British scientific bodies might have paved the way for a fairly uncontroversial development of nuclear technologies in the late 1950s and 1960s. In some ways, the 1956 reports reinforced the enthusiasm of the 1955 United Nations Conference on the Peaceful Uses of Atomic Energy, when the United States seemed genuinely willing to share information, and delegates presented paper after paper about the promise of peaceful atomic energy. Some countries had used that conference as a venue to reveal technical details, as in the case of France, which published its methods of plutonium extraction. It did so in the hopes that others would follow suit and abandon their own practices of secrecy. Others used the conference to highlight their successes, as in the case of the Soviet Union, which drew attention to the fact that it had built the world's first civilian atomic power plant at Obninsk in 1954.[41] Other plants were under way—an American one at Shippingport, a British one at Calder Hall, and a French one at Marcoule. Just prior to the conference, Britain had announced its intention of launching a nationwide effort to produce nuclear energy, which it hoped to make competitive with other sources by 1963.

It had been very difficult to determine from the conference at Geneva the Soviet attitudes toward radiation protection. Although impressed by Soviet work on physics and reactor technology, the Americans observed that the Soviet papers on biology and medicine were unsophisticated. They could not quite discern how health physics and radiation protection standards were organized in the Soviet Union, but they did believe that the apparently high standards in Soviet health physics were "unrealistically restrictive." American delegates doubted that the Soviets really adhered to the stated levels of permissible exposure, and they dismissed them as propaganda. In general, it also was difficult for Americans to assess Soviet biological work, largely because of the persistent adherence of Soviet scientists to government-mandated scientific outlooks. At the time, Soviets were obliged to ignore Mendelian genetics, which had fallen into disrepute in the eyes of the Soviet government because of the influence of agronomist Trofim Lysenko.[42]

Despite cold war tensions, the overall feel of the conference had been positive, which seemed to bode well for international relations and peaceful atomic energy. Even in the confidential supplement to their report to the secretary of state, the American representatives noted how supportive the Soviets were of the conference and how restrained their propaganda had been; as far as they were aware, only an East German broadcast criticized American intentions and belittled American and West European accomplishments (the German Democratic Republic was not represented at the conference).[43] The conference had been a learning process for all of the countries, and nuclear power seemed at a crossroads of several technological and design possibilities. The competition at Geneva was not solely about "firsts"; it was also about technological persuasion. The British and French, for example, were building reactors to use natural uranium, while the Americans and Soviets were using enriched uranium. Whose method would prove most efficient and thus set the standard?

The United States and the Soviet Union saw nuclear power as the technology of the future—the distant future. Both countries were rich in fossil fuels, and the development of financially solvent nuclear power seemed on a far horizon. In both countries, the experimental power reactors were meant to achieve rather than develop, and to demonstrate rather than commit. Peaceful atomic energy could not be separated, in either case, from cold war propaganda campaigns to win prestige in the eyes of the world.

Britain and France, however, saw nuclear power through very different geopolitical lenses. Few events crystallized these differences more than the Suez crisis of 1956. When Egypt seized the Suez Canal and nationalized the Franco-British company owning it, Britain and France colluded with Israel to take control of the canal by force. The geopolitical showdown that ensued led the United States to side with the Soviet Union and Egypt against its closest allies, convincing Britain and France to back down.[44] In the broadest sense, the Suez crisis demonstrated the decisive end to British and French claims of great power status. Despite their far-flung empires, they could not hope to implement major foreign policy actions without some American backing. In a more specific sense, the Suez crisis sent a panic through both Britain and France that the clock was ticking on their own source of fossil fuels in the Middle East. The Suez crisis pushed both Britain and France to commit to the large-scale development of nuclear power. In 1957, Britain announced that it was going to triple its investment into nuclear energy, hoping it would produce a quarter of the country's power by 1965. "The panic that followed the Suez crisis," French atomic energy official Bertrand Goldschmidt later wrote, "led to a momentary feeling that energy must be produced at any price."[45] The Suez crisis convinced the French government of Guy Mollet even to reverse its plan to renounce nuclear weapons. Now, it had ample reason to believe that French freedom of independent action could only be maintained by joining the nuclear club, and it pressed on with the military research it already had begun.[46]

These nuclear ambitions were reinforced by studies in the mid-1950s that painted a bleak picture for energy in Europe. Energy demands were doubling about every ten years, but supplies of conventional fuels such as coal were not rising to meet demand. Western Europe's annual consumption of crude petroleum jumped from 32 to 120 metric tons between 1937 and 1955, and by the mid-1950s over 90 percent of the imported petroleum came from the Middle East. To some, nuclear energy reactors offered the only viable solution, and cooperative bodies such as the European Atomic Energy Community (EURATOM) seemed well positioned to address pressing energy problems. EURATOM was France's brainchild, designed as a way to catch up to the three atomic leaders who refused to share information with it. Some European politicians had toyed with the idea of creating a nuclear-free Europe, with an organization devoted to advancing only peaceful uses of the atom. Thus EURATOM's creation had been delayed by the internal debate over nuclear weapons. Now it was a dead issue because France no longer wanted EURATOM nations to abandon efforts to make bombs.[47]

The path seemed wide open for a major growth period in nuclear energy, which meant that the waste disposal problem might become acute earlier than expected. EURATOM was born in 1957, with the goal of promoting the nuclear energy industry in Europe. Its members were France, Belgium, Germany (the Federal Republic), Italy, Luxembourg, and the Netherlands. The premise of EURATOM, as reflected in the treaty itself, was that "nuclear energy is the essential factor for ensuring the expansion and renewal of production" and for allowing progress along peaceful lines. These countries were convinced that the best way to achieve this was through common effort, and they resolved to create the conditions necessary to develop a powerful nuclear industry to provide enormous supplies of energy. With EURATOM, they hoped to do so with uniform standards of safety and with a common vision. It had the specific goals of accelerating the process in Europe by ensuring dissemination of technical information, establishing standards of safety and supervising their application, pushing companies to invest in new projects, providing for fair distribution of uranium ores, establishing a common market of radioactive materials, and allowing the freedom of capital and job mobility by nuclear specialists within the community.[48] All these European countries, by committing to nuclear power, also guaranteed that the waste disposal problem soon would be widespread among small, industrialized countries on that continent.

While Europeans were collaborating amongst themselves, a more significant international body finally was also born in 1957: the International Atomic Energy Agency (IAEA). This organization had been the backbone of Eisenhower's Atoms for Peace proposal, when he proposed an international body to act as a clearinghouse for materials for civilian nuclear power. It also seemed to be the natural body to establish international standards, or even regulations, for disposing of radioactive waste. At first, however, the IAEA was far less specific in its aims. It wanted to "accelerate and enlarge the contribution of atomic energy to peace, health and prosperity throughout the world." This was explicitly nonmilitary, and it mirrored EURATOM's strategy of fostering communication, encouraging research, and even establishing safety standards. Its actual power would not be in compulsory, concerted action, but rather in the diplomatic power of world consensus, as in the case of any organ of the United Nations. It would write reports, attempt to back United Nations policies, try to establish equitable distribution of radioactive materials, and monitor those materials to ensure their peaceful use.[49] One crucial difference between it and EURATOM was that the latter was more interested in promoting nuclear energy in its member states than in policing their activities, and although EURATOM did want to ensure equitable access to materials, it was not obliged to address the needs of developing countries.

The other major difference was that the IAEA's membership crossed cold war lines. The IAEA became a vehicle for airing complaints, disseminating propaganda, and clarifying cold war divisions. The election of the first director-general was the first of many negotiations between American and Soviet delegates. President Eisenhower wanted W. Sterling Cole, a Republican congressman who had been chairman of the Joint Committee on Atomic Energy in the U.S. Congress, to take the post. Critics hoped for a candidate from a neutral nation who would show the agency's international character rather than remind its members of American dominance. But some, like conservative commentator Holmes Alexander, balked at giving the job to "one of the loincloth nations."[50] The Soviets accepted Cole in return for promises to support Soviet candidates in key subordinate positions. Ultimately, the United States' role in creating the agency, as well as its technical competence, persuaded the members to elect Cole.[51]

Such high profile events in the developments of nuclear power drew attention to other contentious issues, such as the ultimate fate of unwanted by-products. Newspapers highlighted some of the uncertainties, as when they reported in July 1957 some canisters of radioactive waste off the eastern coast of the United States that unfortunately would not sink. A little less than two hundred miles from shore, the U.S. Coast Guard sighted a large steel cylinder floating at the surface. Upon inspection, they realized that this was a canister of twenty tons of radioactive sodium, dumped earlier in the week with twenty-four others just like it, from the tanker *Otco Bayway*. The Coast Guard had escorted the tanker out to sea to dump radioactive materials from the construction of the reactor in the United States' second nuclear submarine, the *Seawolf*, at the General Electric facility in Schenectady, New York. All of the canisters, or drums, had been rigged with explosives so that they would sink with their contents below the surface. But in this case, the explosives had failed, and the giant canister remained afloat, keeping its consignment of radioactive waste on the surface. The Atomic Energy Commission acknowledged that this was a danger, but only to navigation because of its large size and the possibility of explosion—the radioactive materials, it assured the press, posed no risk. Of course, no one wanted to get close enough to retrieve it because of the chance of explosion. So the Navy sent a plane in to attempt to sink it, but it failed to rendezvous with the Coast Guard vessel watching over it, and the dark of night obliged it to give up the chase. The next day, the Navy sent more planes and, after a day's effort shooting at it, finally managed to sink it.[52]

The story made front-page headlines, and the *New York Times* ran a long piece during this incident drawing attention to the problems associated with peaceful atomic energy. It was not just the government's safety record that should concern people, journalist James Reston argued, but also the activities of the four thousand other establishments who were licensed to use radioactive materials. Reston observed that "the use of dangerous radioactive materials is no longer solely in the hands of official establishments operating under a uniform set of security and safety regulations, but is now spread throughout the length and breadth of the nation among private organizations that operate under varying rules and regulations." This statement was probably somewhat misleading—each of these establishments was required to conform to standards set by the AEC or else face the possibility that their licenses would be revoked.

But Reston's general point was sound. With the proliferation of radioactive materials in the private sector and internationally, radioactive wastes would be more difficult to control, possibly involving conflicts between nations and governmental bodies at the federal, state, and municipal levels. This could make the problems of radioactive waste as problematic as those of nuclear fallout. He commended the AEC and the National Bureau of Standards for trying to set guidelines, but "they cannot always compel compliance." Although most experts agreed that these considerations did not need to hamper the rapid development of peaceful atomic energy, Reston warned, closer liaison was going to be needed if the side effects of that development were going to be controlled.[53]

Already in 1957, reactors found stiff resistance in local populations because the great promise of nuclear energy seemed to bring equally great uncertainties about health effects. For example, the company Detroit Edison built a breeder reactor in Lagoona Beach, Michigan, despite the protests of local labor unions that it would be an unsafe facility for workers. This was the reactor, as journalist Gene Smith observed, "that science fiction has portrayed to the lay public." In addition to producing heat for power, such reactors used the neutrons from fission to bombard depleted uranium, which made plutonium. The name breeder came from the idea that they would create more fuel than they used. Yet the enthusiasm for such innovations was not shared by apprehensive locals who did not feel safe living nearby, and they bitterly opposed its construction.[54] Similarly, in Linkenheim, Germany, voters boycotted national elections as a protest against the government's plans to put a reactor nearby. They were reportedly afraid of pollution of crops and water.[55] Although the German boycott might have struck some observers as extreme and irrational, the residents of Linkenheim must have felt vindicated when, a month later, world headlines carried the story of a serious fire at a British nuclear plant at Windscale.

Few events tempered the enthusiasm for nuclear power or heightened anxiety about radiation like the Windscale fire. The event added a new surge in the antinuclear movement, particularly in Britain, where the government was about to test its first thermonuclear weapons, sparking the creation of the Campaign for Nuclear Disarmament.[56] On the afternoon of 10 October 1957, during a routine procedure called Wigner release, some of the uranium in Windscale's Pile Number 1 became so hot that it caught fire. The fire did not spread, but some of the radioactive material converted to gas form and rose up through the plant's large chimneys. Most of this was stopped by the filters, but some escaped. Workers were instructed to stay inside and put on gas masks. Not knowing quite what to expect, the British Atomic Energy Authority arranged to have the local police set up a bus service in case the local inhabitants needed to be evacuated. The next morning, workers hosed down the reactor with water and gradually the fire came under control. After some thirty hours of saturation, they turned off the hoses. The workers had controlled the fire, averted disaster, and had to pump contaminated water into local ponds.[57]

Although the most frightening part of the accident passed quickly, the public relations nightmare had only begun. Twice during the crisis there were major releases of radioactive material from the Windscale chimneys, once around midnight as the uranium burned, and once during the morning when they first put the hoses on the fire, creating a lot of radioactive steam. Windscale's resident manager of health and safety was Huw Howells, who immediately realized that none of the radiation protection literature seemed appropriate for this kind of accident. The recommendations of the ICRP were for lifetime doses, not acute accidental exposures to the population. The only comparable studies were those that dealt with the theoretical possibility of a nuclear attack. His immediate problem was to determine whether the local populace should stop eating local foods, which might contain dangerous amounts of radioactivity. After ordering a survey of local food, Howells found that local milk from Friday afternoon contained between 0.4 and 0.8 microcuries per liter from the isotope iodine-131. Scientists preparing for a nuclear attack had recommended that people avoid drinking milk at any levels above 0.3 microcuries to avoid damage to infant thyroids.[58]

The subsequent milk ban made public controversy unavoidable. This became particularly true when, on 15 October, the ban coverage extended from eighty to two hundred square miles. This delay compounded the consternation already in place about the first one—why were the local inhabitants not warned on Thursday, 10 October? At the very least, commentators wrote in the pages of the *Manchester Guardian*, they might have wanted to stay indoors. The AEA's official response was that they were prepared to warn the public any time it (the AEA) determined that limits had exceeded safe levels. One of the commentators was a Windscale scientist, Frank Leslie, who sent the *Guardian* a letter with fallout levels measured from his own garden. The AEA acknowledged that these levels were higher than normal, though still below those deemed acceptable by the ICRP. Still, Prime Minister Harold Macmillan was outraged that any critical comments came from within the AEA ranks, but he was now helpless—due to the press's desire to speak further with Leslie—to restrain the scientist without provoking further public relations problems.[59]

British efforts to obscure the facts of the Windscale disaster came straight from the top of government, not from the AEA itself. It had been horrible timing for the Macmillan government. Less than a week earlier, the Soviet

Union launched the first artificial earth-orbiting satellite, *Sputnik*, promising an intense technological competition between the United States and the Soviet Union. Macmillan wrote to Eisenhower on the very day of the accident suggesting that, in light of *Sputnik*'s launch, the two countries should pool their efforts on a wider basis, and Eisenhower soon responded favorably by inviting him to Washington for a meeting. On 23 October, this resulted in a Declaration of Common Purpose, in which the president expressed his willingness to try to have American legislation amended to allow closer cooperation with the British. When Macmillan returned, he had the report of the official Windscale inquiry by Sir William Penney waiting for him. Knowing what to do would be easy, he wrote in the margins, but "What do we say? Not easy." First of all, the Penney report was certainly not going to be published. It explained all of the problems and did not pull any punches, assigning collective responsibility for the accident to the entire atomic energy establishment rather than naming a scapegoat. Certainly this appeared to be an admirable choice, but the level of technical detail would provide too much ammunition to critics of Britain's atomic energy program. Instead, a separate white paper would be written. Macmillan wrote in his diary that the report "had been prepared with scrupulous honesty and even ruthlessness. It is just such a report as the Board of a Company might expect to get. But to publish to the world (especially to the Americans) is another thing. The publication of the report, as it stands, might put in jeopardy our chance of getting Congress to agree to the President's proposal."[60]

Macmillan shelved the full Penney report and had an abridged version of it put into the official white paper. Then, after the white paper was released, the Atomic Energy Authority invited a team of scientists from the AEC for a conference in Britain. There Cockcroft and others gave the Americans secret information about some of the technical problems, and the Americans discussed their own experiences operating reactors, including the plutonium facility at Hanford. This marked a change for the British, as they had not been privy to many discussions about Hanford until after the Windscale fire. The Americans were particularly forthcoming about how they handled problems of Wigner release, the procedure that had led to the fire. The British also helped to reassure other nations that had committed to building nuclear reactors—such as Italy, Japan, and Belgium—by sending some technical details of the accident directly to them. Even the French Commissariat à l'Energie Atomique, which had not bothered to ask, was brought into the loop through Britain's initiative.[61] This discreet sharing of information suggested that the British increasingly saw public relations as a problem common to all atomic energy establishments.

Although the milk ban around Windscale made controversy about air contamination unavoidable, a much quieter dispute brewed up behind closed doors about what to do with the radioactive waste. Putting out the fire gave rise to an enormous amount of contaminated water, which was temporarily pumped into a nearby pond. Although the immediate crisis was over, the authorities at Windscale now were faced with a dilemma about what to do with the liquid waste sitting in their ponds. Here the AEA interpreted the notion of threshold levels of safety quite strictly, increasing the amount it discharged from Windscale to the maximum allowable by law. Over the next few months, the AEA pumped the contaminated liquid from the fire into a pipeline heading toward the sea, working as close to the limit of its regulations as possible. But it was not long before Windscale exceeded its authorized levels for discharge into the Irish Sea (it did so in March 1958, by about 5 percent). As one official put it, discharging waste from the ponds "had made it necessary to work close to the limits and this cannot be done without there being some chance of exceeding them." A flurry of correspondence ensued between authorizing ministries and the AEA, in which they expressed less concern about the infraction than they did about news of it spreading. They explicitly acknowledged the need to avoid undermining the integrity of Britain's authorization procedure, in which the British government essentially regulated itself. The infractions were kept quiet, and AEA scientists played with the numbers a little to try to make them approximate the authorization for monthly discharges by averaging the discharges over three months.[62]

Despite the Windscale fire, or perhaps because of the focus on the milk ban, radioactive waste disposal had yet to become an international controversy, taking a backseat to fallout and nuclear disarmament. After all, the leading scientists in the United States and Britain seemed to bless the practice, promising more research to make sure of its safety, and the United Nations voiced no objections in its own ostensibly independent assessment. Even had the brief infraction of Windscale's discharge limits been discovered and publicized, it is difficult to judge what level of consternation would have been aroused. Although scientists did not always agree about Windscale's discharges into the Irish Sea, neither health physicists nor marine scientists had taken strong positions against American or British practices of putting radioactive waste there. Even at Windscale, the world's leading point of radioactive effluent discharge into the sea, scientists and engineers did not worry obsessively about criticism. For example, a year after the Windscale fire, storms on the Cumberland coast destroyed part of the concrete shielding on the effluent pipeline, where it crossed the beach. The task of repairing it to prevent further erosion by the tides presented a major engineering difficulty; consequently, the health physicists at Windscale decided that it would not be necessary to make the repairs. Certainly the permissible exposure to man would be exceeded by almost double, if someone were to sit on the pipe for as little as two hours without the shielding, but the odds of someone wanting to sit for that length of time on a nuclear waste pipeline seemed remote. Huw Howells, the general manager at Windscale, was not sure how to proceed, and he wrote to health physicist John Dunster for confirmation, "since there could be obvious public relations repercussions if these levels were measured by, say, an enterprising newspaper man."[63]

It was an intriguing idea, but such a stunt seemed improbable to atomic energy officials. Dunster agreed that such journalistic entrepreneurship might present a risk, since measuring the pipeline would be a natural first choice for someone looking for a sensationalist story. He passed the question on to a public relations officer, J. A. Dixon, who passed it right back, dismissing the significance for public relations: "So far as the public relations aspect is concerned, my view is that the risk of an enterprising newspaper man taking measurements on the pipe is extremely small. I base this on the fact that so far as we know this has never been done, the pipeline has been in place for about seven years and its history has been so uneventful that there seems little possibility of any reporter thinking at this stage that there might be a story in it."[64] Such was the attitude in 1958, that despite a few controversies about radiation effects, waste disposal options, and the Windscale fire, there was nothing in sea disposal to attract enough media attention to make a good story. It was not long, however, before these attitudes changed dramatically. In a few short years, radioactive waste disposal at sea became a rallying point for an assortment of strange bedfellows: marine scientists, antinuclear activists, politicians, and a vicious propaganda campaign from the Soviet Union.

Chapter 5 No Atomic Graveyards

WHEN OCEANOGRAPHERS CAME TO Göteborg, Sweden, in 1957 to plan the International Geophysical Year of 1957–58, they knew they were on the cusp of a scientific opportunity about which generations of their predecessors could only have dreamed. For the first time in history, the IGY gave oceanographers a chance to pursue a scientific ideal: global studies involving simultaneous multiple observations in various locations around the world. Some sixty nations were to take part, including the Soviet Union and the United States, with full cooperation and funding from their governments. Lev Zenkevich, one of the Soviet Union's leading marine biologists, pointed out that the IGY would allow them to develop a truly global picture of the biggest problems confronting marine science, such as the circulation of water masses and the biological structure of the oceans. Everyone was excited: not only was it an unprecedented opportunity for scientific study, it was also a powerful sign that the scientists might play a role in transcending cold war divisions. Yet at least one American oceanographer, Roger Revelle, wanted to focus this enthusiasm toward the future. After all, the money and international opportunities would dry up after 1958. Their only hope for keeping a viable international oceanographic community alive would be to work together, targeting practical problems that would serve their mutual interests as scientists and attract funding from governments. He suggested that they attach each of their scientific goals to a problem facing humanity: instead of talking about the circulation of water masses, for example, they might speak of the effects of radioactive wastes in the oceans.[1]

It seemed like a brilliant idea, and the strategy of tying science to practical aims would form the basis of future international cooperation in oceanography. But for all its wisdom in attracting government interest in research, incorporating radioactive waste disposal into the IGY proved the catalyst for a sharp divergence of views about the use of the oceans as sewers for the atomic age. Whereas Revelle innocuously hoped that radioactivity might become an oceanographic tool, scientists from other countries—notably the Soviet Union—took the IGY as an opportunity to scrutinize the deep ocean for evidence of water circulation. And they found it. The fact that even the deepest parts of the sea showed evidence of circulation seemed to imply that there were no dead zones into which high-level radioactive wastes could be put with any expectation of long-term isolation. In fact, Soviet scientists in particular began to question whether the oceans should receive any waste at all. The IGY thus became a vehicle to overcome the existing international consensus, formed by American, British, and United Nations reports, that the development of atomic energy should proceed, in a rational and controlled way, with the seas as the best place to put dangerous waste materials.

Compounding this criticism were new reports coming out of the United States, under the auspices of the National Academy of Sciences. The high-level support for the IGY gave American oceanographers the means to assert an even greater role in atomic energy affairs, beyond their panel for the BEAR committees. Now there was a permanent committee under the academy devoted explicitly to the oceans: the National Academy of Sciences Committee on Oceanography (NASCO). These reports not only pointed out that high-level radioactive waste should not be put in the sea (following the IGY results), but also took issue with the attitudes of the (American) Atomic Energy Commission and the policies of the (British) Atomic Energy Authority. One study on nuclear-powered ships took aim at the British discharges at Windscale, provoking consternation and outrage from the British health physicists who already had blessed the practice. Another study on disposal sites caused a political uproar all along the Atlantic coast of the United States and, in a withering blow, undermined not only the academy's credibility but also led ultimately to the United States abandoning the oceans as repository for even its lower-level packaged wastes. A third report, on waste disposal in the Pacific, was hotly contested—the deliberations over this report, discussed here, revealed growing frustration at oceanographers' opportunism and the politicization of an issue that atomic energy establishments seemed helpless to overcome.

No Evidence of Dead Zones

The IGY was the first international scientific project to include both the United States and the Soviet Union, following the goodwill of the 1955 Geneva conference on the peaceful uses of atomic energy. On the surface, geophysics appeared to be a fairly noncontroversial topic (unlike atomic science), but it was not long before the participating scientists realized that the earth, sea, and atmosphere—all subjects of inquiry in the IGY—could be even more politicized than laboratory sciences. The launch of *Sputnik* in October 1957, as part of the Soviet Union's atmospheric program, made this abundantly clear. The quest to discover whether deep ocean water was stagnant or not became an object of the IGY oceanographic program, and the answer would hold the key to understanding the role of the oceans in disposing of nuclear waste.

Some American scientists seemed to suggest that there was no harm from sea disposal at all. In December 1957, Scripps Oceanographer Theodore Folsom concluded this to an international audience at the Pacific Science Congress in Bangkok, Thailand. Reviewing American waste disposal practices in the ocean off San Francisco, he pointed out that some ten thousand concrete barrels had been dropped there since 1946. The *Los Angeles Times* quoted him saying that each barrel contained one curie of radioactivity. "That is pretty hot," he said. "You would have to stand off and stir that with a long pole." Yet the studies of the area by scientist at Scripps had found no evidence of contamination in their surveys of the area. This seemed to point up the power of the ocean to render even "pretty hot" wastes harmless. There were certainly higher than expected levels of radioactivity there, but he credited these to the nuclear weapons tests in the Pacific, not to waste disposal.[2]

Overall, however, the IGY proved disappointing to those hoping to use the sea for the disposal of high-level wastes, because the assumptions about stagnant waters seemed to disintegrate as the eighteen-month "year" wore on. Early evidence of deep circulation came from Japanese and French scientists. Working jointly, they used a bathyscaphe, a manned submersible craft recently invented by Swiss scientist and explorer Auguste Piccard. Some 120 miles off the coast of Japan, after descending to a depth of over nine thousand feet, the two-man *FNRS III* measured slow currents. A Japanese scientist who went down with the bathyscaphe, Tadayoshi Sasaki of the Tokyo Fishery College, reported water movement at slightly less than an inch per second. He concluded that the minute movement was probably caused by ice melting at the poles. Despite the widely held belief that the world's deepest waters were stagnant, here was evidence to the contrary. To Sasaki, the implications were crystal clear. "Considering the length of half-life of radioactive waste," he said, "this sluggish flow of deep sea water would make the sea unsafe as a dumping place for atomic reactor waste."[3] Thus a Japanese scientist once again led the way in pointing out the dangers from American nuclear activities.

Soviet scientists came to similar conclusions. This was no surprise, as Western intelligence agencies had presumed from their translations of recent Russian literature that the Soviet Union was going to play up the differences of scientific opinion about the deep waters. A November 1958 article in *TASS*, for example, told readers that the research vessel *Vityaz* had shown that there were no stagnant waters in the deep Pacific, and that the circulation provided ample oxygen supplies for creatures to live at such depths. "These explorations," the article stated, "have shattered the contention of some foreign specialists that deep water depressions can be used as a 'dumping ground' for atomic fission waste products."[4] American newspapers put the disagreement more mildly. "Huge canyons in the oceans, far deeper than the Grand Canyon, are not good ash cans for the nuclear age, a Soviet scientist said today." Thus the *New York Times* heralded the Soviet findings of the IGY. Quoting Lev Zenkevich, the newspaper noted that water was not stagnant in the deepest parts of the ocean, as some oceanographers had hoped. Instead, the water circulated, meaning that radioactive wastes could poison nearby sea life and, ultimately, man.[5]

The *New York Times* followed up these revelations with a major piece by journalist Walter Sullivan entitled "Sea Depths Yield Secrets in I.G.Y." Sullivan painted a bleak picture of the deep ocean. The region of the western Pacific that the Soviets had measured, the Marianas Trench (sometimes called the Nero Deep), Sullivan described as "perhaps the blackest and most inhospitable place on the face of the earth," with no signs of life. But there was evidence that there was indeed life in most of the deep trenches, albeit a meager amount. There was enough of it, however, to lend credence to the claim that there was circulation in the area to provide oxygen and thus sustain life.

Soviet scientists took an interest in deep circulation, partly because the Black Sea was mentioned most often as a potential international atomic graveyard. In Sebastopol on the Crimean peninsula, the Academy of Sciences of the USSR managed a marine biological station that conducted studies during the IGY of the concentration of fission products (Sr^{90}, Cs^{137}, and Ce^{144}) in certain organisms such as algae, mollusks, and crustaceans, as well as transport of water from the bottom. They determined that it took between 60 to 130 years for the deep water to rise to the surface, which led them to conclude that the concentration of dangerous isotopes—particularly strontium-90—would diminish between five and thirty times by the time the water reached that level. Still, it was not a stagnant sea.[6]

The Soviet Union faced many of the same technical problems about waste disposal as did the United States and Britain, but it was far less forthcoming about its practices. The country's first plutonium production plant, called

Mayak (known to Westerners as Kyshtym, after a nearby town), was built in 1948 some seventy kilometers north of Chelyabinsk in the region of the Ural Mountains. Although far from the ocean, the Mayak complex was located on the Techa River and was very close to lakes and marshes, with groundwater at shallow depth. Much of the high-level waste was stored in tanks, but the Soviet Union also released radioactive materials directly into the nearby Lake Karachai, beginning in 1951. Some other major facilities for producing weapons-grade materials and for reprocessing nuclear fuel were built near cities farther east in Siberia, such as the Siberian Chemical Combine at Tomsk (also known as Tomsk-7) and the Mining and Chemical Combine at Krasnoyarsk (also known as Krasnoyarsk-26); all of these major weapons complexes used local ponds, reservoirs, or lakes for discharge of liquid wastes and also released some radioactive materials into rivers, the Ob (carrying waste from Chelyabinsk and Tomsk) and the Yenisey (carrying waste from Krasnoyarsk) rivers, northward toward the Kara Sea. The Soviet Union also dumped solid radioactive waste in the Arctic Ocean, and in the 1960s it would even dump whole reactors with their spent fuel.[7] But the details of Soviet activities were not known publicly in the 1950s and 1960s, and in fact the Soviet government denied putting any radioactive waste in the sea; consequently, the primary dumping nations appeared at that time to be the United States and Britain.

Soviet scientists made their negative views of sea disposal known in 1958, at the Second International Conference on the Peaceful Uses of Atomic Energy, held again in Geneva. In some ways the meeting reflected the cooperative spirit of the first conference three years earlier. For example, the scientists discussed the need to break down the barriers of secrecy that hampered cooperation in the field of atomic energy. They urged following the spirit of the IGY, ongoing since mid-1957, which emphasized the free exchange of data. Some previously secret areas of research, such as controlled thermonuclear reactions, now were shared at the international conference. They hoped that a third conference, projected in 1961 or 1962, would ensure that turning their backs on secrecy was not an ephemeral step. American physicist Isidor Rabi used the conference as a means to voice the position of the Eisenhower administration, that these conferences would serve to bind the nations of the world together in a commitment to use atomic energy toward peaceful rather than destructive ends. The chairman of the Soviet Union's State Committee for the Utilization of Atomic Energy, Vassily Emelyanov, echoed this sentiment and added that perhaps peaceful atomic energy would stimulate popular movements for peace among the world's peoples.[8]

Still, the 1958 Geneva conference was more tense than the first one had been. Delegates from the United States and Soviet Union were eager to outdo each other. Emelyanov gave a talk in which he predicted that ships would one day be powered by thermonuclear energy, based on the same kinds of fusion reactions that powered megaton-range weapons and the sun itself. The *New York Times* observed that Western scientists dismissed the Soviet claim as fanciful; any fusion power facility would have to be housed in a huge power station, not aboard a ship.[9] But the Soviets had progressed far in the construction of a (fission) nuclear-powered icebreaker, the *Lenin*, which the representatives at Geneva described in some detail.[10]

The American delegates were certainly proud of the impact they had made at Geneva, particularly through its technical exhibit—which included two operating reactors. Scientists returned to the exhibit again and again; the American delegation's confidential report to the State Department quoted foreign scientists with phrases such as "beyond belief," "can scarcely believe it," and "have stolen the show." Turkish delegate B. Kurzunoglu commented, "The difference between the United States and the Soviet exhibit shows the real difference between the people. Am I glad we Turks are so near to the Americans."[11]

The goodwill of the conference dissipated when the Americans raised the possibility of using nuclear explosives for peaceful purposes. The United States already had begun its Plowshare program and planned to conduct an experimental nuclear excavation in Alaska to build a harbor.[12] Emelyanov pointed out that the Americans could use such ostensibly peaceful activities to cover up military activities. Particularly if the Americans were planning to cease military nuclear testing for political reasons, such activities seemed questionable. Emelyanov accused AEC Commissioner Willard F. Libby of trying to find a way to continue testing, and he said that the Soviet Union had no intention of developing nuclear explosives for peaceful purposes, as they would have only political purposes. Reporters at the conference challenged this point, noting that a previous Soviet foreign minister, Andrei Vishinsky, had said the opposite in 1949—that the Soviet Union would use such explosive to dig canals, level mountains, and change the course of rivers. But Emelyanov maintained that the Soviets had never done it and never planned to do so.[13]

The day after Emelyanov's remarks, the *New York Times* editorialized that all men of goodwill should regret the taint that the Soviet delegate brought to the conference. It lamented that Emelyanov's "political instructions seem to have overridden not only his scientific courtesy but also his common sense and his memory." It was obvious to anybody, the newspaper continued, that if nuclear explosions could be used for engineering projects, they ought to be used; besides, the Soviet already had said as much in the past.[14]

Emelyanov followed up his accusations by announcing the inauguration of the largest-yet atomic power station, designed to produce power on the order of 100,000 kilowatts. According to the Soviets, it was the first of six to be built in an undisclosed location in Siberia. He unveiled the facility by showing a color movie to the scientists attending the conference, complete with ballet music as a backdrop to scenes of the plant, its workers, and their families. The output was far more than the American prototype station in Shippingport, which was a 60,000-kilowatt power plant (though an upgrade to 100,000 kilowatts was in the works), and even more than the three British plants, which operated at a comparatively meager 40,000 kilowatts.[15]

As the Soviets, Americans, and British compared the sizes of their reactors and debated the technological implications (i.e., the possible dual uses of huge reactors to produce plutonium), other countries voiced concerns about the dangerous by-products. The Dutch delegation issued a statement asking for international regulations to protect public health. Especially in Europe, where the countries were industrialized but small, the danger of one country's radioactive waste reaching another seemed very high. Although many of the delegates did not want to discuss the issue, seeing it as the province of politicians, the Royal Netherlands Academy of Sciences backed early plans for international regulation. Even if they were not supposed to negotiate with each other, Dutch delegate Johan H. de Boer and others argued, they could at least urge the responsible bodies to do something.[16]

After the conference ended, more superpower confrontation emerged, this time at an international conference assessing the first year of the International Atomic Energy Agency, in late September 1958. John McCone, who recently had replaced Lewis Strauss as AEC chairman, tried to continue the spirit of his country's Atoms for Peace initiative, promising research money to the IAEA. The United States, he suggested, could identify problems for the agency and provide money so that scientists from all over the world could work on them on a contract basis.[17] But Emelyanov renewed his criticism of the United States, saying that the Americans were not really interested in seeing the agency succeed. Again he alluded to American desires to find ways to continue testing nuclear weapons. The Americans, in turn, dismissed his critique as irrelevant and politically charged.[18]

The bickering at these conferences heralded a serious deterioration in superpower relations in the arena of peaceful nuclear energy. In addition to its criticism of nuclear testing, the Soviet Union made it clear that it stood against ocean dumping of radioactive waste, and this was repeated routinely by political representatives and scientists alike. Soviet scientists writing on the subject implied that the principal factors in making disposal decisions in Western countries were more economic than scientific. One of these scientists was Evgeny Mikhailovich Kreps, a physiologist who had studied under Ivan

Pavlov and was known primarily for his work on the physiology of marine life. His perspective was an ecological one, and he focused on the uptake of fission products by organisms. In his discussion of waste disposal at the 1958 Geneva conference, Kreps noted that decisions about land and ocean disposal should be made from consideration primarily of the safety and security of the population and only secondarily of economic factors. Storing radioactive waste in containers buried in the ground was expensive, thus "the atomic industry in the capitalistic countries is consistently on the lookout for cheaper means of disposal." They did this by scattering their waste into the natural environment, he said, to achieve dilution in the ocean and other waters.[19]

Kreps pointed his finger primarily at the United States and Britain, both of whom he said dumped waste in the ocean purely to avoid the cost of storage. This was dangerous, he argued, because of scientists' insufficient knowledge of oceanic circulation. Ambiguity about circulation called into question both the possibility of adequate dilution and the feasibility of deepwater isolation. In addition to these questions of physical oceanography, there was the issue of radioactive concentration, Kreps's own area of expertise. Plankton and other organisms could concentrate such materials by "hundreds of thousands of times," and the migration of such organisms complicated scientists' understanding of the exchange of radioactive elements in the ocean. These factors, Kreps argued, "created a serious threat to all humanity," hence the recent UN agreement in 1958 "forbidding the radioactive contamination of marine waters."[20]

The greatest danger, Kreps argued, resulted from the fallout after nuclear weapons tests. In a nod to his own political context (the Soviet Union also conducted tests), he said briefly that some of the long-lived isotopes from such tests probably would not show obvious biological effects. Then he stated that the American activities in the Pacific were much more serious, raising the degree of radioactivity in the ocean, atmosphere, and in marine organisms. On the *Vityaz*'s recent Pacific cruise during the IGY, he wrote, "the sea water contamination sometimes exceeds all permissible doses and renders some of the marine products unusable by man."[21]

American and British scientists at the conference felt unable to reply. For one, they were under instructions to avoid political entanglements of this kind. But, more important, Kreps was trying expose the American and British practices without discussing Soviet practices in detail. Any kind of direct comparison of operations proved impossible. "Since the Soviets have said almost nothing about their own waste disposal methods," the American delegates reported to the State Department, "they could not be counter-attacked on this point." In fact, the Soviets deftly avoided criticisms of any of their papers, simply by holding them back and only submitting them for general consumption just prior to presentation. Even the list of Soviet participants was unavailable until after the scientists arrived in Geneva.[22]

The UN agreement to which Kreps referred was the United Nations Conference on the Law of the Sea (UNCLOS), which met from February through April 1958. This conference added renewed urgency to the problem. Scientists needed to find answers quickly in order to inform future agreements. United Nations signatories agreed to prevent pollution of the seas from the exploitation of offshore oil and other resources, and to take steps to avoid contamination of the sea by radioactive waste, according to principles to be set forth by a competent international organization. Because that organization would be the newly created International Atomic Energy Agency, the handwriting now was on the wall: oceanographers and other scientists would have a limited amount of time to gather data and bolster their professional judgments about international regulation of dumping, because the IAEA would see the formulation of specific regulations as one of its principal mandates.[23] In fact, it began to act on this shortly, convening in Monaco a conference of oceanographers, geologists, and atomic energy representatives. It also arranged to have a study made to assess sea disposal, under the leadership of Swedish atomic energy official Harry Brynielsson. Both of these efforts, discussed in the next chapter, would do more to exacerbate disagreements than to settle them.

In the Hands of American Oceanographers

With more international attention focused on waste disposal at sea, oceanographers continued to rise in prominence. Their work seemed to hold the key to answering the contentious questions raised by scientists and politicians everywhere. In the United States, oceanographers used the wide-ranging activities of the IGY to persuade the National Academy of Sciences of the need to establish better coordination in marine science. The academy established the Committee on Oceanography (NASCO) to formulate recommendations to the government to help it establish national ocean-related policies. This body acted as the springboard for the BEAR oceanographers to turn their scientific conclusions into policy recommendations. Two of the BEAR oceanographers became members of NASCO, Roger Revelle and Milner Schaefer, and their voices—especially Revelle's—soon dominated national discussions about America's use of the oceans. The NASCO chairman, Harrison Brown, wrote to Bronk about the areas needed for policy recommendations, as stated but not yet acted upon by the BEAR oceanographers, such as incorporating more scientific work into weapons tests, relaxing security restrictions on research publication and data exchange, charging a national agency with regulating ocean disposal, and entering into international agreements. Bronk agreed, paving the way for the BEAR oceanographers' voices to be heard not merely as detached experts but now as policy advisors.[24]

The Atomic Energy Commission asked NASCO to conduct a number of studies in the late 1950s. One involved a precious Navy innovation—the nuclear-powered ship. To the Navy, nuclear fuel was a glorious new paradigm in strategic power. Vessels soon would avoid the time-consuming tasks of refueling, and submarines could stay submerged as long as their oxygen supplies held out. This meant that each ship was going to have a reactor, and each reactor would produce waste. Although the AEC was content to leave this issue to the Navy, it knew that part of Eisenhower's Atoms for Peace initiative was to see merchant and cargo vessels carry nuclear reactors of their own. There would be two options: keep the waste aboard or allow the ships to release it at sea. The former would present no problem to oceanographers, but shipbuilders and the AEC agreed that it would place a heavy burden on the cost of operation and would decrease the value of these ships as cargo carriers. In order for them to be cost efficient and thus attractive commercially, these ships might need to dispose of their waste by discharge. The first major civilian surface ship, the NS *Savannah*, was being built by General Electric and was scheduled for sea trials in 1960. Although it was being designed to collect and retain its wastes, without discharges to the ocean, this was not intended to be a permanent feature of all future vessels of the kind.[25]

Because there were less than a hundred merchant vessels in the world with displacements greater than 20,000 tons, General Electric assumed that one could conservatively estimate that one hundred was the maximum number of nuclear-powered vessels to be built in the foreseeable future, even with a massive campaign to convert the world's ships. The probability of there being a thousand such ships across the world over the next few decades was, to General Electric radiological engineer James M. Smith Jr., "vanishingly small." Each would produce both solid and liquid waste. Smith envisioned that all future ships would include waste concentration systems, which would decrease the amount of space needed for storage, to allow some flexibility about whether to keep them aboard or simply release them at sea.[26]

Because of the urgent need to provide data in support of these developments, AEC sanitary engineer Arnold Joseph asked NASCO to arrive at some conclusions about safe discharges that they could use for actual operations. But there was more to the study than that. Judging the safety of disposal from ships would necessitate a definitive statement about discharges from any source into any part of the open ocean. Since the IAEA was putting together recommendations about disposals and discharges, the academy was being asked to formulate the consensus opinion of American oceanographers about how much the ocean could receive—a threshold value—without tailoring it to the specific environment. In other words, the AEC needed an "American" position, rather than just the AEC's or oceanographers' position, upon which American representatives at the IAEA could rely. The American position on the subject needed to be established early, AEC scientist Douglas Worf told NASCO, "before it becomes an acute issue within international political bodies."[27]

The American oceanographers undoubtedly were keenly aware that more was at stake than just civilian nuclear-powered ships, or even waste disposal policies. The United States had begun to build a nuclear navy—thus ocean disposal had to be defended to some degree. Declaring the safety of radioactive waste discharges from the merchant fleet was one way to do this without inviting criticism from the Soviet Union that the American military/naval arsenal was contaminating the oceans. Nuclear-powered vessels would release such wastes in liquid form very near the surface. They might take the form of reactor coolant, corrosion products, gases, leakage from the core into the coolant, and contaminated water. The United States would not be the only country developing such ships, so the oceanographers needed to imagine a scenario in which entire fleets of nuclear ships continuously released radioactive materials into the sea. Worf estimated that several countries over the next ten years would have nuclear-powered cargo ships operating in the major sea lanes.[28]

As NASCO continued its work, the close involvement of the AEC left a bad taste in some oceanographers' mouths. The money for these studies was coming from the pot allocated to the BEAR study, which meant that the Rockefeller Foundation was paying for them. It dawned on the National Academy (particularly Charles Campbell, who managed the BEAR committees) that the AEC might be requesting these studies purely for the benefit of being able to say that they were conducted by the academy, through independent funds. Yet they were anything but independent of the AEC's interference. NASCO executive secretary Richard Vetter cautioned Revelle that perhaps they had taken the wrong course: "I discussed this arrangement with Charlie Campbell. He questioned the appropriateness of using Rockefeller Foundation money . . . if the main motivation for so doing is to avoid any possible prejudice that might be connected with a closer identification with AEC support." However, another view won out: any of the sponsors of NASCO should expect to receive policy advice from it. And the results would contribute to the goals of the BEAR group because the study of nuclear-powered vessels would have worldwide ramifications.[29]

Because the academy gave the nuclear ship study to the BEAR Committee on oceanography and fisheries, Roger Revelle, the chairman, again had an opportunity to carve out a place in atomic energy matters. The report, for which Revelle wrote an introduction, was based on the work of a smaller group led by Donald W. Pritchard, a scientist at Johns Hopkins University's Chesapeake Bay Institute. Of the twenty scientists listed as contributors in the report's introduction, one was an AEC sanitary engineer (Arnold Joseph, listed only as a consultant) and another worked for General Electric (James M. Smith Jr.), the company building the *Savannah*. The rest were marine scientists, some government-employed but most working for universities and oceanographic institutions.[30]

Revelle's and Pritchard's committees made some fairly strident remarks in their report, making critical comments about the AEC and AEA practices of working as close to the permissible levels of discharge as possible. Although ostensibly about nuclear-powered ships, their report actually was conceived more broadly, describing the general considerations for judging permissible discharges into the sea. The report cited the methods used in Britain by health physicist H. J. (John) Dunster, and those used in the United States by the Carritt committee, both of which began with the maximum human exposures recommended by the International Commission on Radiological Protection. If one looked at the environment as a source of exposure, then one could limit the introduction of radioactive material to the sea accordingly. Commending this method as a "reasonable" procedure, the academy committee tried to tackle the misleading notion of thresholds by drawing a distinction between "permissible" rates of discharge and "acceptable" rates of discharge, the one not necessarily determining the other. The acceptable discharge always should be less than the permissible rate, the report stated, to be reduced until it should become too costly to do so. Given the evidence that there were no threshold levels of danger from radiation, the oceanographers concluded, they must reduce the levels if at all practical to do so. In some cases in which alternatives existed at low cost, even with existing permissible levels, "*the acceptable discharge of radioactive waste may be zero.*" The report added: "This is emphasized because, although it has been repeatedly point out by other committees dealing with such problems, it has often been ignored."[31]

Certainly the tone of the report reflected a new confidence from the oceanographers. They were no longer asking to be invited to the nuclear table, but instead were making brazen authoritative judgments. Still, neither Revelle nor his colleagues saw themselves as turning against ocean disposal; indeed, most criticisms pointed toward the need for more research, and thus one could regard the deprecating comments not as general disparagement of the AEC but rather as a patronage strategy, revealing gaps in knowledge needing to be filled. Were the AEC to forgo filling them, it might mean the end of sea waste disposal—and it certainly would mean the end of oceanographers' blessing. This was not outright blackmail, but it did put the pressure on. Scripps oceanographer Milner B. Schaefer spelled it out explicitly in a presentation at the June 1960 annual meeting of the Health Physics Society and later in the journal *Health Physics*. He pointed out that the incompleteness of scientists' understanding of the sea meant that "waste disposal practices must be more restrictive than would be the case if we knew more. This, of course, entails economic costs which can in the future be very large, and which may be reduced by the results of further research." He then referred to the research requirements outlined by NASCO in 1959 and by himself and Revelle at the Geneva conference in 1958.[32]

Pritchard, fresh from his experience with the nuclear ship study, told the same audience of health physicists that radiation policies did not stop with setting exposure levels. There were many other avenues to explore as well, and Pritchard relegated the total allowable dose rate to humans, certainly the most familiar subject to the health physicists in attendance, to a very small role. In addition were the routes to humans through the marine environment—eating seafood, use of contaminated fishing gear, contact with contaminated sand, etc.—all of which necessitated different dose figures tailored to the maximum permissible concentrations in various parts of the marine environment. One had to know how humans received these doses from the environment, and of course one also had to know how radionuclides concentrated in seawater, sediments, and biota. All of this required more money for research. Still, Pritchard was careful not to set himself up as opposing the practice; he observed that in the meantime even conservative estimates should allow the oceans to be utilized to receive limited amounts of radioactive materials. This followed the BEAR report of 1956 and the preliminary conclusions of other studies by the academy and by the IAEA.[33] Pritchard and others did not oppose threshold values; rather they wanted a greater role for oceanographers in setting them.

In addition to the growing divide between oceanographers and health physicists in the United States, there were increasingly sharp contrasts between the American and British views. The latter were based not on oceanographers' research but on work by government scientists at the Atomic Energy Authority and at the Ministry of Agriculture, Fisheries and Food. In 1958 John Dunster and F. R. Farmer published an internal report for the British AEA's Health and Safety Branch detailing its studies of effluent discharges from Windscale. They provided an overview of the early dye experiments and gave a general description of the importance of concentrated radioactivity in organic material, particularly edible seaweed, in limiting the amount of effluent discharged. They also insisted that even now the discharges were still considered experimental. When Windscale began regular discharges in 1952, AEA officials believed that they could call them experimental, because any studies without actual discharges were unlikely to be conclusive. Ever since the initial discharges, they noted, the discharges had always been kept below authorized levels, but they felt compelled to keep them high enough to allow measurable radioactivity levels in fish—all necessary, they said, for an experiment to yield conclusive results. It was fairly easy to take samples of possible concentrations of radioactivity in the sea off Cumberland, the location of Windscale, because there were masses of suspended solid materials, including sand, mud, and organic debris, making up the local biological habitat, not to mention the seaweed and fish such as plaice. These measurements were used not only as data in the ongoing experiment, but also to "monitor" the safety of effluent discharges.[34]

The report laid out British assumptions on permissible levels, which were based entirely upon the recommendations by the International Commission on Radiological Protection. This body's goal was to protect workers, and thus it confined itself to defining permissible exposures to humans, without reference to the ocean.[35] The British health physicists working on ocean discharge adopted the same approach. They did not monitor levels of radioactivity in the seawater, for example, but instead focused their work on those organisms that presented the most likely pathway for dangerous levels of radioactivity to reach humans. This necessitated studies of how different radioactive elements concentrated in organic material and the seabed. By 1958, the British identified different discharge levels for beta activity (15,000 curies per month) and the most dangerous alpha activity from strontium-90 (3,000 curies per month), ruthenium-106 (5,000 curies per month), and plutonium-239 (200 curies per month)—all of these levels they considered extremely conservative, with an overly cautious margin of safety. From all of these studies, the health and safety officials at the AEA determined that seaweed was probably the most likely pathway to reach humans, because most of Cumberland's harvestable seaweed grew within twenty kilometers of the pipeline's outlet. Permissible doses thus were based on calibrating the concentration of radioactivity in seaweed with expected consumption, to approximate the safety levels published by the ICRP. This technique of setting discharge levels has been touted as the "critical pathways" approach, and it placed no significance upon the effects of radioactivity on the ocean environment per se. Instead, it was based on one chosen route by which radioactivity reached humans.[36]

These policies bore the mark of health physicists, not of oceanographers. The difference in outlook illuminated an uncomfortably obvious rift between two sets of professionals, the academic oceanographers and the scientists who worked within atomic energy establishments—sometimes called oceanographers but more typically identifying themselves as radiobiologists, biophysicists, health physicists, and sanitary engineers. Health physicists, taking their cues from national and international radiation protection bodies, often resisted the efforts of outside scientists to assert roles as experts in formulating policies; this meant that they embraced an all-encompassing definition of their own field. In the first issue of the journal *Health Physics*, in 1958, Walter D. Claus defined his field as the "science or profession of radiation protection in all its aspects." This meant that health physicists had to familiarize themselves with an array of disciplines: physics, electronics, biophysics, chemistry, biochemistry, biology, physiology, genetics, toxicology, and ecology. Moreover, the health physicist needed to be keenly attuned to the legal and public relations aspects of his work. Claus pointed out that waste disposal, with most of the high-level waste stored in temporary tanks, was not yet solved, and it might be necessary to put such wastes into the sea. "Without doubt," he admitted, "the whole question of waste disposal is the knottiest problem which the health physicist has to handle." Claus did not suggest, however, that this was a problem for others to solve; he clearly saw it as the province and responsibility of health physicists.[37]

The health physicists in the United States and Britain who had developed the methods to judge safety levels of sea dumping were quite disturbed to see repeated references in the statements of oceanographers (particularly American oceanographers) that radioactive waste disposal at sea ought to be limited by the potential harm to humans *and to the marine environment*. In fact, scientists at the Atomic Energy Authority were bemused at how American health physicists had conceded so much authority to marine scientists. They were surprised at the scale of work being done in the United States on what they considered to be nonessential issues for developing sensible operating policies. Reporting on a visit to nine American and Canadian research sites in 1959, British scientist W. L. Templeton told colleagues at Windscale that the Americans "are carrying out an enormous programme by our standards on theoretical assessments of the effects of radioisotopes discharged to the sea." Part of his visit was devoted to the laboratories near the major production facilities in Oak Ridge, Tennessee, and Hanford, Washington; at these sites, Templeton learned of extensive work in ecology, aquatic biology, and other studies on ground and freshwater contamination. But most of the other sites dealt specifically with issues directly related to the sea. For example, American scientists were conducting research on the effects of nuclear powered ships, financed by the AEC's Reactor Safety Division, to advise on the operations of the reactor in the *Savannah*. Other projects on the effects of radioactive waste disposal on fish, the use of radioisotopes in fisheries laboratories, and the marine biological surveys of weapons tests areas were paid for by the AEC's Division of Biology and Medicine, the U.S. Fish and Wildlife Service, and the U.S. Navy. Aside from the known isotopes such as strontium-90, the discovery of large amounts of radioactive zinc, cobalt, iron, manganese, and copper in marine organisms following the hydrogen bomb tests in the Pacific added a renewed urgency to the problem of biological concentration, and it seemed that American scientists had succeeded in getting various divisions of the AEC and other government agencies to pay for such studies.[38]

During his visit, Templeton was urged frequently by the Americans to pay more attention at Windscale to making surveys of all the different isotope concentrations in as many species as possible. Scientists at the Fish and Wildlife Service's radiobiological laboratory in Beaufort, North Carolina, W. A. Chipman and T. A. Rice, were particularly critical of the British. They told Templeton that the American public would never accept the levels of isotope concentration that they produced in their marine organisms around Windscale. They discussed the fact that the British discharge policies were based primarily on the critical pathway to humans, particularly the concentrations of radioactive isotopes in seaweed. What the British allowed in their seaweed was far higher than the Americans ever would allow, they claimed. In addition, the scientists at Beaufort pointed out, the British had little information about other organisms, and they paid almost no attention to isotopes other than strontium-90 and Ruthenium-106. The Americans, by contrast, were doing more inclusive studies of a wider range both of organisms and of radioactive elements.[39]

In addition to all this, oceanographers at Woods Hole Oceanographic Institution had begun a major project on the distribution of radioisotopes in the sea, partly financed by the AEC's Division of Biology and Medicine. The goal of this work was to judge the probable effects of dumping large amounts of high-level radioactive waste into the deepest parts of the ocean. Under the influence of Bostwick Ketchum and V. T. Bowen, these studies emphasized not just current systems and mixing processes, but also biological and geochemical transfer of isotopes. One special focus of theirs was the vertical migration of organisms, which could affect the transfer of dangerous material from wastes sitting on the sea floor toward the upper layers where fishing boats trawled. The marine scientists at Woods Hole were troubled by the possibility that the most likely vertical transfer of isotopes would be from biological sources, not from the ocean's physical systems.[40]

British health physicists like Templeton found the vast American program baffling. American scientists had acquired funding in every subject that could be construed as touching on the AEC's waste disposal operations. These scientists had succeeded in dipping into the deep pockets of the federal government to an extent that would have seemed ludicrous in Britain. This was partly due to the American government having more money to spend, but it also was due to the fact that American oceanographers' success in breaking into the atomic energy field had no equivalent in Britain. The British government certainly supported research, not just at Harwell but also at other locations, such as the radiobiological laboratory operated by the Ministry of Agriculture, Fisheries and Food. Most of the studies, however, were operational in nature and were designed to inform the AEC's operations by studying ecological pathways to humans. The same could be said of some of the American studies, to be sure; but many of the studies seemed superfluous to waste disposal. Why, for example, was it necessary to study every isotope in every major organism? Were such studies designed to help the AEC refine its policies or simply to pay for scientific research? In British atomic energy circles, American marine scientists were gaining a reputation for second-guessing American health physicists; they appeared as opportunists whose only conclusion about waste disposal was that more research needed to be funded.

The Atlantic Coast Debacle

What the British perceived as bizarre concessions to marine scientists by the AEC actually reflected a mounting contest for scientific authority between health physicists and oceanographers in the United States. This began to manifest itself most clearly after the AEC sought advice from the National Academy of Sciences Committee on Oceanography in January 1958 about using the Atlantic coast as a major dumping area. Already waste disposal had become something of a routine there. In September 1957, the *New York Times* heralded the five hundredth waste disposal voyage of the *Irene Mae*, a cargo vessel that dumped radioactive waste less than thirty miles off the coast of Massachusetts. The ship's owner, Crossroads Marine Disposal Corporation, had been in the radioactive waste disposal business since 1946, with research centers and the AEC (beginning in the early 1950s) as its main customers. When it set up an advertising booth at a nuclear-related conference in Philadelphia in 1957, its clientele expanded dramatically. A similar firm, Nuclear Engineering Company, dumped waste from California into the Pacific. Licensed by the AEC to dump at sea, Crossroad Marine Disposal Corporation was profiled by the *New York Times* as "one of those unusual satellites that have grown up around the atomic age."[41] Not everyone saw these developments as benign. One New York assemblyman, Leo A. Lawrence, made headlines by asking about the possible harm to the seafood industry as a result of radioactive waste disposal, nearby nuclear plants, and more recently, the operation of nuclear-powered ships by the Navy.[42]

The AEC wanted to permit the dumping of low-level radioactive wastes closer to shore than it then allowed—that is, closer than one hundred miles out. "Time is of the essence," AEC sanitary engineer Arnold Joseph told NASCO, because the AEC wanted to allow the use of close-in areas by May. Joseph was worried about the public relations problems that might come with increased use of these areas. With the blessing of Joseph and other health physicists (or sanitary engineers, as those directly concerned with public health often called themselves), the AEC had been allowing a commercial firm to dump its low-level radioactive wastes in shallow water (fifty fathoms), less than fifteen miles from shore. That company, unnamed by the AEC but later revealed to be Crossroads Marine Disposal Corporation, was profiting from the fact that it did not have to incur the expenses of getting the waste farther out to sea. The problem, however, was that the company was getting bigger and bigger. According to Joseph, the AEC "has not strenuously objected to this shallow water disposal in the past because the amounts of radioactivity so disposed were not large (equivalent roughly to what is allowed to be discharged into sewers) and because there was no valid scientific reason for disallowing it." But with increased size, the AEC was concerned about three things: public relations, possible contamination of local creatures, and the physical obstruction (by waste containers) of fishing and trawling areas. It seemed to the AEC that the operations were just going to escalate. In addition, the AEC was in the predicament of receiving applications from other companies wanting licenses to the do the same.[43]

From the AEC's perspective, the situation had not gotten out of control—so far. The Navy was dumping its radioactive wastes farther out, and this did not concern the AEC because the Navy had a good track record of following the recommendation of the National Committee on Radiation Protection. In addition, the Navy used weighted containers and always disposed of the wastes in areas specially designated for unused ammunition and hazardous chemicals. The AEC had asked the commercial businesses to use the Navy areas, but they balked at the expense of having to go so far out to sea, which would require the services of ocean-worthy vessels. The AEC did nothing to compel compliance. "There are no strong scientific reasons," Joseph wrote, "for requiring them to do so."[44]

Joseph's purpose in writing to NASCO, then, was to see if the oceanographers could provide some scientific justification (external to the AEC) for moving the recommended dumping areas closer to shore and in shallower water. He was uncomfortable with the AEC requiring distant dumping if commercial firms did not want to bear the expense of it. Joseph called this a "conflict of interest." He did not appear to share the same view of the AEC's relationship with NASCO, which was expected to provide scientific credibility to less inconvenient practices. Already the AEC had made the decision to bow to pressure from commercial firms—Joseph mentioned that "the Atomic Energy Commission feels that as many as 4 or 5 disposal areas can be established along the Atlantic Seaboard," all conveniently located near ports. The areas should, he said, be rather large so that companies would not need to be precise about navigation. Joseph went so far as to specify who should be placed on the NASCO panel to look into the matter, and he included himself. Ultimately, Joseph was not formally attached to the NASCO panel, though most of the scientists he suggested were. The panel's chairman, Dayton Carritt of the Chesapeake Bay Institute, was hand-picked by the AEC.[45]

The AEC wanted to use NASCO to help justify its policies, not to evaluate them, and AEC health physicists became hostile when NASCO raised objections. When Dayton Carritt's group presented drafts to the AEC, Joseph and his colleagues gave it only moderate praise before launching into a major critique, mostly on semantics. One of Joseph's serious criticisms was in Carritt's use of the word "hazard" and the designation of all man-made environmental radioactivity (including fallout) as radioactive waste. Such words were suggestive of danger, he said, and perhaps they were acceptable when discussing nuclear weapons. Civilian nuclear power, he believed, ought to be treated differently. Radioactivity from weapons testing ought to be differentiated from waste disposal, he said; after all, the purpose of the report was to show businesses where to dump, not to highlight dangers. Also, he observed, "'Hazard' according to Webster implies a foreseeable, but uncontrollable possibility of danger, i.e. liability to injury, damage or loss." In places where Carritt referred to "hazard levels," Joseph suggested replacing it with "safety levels." He also objected to Carritt's use of the word "misleading" when "inaccurate" could suffice, or his use of the word "waste" when "concentration of isotopes" might do.[46]

The AEC's efforts to bring atomic waste closer inland—or at least to justify those practices already taking place—sparked consternation in Congress. NASCO did as the AEC requested and, after mulling over the possibility of dumping off the Atlantic seaboard, chose twenty-eight sites that might be suitable. Then the locations were revealed to the public, and a map was soon published in the *New York Times*.[47] Virginia congressman Thomas N. Downing, a Democrat, was shocked to discover that three of these were off Virginia. He wrote to AEC chairman John A. McCone to "register a strong protest to any further consideration of the sites located off the coastal waters of Virginia." While he had great confidence in NASCO's scientific abilities, he wrote, "I cannot see where it is either necessary or practical to dispose of this radio active material in waters so close to our shore." Aside from the possible physical harm, such disposal would be bad for business. The sites, it seemed, were very close to resort areas, so "there would also arise a psychological factor which could possibly be harmful to the economy of this area."[48] Downing received a reply not from the AEC but from the Navy's Chief of Naval Research, Rear Admiral Rawson Bennett. He agreed with Downing that psychological attitudes of the general public were crucial, and that this "can be overcome only by truthfully informing the public of all of the facts."[49] He did acknowledge that the problem was going to become more acute with the spread of atomic power and simply enclosed another letter a Navy official had written to another Congressman, Democrat Bob Casey of Texas. In that letter, he told Casey that NASCO had been extremely conservative in its estimates, that they were only considering low-level waste, and that the NASCO scientists were "among the best the country affords, and that they are also concerned over the adverse effects to life which are implicated in radioactive waste disposal."[50]

The report had followed on the heels of a controversy about issuing a license for a commercial firm to dump radioactive waste in the Gulf of Mexico. That incident had roused not only domestic outcries but also a diplomatic protest from Mexico.[51] Because of the existing controversy, pointing to the NASCO scientists' credentials did not end the matter. Bob Casey grumbled that he had no quarrel with scientists, and he did not oppose dumping in general. "But it will take only one radioactive red snapper, or load of shrimp, to seriously cripple a multi-million-dollar a year industry in my state." The scientists' report had done nothing, he wrote to NASCO executive secretary Richard Vetter, to alleviate his concern.[52] Another congressman, M. Glenn, wrote to the academy that he was "amazed and concerned" that one of the sites was 22.5 miles from Atlantic City. This was very close to the bathing beaches of his district, not to mention the sport and commercial fishing being done in the whole area. A few hundred miles to sea would be much safer, he reasoned. He requested a justification of how the site was chosen—why they must dump here, "rather than in all the sections of the big expanse of the Atlantic that could not possibly cause any concern to a layman."[53]

For politicians in Texas, where AEC dumping licenses in the Gulf of Mexico already had created anxieties, the academy report simply had added fuel to the fire. In an interview with a reporter, Congressman Jack Brooks pointed out that dumping radioactive waste had become a national problem, with dumping sites identified off Boston Harbor, off Rhode Island, off Atlantic City, off Virginia and North Carolina, and off New Orleans. "And the site that was closest to my people, that really got me radioactive, was that one at 19 miles from my district, in the Gulf Coast just out of Sabine."[54]

Such reactions proved embarrassing to both the AEC and the academy. After all, the AEC already had been allowing the dumping practices. And the locations were chosen not to optimize safety but rather because of their proximity to cities, to make waste disposal convenient. The academy tried to control the damage when the report was called into question in hearings before Congress's Joint Committee on Atomic Energy in late July 1959. When the hearings highlighted the fact that the AEC already had been allowing the disposal of radioactive waste in close-in, shallow areas, a storm of media attention rained down on the AEC.[55]

AEC General Manager Alvin R. Luedecke pointed out during the hearings that there was no urgent requirement for these offshore disposal sites. Indeed, given the controversy, the AEC was going to begin requiring Crossroads Marine Disposal Corporation to move its disposal operations farther out to sea, to meet a 1000-fathom depth requirement. That company had been authorized by the AEC since 1952 to dump in water twelve to fifteen miles from shore, in water about fifty fathoms deep (this site had already been used by the Army Corps of Engineers as an explosives and toxic chemical waste dumping site). He added that, despite the daunting and all-inclusive term "radioactive," the wastes dumped at sea were of low and intermediate levels of radioactivity, particularly compared to the high-level wastes produced from production and chemical reprocessing. Most of the waste dumped in package form at sea came not from these major production facilities but rather from research and development activities. Luedecke compared the two: the concentration of radioactivity in packaged waste dumped at sea was "generally in the thousandth or millionth of a curie per gallon range, whereas the liquid high level waste resulting from chemical processing operations at Idaho might have concentrations in the hundreds or thousands of curies per gallon." Moreover, the total amount in gallons was vastly different. Luedecke revealed that in 1957 the AEC disposed off both coasts 686 fifty-five-gallon drums of solidified laboratory waste; but in storage tanks at Hanford, Idaho, and Savannah River, there were about sixty-five million gallons of high-level waste.[56]

Luedecke adopted an attitude that made the AEC look like a model of restraint. He was quick to differentiate the United States from "other countries" who put liquid radioactive waste in large quantities into the sea. He did not mention Britain specifically, but anyone with knowledge of waste disposal policies would have realized that Britain relied heavily on "bulk liquid" disposal from pipelines. The United States did not use bulk liquid discharges into the sea, except for "two minor exceptions" involving millicurie quantities. He added that oceanographers had suggested the AEC might dump their high-level waste at sea using a "dilute and disperse" philosophy, but that the AEC had no intention of doing so.[57]

As a further armament against criticism, Luedecke claimed that the AEC supported an "extensive research and development program" in the marine sciences. In fact, the AEC had been one of the bodies that asked the National Academy of Sciences to create its committee on oceanography, and it had given that group financial support. He announced the expected expenditures on oceanography in 1960 as $250,000 to help understand the fate of radioactive materials in estuarine and coastal environments; he also noted that oceanographers were involved in projects related to weapons tests, and the information could be applied to waste disposal.[58]

At the hearings, the members of Carritt's working group admitted that they had not thought some of the issues through. For example, they should have consulted state and local authorities about the sites. Had any of the scientists done so, the outcry might have been prevented. In any event, they said, backtracking a bit, these sites were chosen for further study, rather than as definitive choices. The committee realized that choosing specific sites, and including precise figures of latitude and longitude, was its greatest mistake. "We suspect that if specific locations had not been given," it observed in a statement for Congress, "there would have been less alarm." The report's major flaw, the committee acknowledged, was that it based its judgments mainly on the degree to which people used the sea as a source for food. It put much less emphasis on the recreational uses of the sea—sport fishing and bathing. Some of the sites, upon further reflection, "will be rejected because they are in important fishing areas." Their primary mode of reasoning in choosing the sites had been economy and ease of monitoring—why insist that the government spend more money to monitor distant sites when it was reasonable to have them closer in? Overall, Carritt's committee presented itself before Congress with humility but insisted that it had done everything on sound scientific and economic grounds, but perhaps without proper sensitivity to the various actors taking an interest in coastal waters.[59]

Embarrassed, the academy terminated Carritt's committee. NASCO chairman Harrison Brown wrote to Carritt thanking him for the "excellent report" and for the "considerable service" he had rendered, but that the group's activities now had "reached a successful conclusion." Academy president Detlev Bronk was less polite and was irritated at Carritt because of the damage his report had done to the academy's reputation. He saw an article in the

Washington Evening Star reporting the hearings before Congress. The article quoted more congressmen who were incensed with the working group's findings, including some from Florida who had discovered one such site proposed three miles off of Fort Lauderdale, a major resort area. The newspaper drew attention to the fact that already Carritt had deleted a couple of the sites because they were in fishing areas, undermining the report's credibility. Bronk enclosed a copy of the article in a letter to academy executive secretary S. Douglas Cornell, drawing attention to the critical remarks made, in Bronk's view, "because of Carritt." He wondered if they ought to make a formal statement to rectify Carritt's missteps and added, "I was never impressed by his scientific quality as are some."[60]

Despite the Carritt report, the hearings in Congress vindicated American policies. In its press release summarizing the hearings, issued in September 1959, the JCAE stated that the scientists had assured them that waste management and disposal practices had "not resulted in any harmful effects on the environment, the public, or its resources." Only low-level wastes were being dumped, because the high-level waste problem had not yet been solved. "Some radioactive wastes have been disposed into the ocean, but it should be emphasized that these were, and will continue to be, of a low level, solid, packaged variety." The AEC announced that it would not dispose of high-level wastes at sea any time in the foreseeable future, and it would require disposal of low-level wastes in water at least a thousand fathoms deep.[61] Perhaps to acquit itself, the AEC soon sponsored a survey of the area off Boston Harbor used by Crossroads Marine Disposal Corporation until August 1959; it concluded that no harm had been done.[62]

Carritt's experience served as an unprecedented warning both to the academy and to the AEC that the public resistance to nuclear power might be intense if it meant putting waste nearby. In addition, strong resistance might come from those who feared economic impacts rather than radiation itself. If it were true that "one radioactive red snapper" could make an industry collapse, politicians and business alike had reason to be afraid. The second round of BEAR studies, now in the works, would need to keep these factors in mind. The minutes of a meeting of some BEAR committee members in October 1959 record Carritt describing "in some detail the violent reaction of the press and the public to the east coast disposal report and the Congressional hearings on the subject, despite repeated emphasis on the low-level characteristics of the wastes and on the fact that this was merely a feasibility study." In general, the lesson to be drawn here was that great care had to be taken in presenting material to the public, through careful wording of all statements. If the recommendations of the new BEAR committees were to change markedly from those of 1956, they would need to be as conscientious as possible in managing public information to avoid outbursts in the press.[63]

In the long run, the Carritt report also drove a wedge between the academy and the AEC. Bronk and others detested the idea of rubber-stamping AEC policies, and here they had appeared to do just that. Other academy committees would be extremely sensitive (and some remarkably resistant) to pressures from the AEC. From the AEC's point of view, the academy had blundered, bringing waste disposal into public view for the first time and seriously threatening AEC policies. Ultimately, the AEC blamed the report for catalyzing public opinion against ocean dumping, leading the AEC to cut its losses in the early 1960s and abandon the sea as a major disposal option.[64]

None of this should suggest that American oceanographers wanted to keep radioactive waste out of the sea; in fact, quite the opposite was true. With Luedecke and others using the AEC's past patronage for oceanography as a way to placate Congress and the general public, it is hardly surprising that oceanographers tried to capitalize on it. The same year that these controversies were brewing, NASCO oceanographers were busy writing and promoting a report entitled *Oceanography 1960 to 1970*, a document that spelled out the country's current and future needs in the marine sciences. NASCO members unabashedly shopped the report around to congressmen, who used it for their own purposes—occasionally to criticize the president for allowing the United States to lag in oceanography while the Soviets gained fast.[65] But essentially the NASCO report was another roadmap for government patronage, and the oceanographers put atomic energy firmly into their realm. Luedecke had mentioned a figure at around a quarter of a million dollars. But in the report's chapter 5, "Artificial Radioactivity in the Marine Environment," the oceanographers envisioned an even more lavish future for themselves and their institutions. They broke down research needs into five categories: coastal and estuarine environments, the open ocean, sedimentation, distribution of radioisotopes, and the effects of radiation on marine organisms. They recommended at least a five-year program that would help determine the processes and effects of radioactive materials in the sea, at a cost of more than $30 million.[66]

"Unacceptable" Is Not a Scientific Viewpoint

While the Carritt study was being prepared, the AEC confronted a different set of problems about ocean dumping, this time on the Pacific coast. Because the depth of the Pacific dropped off considerably closer to shore than the Atlantic, the AEC did not anticipate the kinds of commercial gripes that plagued them in the Atlantic. Taking ships out to the appropriate depths, at least a thousand fathoms, would not be as costly or time consuming. But the AEC did not anticipate the political cost of the Carritt report and subsequent Congressional hearings. Politicians in the California state legislature formally objected to the AEC's methods of dumping radioactive waste offshore. The legislature petitioned the federal government and the armed forces to extend the depth requirement to two thousand fathoms and to ensure that dumping areas were at least sixty miles away from known seamounts (undersea mountains where depth decreases dramatically). It further asked that containers be used that could withstand water pressure at two thousand fathoms, and it wanted—as Arnold Joseph summarized it—"to go on record that they are opposed to the philosophy of unsafe bulk disposal of radioactive wastes."[67]

AEC scientists believed these demands were ludicrous because they were based on the winds of politics, not on any particular scientific findings. To Arnold Joseph, the requirements set down by the National Committee on Radiation Protection, which the AEC dutifully respected, already were cautious enough. Now the state of California expected an even wider margin of safety, despite the fact that the AEC assured it that there were no discernable effects after ten years of radioactive discharge into some areas. Only two areas were being used, one off of San Francisco and the other off of Long Beach.[68]

The AEC's assurances were not enough to satisfy the demands of California politics, so the AEC asked NASCO to conduct an assessment of the effects of radioactivity in the Pacific, once again to justify existing AEC policies with an external evaluation. This new Pacific committee was led by Scripps oceanographer John D. Isaacs, who tried to approach the inflammatory issue with levity. He joked that the severity of their situation seemed to deepen by having some of their meetings at the California Academy of Sciences, "with perhaps an additional sobering element injected by our collaborators and sober spectators, the animal heads of the Auxbury Room." His correspondence to his committee typically was upbeat, humorous, and whimsically conspiratorial. He seemed determined to avoid the fate of the Carritt committee, which meant he might not toe the AEC's line. By mid-December he had penned a draft, which he worriedly felt was "too long and too apologetic," and sent it to the others for comment. He urged them to make any changes they wished, as it was crucial to speak with a united voice. "We must have a report behind which we all unfailing stand in the spirit of the 'Lost Battalion' if necessary, only ready to make changes based on fact." Isaacs was both tongue-in-cheek and serious, and he hoped a quote from Lao Tzu would aid his fellow scientists in shaping up the final report:

The Wise Man says
That only those who bear the nation's shame
Are fit to be its hallowed lords;
That only one who takes upon himself
The evils of the world may be its king.

Lay readers might have found these lines chilling had they realized Isaacs was referring to radioactive waste disposal. But it is likely that he meant to impress upon his committee a sense of responsibility that went beyond a mere scientific report. They needed to appreciate the implications of their work, and to be willing to stand by their conclusions amidst likely political pressures on all sides. Carritt's committee members had bowed to these pressures—first from the AEC and then from politicians—with ruinous consequences for the committee's credibility. This was only one of the verses Isaacs included in a peculiar letter to his panel members, the second being in a more practical vein to help them in arriving at conclusions, deflecting potential criticism, and managing public opinion, not to mention discussing the technical issues of dispersal and drum construction.

A thing that is still is easy to hold.
Given no omen, it is easy to plan.
Soft things are easy to melt.
Small particles scatter easily.
The time to take care is before it is done.
Establish order before confusion sets in.
Tree trunks around which you can reach with your arms
were at first only miniscule sprouts.
A nine-storied terrace began with a clod.
A thousand mile journey began with a foot put down.
The stiffest tree is readiest for the axe.
The virtuous man promotes agreement;
The vicious man allots the blame.

For the tasks at hand, Isaacs wrote, these made for a remarkable group of aphorisms.[69]

At first, Isaacs's concerns that they would need to hold steadfast against salvos of criticism from AEC health physicists seemed unjustified. Isaacs drafted several versions of the report over the next several months. In May 1960, the BEAR committee and AEC scientists met with him and his panel to assess its basic assumptions and to try to resolve the differences between the oceanographers and the AEC. Isaacs was shocked at the result: "The upshot was that they bought *all* our assumptions!" The AEC called it sophisticated and even a potential basis for AEC policy.[70] Isaacs was elated at the AEC's apparent approval and churned out a final draft soon thereafter.

This parting of the clouds did not last long, however, and soon the AEC resumed its critical stance toward what it considered overly enterprising oceanographers. As in the case of Carritt's committee, the principal critic was sanitary engineer Arnold Joseph. Reading a preliminary draft, he disliked the tone. He wrote to Isaacs, "you are really plowing and discing for an emotional approach, contrary to the words which state that the Committee's recommendations are realistic, rational and unemotional." Joseph wanted Isaacs to "let your technical considerations in the report speak for themselves." The draft seemed to imply that radioactive waste disposal was going to limit man's use of the sea. "If true," Joseph complained, "AEC perhaps should have curtailed sea disposal some time ago." Some of the conclusions seemed ungrounded in evidence. For example, the oceanographers argued that biological communities could form around the container areas at the bottom of the sea. These might attract fish, which would then eat highly concentrated radioactive material and be caught in fishing nets. Joseph disputed this, annoyed that in the report "it appears to be a fact, whereas in reality this is still largely hypothesis." In addition, there was too much discussion of the administrative and logistical issues related to disposal, which struck Joseph as "axe grinding."[71]

Isaacs attempted to avoid the problems that plagued Carritt's Atlantic report by talking to local organizations about the issues. He later told the *Houston Post* that he consulted over thirty different bodies, including sportsman's clubs, commercial fishermen's groups, and antipollution leagues. He targeted laypersons because he recognized that the main source of resistance would not be scientists, but rather the public at large. He kept this audience in mind when drafting the report. "Thus there are no vital steps of erudition that an audience must take on faith, but, rather, each step in our picture can be considered and criticized by any intelligent 'natural naturalist,' such as a crab fisherman, as well as by the formal scientist." All along his group had felt, Isaacs wrote, that explaining the information as clearly as possible was the best approach, because it would lead the public—should it take issue with the conclusions at all—to make "constructive rather than emotional criticism."[72]

The AEC interpreted this approach as manipulative and opportunistic pandering to public perceptions, often at the expense of health physicists' authority. Joseph recognized that Isaacs's group faced "a difficult task, politically as well as technically." Still, he maintained serious objections as late as July 1961, largely because of what the report implied about the AEC's negligence. "We, too, are very sensitive to reactions by public, civic, political and business interest groups." In Joseph's view, the report's fundamental flaw was that "the numbers say one thing and the words something else." These should coincide in a scientific report, he argued to Roger Revelle, Isaacs's boss at the Scripps Institution of Oceanography: "Or am I too naïve? Most people—lay and scientific—read the words to get the writer's evaluation—not the numbers. Any disparity will create difficulties and further misunderstandings not only in this country but wherever the report goes. In any event, the words modify, amplify, clarify and sometimes cancel out what the numbers say." Some clarifications might be necessary, Joseph argued. For example, one might mention that sea disposal was not a new phenomenon, and in fact the University of California's Radiation Laboratory began to practice it in 1946, and since that time it had been the primary means of disposal for radioactive laboratory materials. The report implied that this was a new and risky proposition. Joseph deplored the mixing of fact with speculation throughout the report, and it seemed far from impartial. "As presently worded," he argued, "the report kindles as many or more public relations fires than it puts out."[73]

The AEC had requested the report to put out such fires. But in Isaacs's hands, the report seemed poised to paint a negative portrait of atomic energy establishments everywhere. It indicated a widespread feeling of complacency and an unwarranted confidence that problems would solve themselves. Joseph countered: "Has AEC exhibited 'complacent confidence'? Does not the fact that this study was requested mean anything?" The draft report "calls the AEC, collectively from the Commissioners to the janitors, a bunch of untrustworthy people."[74]

To Joseph, the oceanographers had introduced a hypothetical possibility about biological transfer from waste drums and used this hypothesis to recommend very deep disposal. Focusing on isotope concentration was not new. Indeed, it was the basis of Soviet physiologist E. M. Kreps's criticism of American and British practices. In the context of California, deep waters were close to shore, so the report did not amount to recommending a distinct break in policy. But Joseph recognized that if the principle were applied to other environments than the Pacific coast, waste disposal typically would have to be done very far from shore. He was worried that the report might be criticized by the British. "As worded," Joseph noted, "this is a real slap at the British in their pipeline disposal only a few miles off shore. It means all the work of their Ministry of Food, Fisheries & Agriculture is meaningless."

The AEC saw the Pacific group as a scientific advisory body. Since the AEC had requested the report, Joseph felt that the committee should consider itself as advising *the* AEC. But instead, the committee had taken a different tack and saw themselves as reporting to the country, which allowed them to take the position that their duty was to report to the public. This was not what scientists in the AEC wanted, especially now that it seemed clear to them that their voices were being drowned out by the editorializing of oceanographers. The request had been made for the purpose of advising the AEC, Joseph lamented, but "outside of seeing, in a round-about fashion, copies of the drafts of the group, our office, which originated the request, was not consulted for data or information and did not meet with the study group to discuss the sea disposal problems nor the text of the report."[75] But Joseph was misrepresenting the origin of the report. Like the Carritt committee, the Isaacs committee was formed to lend scientific credibility to existing policies. The problem was that Isaacs now was asserting oceanographers' authority to set levels of safety—threshold values—that differed from the AEC's.

Because Isaacs's group also called for more research, some at the AEC believed that oceanographers were manipulating public apprehensions to compel the AEC to spend more money on oceanography. But the AEC already was financing quite a lot, certainly more than the British were, despite their much higher reliance on sea disposal. "Contrary to the opinions held by some," Joseph complained, "AEC is limited in the funds it can spend for research." How much research was really justified on the basis of comparatively limited use of the ocean? The oceanographers had not spent much time asking the AEC for its expertise but spent a disproportionately large amount of time gauging the views of the public. The oceanographers were making their own conclusions about what the public could or could not handle while clamoring for more research money. What did they mean when they used terms like "unacceptable levels"? This was not a scientific viewpoint, Joseph held. "There will always be 'unacceptable' levels of radioactive substances to some people." They ought to stick to words like "injurious," which lent a more specific, technical weight to their judgment.[76] In his view, the oceanographers were pandering to public perceptions to augment their own authority by second-guessing experts who had been studying these issues for years.

Joseph's attitude represented a fundamental difference in opinion about the role of the general public in decision making. From his point of view, Isaacs was playing to laypersons' visceral reactions. Joseph acknowledged that most people did not like the idea of ocean disposal, but this was due to their lack of appreciation of the scientific and technical issues. And besides, who should be trusted to make such decisions—the ill-informed, prejudiced masses, or the technical experts who dealt with these issues on a daily basis? Joseph argued that they should not base recommendations upon what the public would find repugnant—the public was not informed about the issues. This attitude stood in sharp contrast with that of Isaacs's panel, which embraced political necessities and was quite willing to compromise and negotiate before producing their scientific assessment. This reflected the point of view stated explicitly in the other academy report on nuclear-powered ships, that there was a difference between "permissible" levels and "acceptable" levels. Isaacs's group seemed interested in getting at the "acceptable" levels, which meant taking nonscientific issues into consideration.

The sense of outrage within the AEC was perhaps best captured by a comparison, used by Joseph himself, to racial integration. After World War II, President Truman had ordered the armed forces to integrate their units, and in 1954 the U.S. Supreme Court had ruled that racial segregation in public schools was unconstitutional. But there was still widespread, often violent, resistance to government-mandated integration. Surely this was a clear case, Joseph argued, of experts needing to stay their course and do what was right, rather than accommodate public opinion. The widespread fear and visceral opposition to racial integration had to be overcome for the common good. The same was true of radioactive waste. Joseph challenged the oceanographers to resist equating majority views with correct ones and, more important, to avoid appealing to the emotions of laypersons. Issues leading to feelings of "'repugnance or apprehension' like race integration problems will probably be with us for a long time," Joseph argued. "Is it proper for a scientific community to sway in its scientific judgment because these states of mind 'might cause' rejection?" The only way to quiet people's fears was to inform and to educate, not to finesse the findings and present them according to what the public might find palatable. Joseph's allusion to racial conflict highlighted AEC officials' indignation at oceanographers who not only second-guessed their scientific decisions but also did so by exploiting negative public attitudes.[77]

To the AEC, there was a "scientific" viewpoint and there was a wrong viewpoint. But the AEC's unwillingness to compromise or even to consider the political implications was self-defeating. It compelled scientists in the AEC to take dim views of the general public and to assume that they knew what was best for it. This made them intransigent against what they considered irrational and illogical actions by critics, which in turn alienated not only the oceanographers, but also the politicians upon whom they depended for support and even other government offices whose jobs required far more diplomacy than AEC scientists were prepared to offer. The AEC learned this lesson the hard way when it butted heads with the State Department.

The AEC perhaps should have expected such a collision, given the fragility of relations with the Soviet Union and its strident opposition to waste disposal. But the catalyst for conflict with the State Department came not from the Soviet Union, but from much closer to home. The government of Mexico lodged a complaint against the Atomic Energy Commission's decision to grant a license to a Houston-based company, Industrial Waste Disposal Corporation, to dump radioactive waste into the Gulf of Mexico, about equidistant to U.S. and Mexican shores. The Mexican embassy in Washington tried, through conversations between embassy officials and the Atomic Energy Commission, to prevent the license from being granted, but the Mexicans failed to make any headway. The AEC invited Mexican delegates to its own internal hearing about whether the operations were safe, but the Mexican view during the hearing had no effect. In June 1959, Mexico made its discord more official, saying to the U.S. State Department that, for technical and scientific reasons, and "because of the unfavorable impression that the project can have on Mexican public opinion," it was obliged to lodge a formal disagreement.[78]

Milton Eisenhower, the president's younger brother (and himself president of Johns Hopkins University), visited Mexico in late August 1959. Milton Eisenhower was deeply involved in advising on Latin American affairs during the Eisenhower administration, and he often served as the president's personal representative. When he brought up the subject of waste disposal to Mexican officials, he was surprised to find that the Mexican objections were quite serious. He had assumed that they were simply formal, but when he inquired about it he found real and strong resistance. He credited it to fear of the unknown and an adverse reaction to nuclear weapons testing. But he also did not wish to antagonize the Mexicans. "Mexican officials almost pleaded for another solution," the internal report of his visit stated, "and Dr. Eisenhower said he hoped the Department of State would use all its influence to persuade those responsible to find another hole for the disposition."[79]

The State Department chastised the AEC for ignoring the Mexican protest. Surely, it stated, the Mexicans would rightly regard this as an arbitrary and unilateral act, since their voices were not being heard. As the secretary of state wrote to AEC chairman John McCone, inviting Mexicans to attend an internal hearing "does not give the Government of Mexico the effective voice in the final decision that the United States would expect if the situation were reversed." The State Department, advised by the U.S. Embassy in Mexico City, made it clear that the AEC's unilateral decision would elicit uniformly adverse reactions from the Mexicans and would harm the two countries' relations, "regardless of any explanations that might be given." It also would inflame relations with other countries in the Western Hemisphere, who probably would agree with Mexico that the United States was acting arbitrarily. This was particularly so, the State Department declared, in light of the lack of certainty regarding the long-term effects of radioactive waste disposal at sea.[80]

AEC officials had not acted illegally—and, after all, theirs was the "scientific" view—but they had acted imprudently in the diplomatic arena. While it was true that there were no international laws regulating dumping radioactive waste in international waters, that did not mean, in the State Department's view, that such an arrogant legalistic approach was in the best interest of the United States. As Robert Lowenstein, of the AEC's Office of the General Council, put it, the AEC ultimately "was obliged to accept the view of the Secretary of State, as the President's spokesman, with respect to the impact on our foreign affairs of issuance of the proposed license."[81]

AEC Commissioner John F. Floberg was dismayed at the decision to bow to diplomatic and public pressure, writing a separate opinion: "I consider this decision to be principally based on grounds other than supportable scientific or technical conclusions that either the integrity of the containers or a change of site is essential or, indeed, even relevant to the health and safety of the residents of the land areas surrounding the Gulf of Mexico. Rather I believe it to be based on the eagerness of the Commission to allay all anxiety, however unreasonable, unfounded, and scientifically unsound, of the residents of those areas with regard to safety."

Floberg added that he was "completely satisfied" that the planned waste disposal operation posed no threat to public health in either the United States or Mexico. Lowenstein, discussing the issue at a conference of the Health Physics Society, tried to cheer the health physicists, most of whom undoubtedly shared Floberg's view. In atomic energy, he said, "as in all other activities in a democratic society, it is not enough merely to be confident we are right." The lay public, it seemed, also had to be persuaded. Public pressures from laypersons should not be shunned, he said. Instead, one should view public pressure as a mark of the atomic energy industry's successful growth.[82]

Ultimately, the AEC gave up trying to satisfy either oceanographers or the lay public on the question of sea disposal. In 1960, AEC General Manager Alvin R. Luedecke wrote to Chairman McCone that land burial of waste was turning out to be more cost effective in any event. Coastwise Marine Disposal Corporation, a company that operated out of Long Beach, California, charged the AEC $11.00 per fifty-five-gallon drum; taking into account that the concrete in the drums made up for a great deal of its weight, that meant $22.00 to $33.00 per fifty-five gallons of waste. The Army and Air Force were paying even higher prices for dumping chemical wastes at sea, up to $48.75 per drum. By contrast, the AEC could bury on land for $5.15 per drum. When transportation and handling costs were factored in, differences between land and sea disposal became even greater: for sea, $23.31 to $69.00 per drum; for land, $9.00 to $18.84 per drum. Although some particular situations might render sea disposal more attractive, Luedecke concluded, land burial usually made more sense from an economic point of view.[83] This, combined with the political problems now associated with sea disposal, convinced the AEC in June 1960 to stop issuing new commercial licenses for sea disposal, pending the outcome of some future research program to ensure the integrity of the drums en route and at the bottom of the sea. Thus the AEC ended its reliance on ocean disposal but left the ocean disposal option for the future.[84]

The AEC tried to delay a decision about the controversial Gulf of Mexico dumping license until after the upcoming conference on sea disposal to be convened in Monaco by the IAEA. But this conference did more to exacerbate uncertainties than ameliorate them. Despite its scientists' frustrations, the AEC decided that the apparently unscientific attitudes of politicians and foreign governments, not to mention the exorbitant expectations of patronage by oceanographers, were causing them too much trouble.[85] In 1961, AEC chairman Glenn T. Seaborg wrote to his counterpart in Mexican Nuclear Energy Commission, Nabor Carrillo Flores, that the material had been disposed of in a different manner.[86] The AEC had capitulated to domestic and international pressures, denying the permit to dump into the Gulf of Mexico while claiming publicly that it now would focus its disposal on land.

Chapter 6 The Environment as Cold War Terrain

WHEN ARNOLD JOSEPH, an Atomic Energy Commission sanitary engineer, feared that John D. Isaacs's report would rattle the British Atomic Energy Authority (AEA), he was right to do so. Already the report on nuclear-powered ships implicitly criticized Britain's high levels of discharge at Windscale. Now the American oceanographers were taking aim at dumping packaged waste at sea, throwing doubt on the British practice of putting most of theirs into the relatively shallow English Channel. The AEA had worked hard over the years to convince other British government offices to authorize ocean dumping. The American oceanographers' focus on biological absorption and subsequent transfer through the food chain stood against health physicists' conventional wisdom that dilution and dispersal would take care of most of the dangers in radioactive waste.[1] Isaacs's report was not officially published until 1962, largely because of repeated draft criticisms by the AEC, but its outlook was no secret. When finally released, the report indicated that packaged waste should only be dumped into very deep water, on the order of 1,200 fathoms.[2] If accepted, the British would have to abandon their most convenient dumping ground, the Hurd Deep in the English Channel. AEA health physicist H. J. Dunster wrote to Roger Revelle, Isaacs's boss at the Scripps Institution of Oceanography, that the report ignored the physical and chemical properties of the ocean—the power of dilution—in favor of a biological approach. This led to some pretty restrictive recommendations, ones the British were not prepared to accept.[3]

Unfortunately for the AEC and the AEA, the difference manifested itself at an inopportune moment, when the International Atomic Energy Agency was hoping to lay the scientific foundations of its policy recommendations for waste disposal at sea. It convened a conference in Monaco on waste disposal, inviting oceanographers, geologists, and health physicists. But the conference only widened the differences of opinion, and it strengthened atomic energy officials' resolve that marine scientists were more interested in patronage than in solving the waste disposal problem. The tensions worsened right after the conference when France's Commissariat à l'Énergie Atomique (CEA) began to plan waste disposal operations in the Mediterranean. The French found that the oceanographers in France and Monaco were far more vocal in their criticisms than any Americans had been, going as far as actively stirring up resistance and bitterly criticizing the CEA. By the end of 1960 waste disposal at sea was a full-blown international controversy.

In the meantime, the Soviet Union seized upon waste disposal at sea to launch a major propaganda campaign against Western nuclear powers, claiming that the United States and Britain had poisoned the oceans. Quite aware of their hypocrisy—the Soviets falsely claimed to discharge nothing into the environment—Western governments struggled to find a way to respond without relying on secret sources of information. Atomic energy establishments cooperated with each other on public relations issues, worked in concert in international meetings, and the United States even put solidarity about waste disposal at sea on the North Atlantic Treaty Organization's (NATO) agenda. If the Soviet Union wanted to make it a cold war issue, then the United States reasoned it could call upon its cold war allies. But ultimately officials in these countries decided that the most effective means of combating Soviet propaganda was to be seen supporting scientific research. Though many saw marine scientists as opportunists, support for oceanography became an important part of the West's counterpropaganda campaign against the Soviet Union.

Revolt of the Oceanographers

Because it depended so much on the sea for waste disposal, Britain had the most to lose in any international negotiation. Consequently, its representatives came to the scientific conference at Monaco ready for a fight. Although the Soviet Union's increasingly shrill outcries against ocean disposal loomed as a sticky political issue, the AEA was far more concerned about the effects of an international group of oceanographers entering into atomic matters. These oceanographers seemed to be in revolt against the waste disposal policies of atomic energy establishments, particularly against the British Atomic Energy Authority. AEA officials assumed an authoritative stance to ward off any impression that a scientific negotiation was taking place. Two of the British representatives, health physicists John Dunster and A. H. K. Slater, agreed that their main object would be "to give oceanographers and geologists an idea of what was involved in waste disposal problems," and to discuss the general difficulties, "but not their detailed solution." In other words, the AEA was to instruct the oceanographers, not vice versa. They saw the nongovernmental scientists as special interests with a stake in the outcome, and it would be crucial "to make sure that these other interests did not seek to divert money from atomic energy projects for their own particular problems." The British delegates to Monaco were thoroughly warned to beware of oceanographers and geologists who wanted to give their research "an atomic twist merely in order to divert funds to them for their purposes."[4] Dunster, the head of the delegation, belittled the "marked tendency" of oceanographers to "batten on to waste disposal" as a way to obtain funds, without any genuine interest in solving the waste disposal problem.[5]

With such attitudes in mind, the absence of an oceanographer on the British delegation to this scientific conference should not have been a surprise. AEA scientists and officials went to the conference to defend their policies, and the last thing they wanted was an elaborate research program (for which they would have to pay) that might question them. Instead, the AEA sent people knowledgeable about their practices near the coast and the deep sea, whose primary purpose would be to speak about the "evidence that these are non-hazardous."[6] Still, in the interest of a balanced delegation, Dunster suggested asking someone from Britain's National Institute of Oceanography to come to Monaco too. But the AEA offered no travel funds, and the institute politely declined.[7] Taking leadership of the British delegation to Monaco, Dunster wrote that he expected he would have "to attempt to sell our approach to waste disposal at sea in the teeth of what might well be considerable barracking from other countries."[8]

Few oceanographers saw the Monaco conference in this way. They were not there to be instructed, but rather to discuss the science and, in doing so, to help the IAEA develop international regulations. The IAEA's director-general, Sterling Cole, had written that it was to be a discussion between atomic energy officials, on the one hand, and oceanographers and geologists, on the other. In his form letter to member states, Cole explicitly suggested that the oceanographers and geologists might contribute to the solution of the waste disposal problem.[9] But when they tried to do so, atomic energy officials resisted, widening divisions that were more institutional and disciplinary than national. As one British official later chastised his own delegation's attitude, it was likely that most countries saw the Monaco conference as a step toward consensus about standards and procedures, and that the representatives of the AEA should not have been so indifferent to what others had to say.[10]

British health physicists targeted as opponents the oceanographers and geologists who presumed to advise on atomic energy policy. On ocean disposal, they were leery of American scientists who made up the National Academy of Sciences Committee on Oceanography (NASCO), then in the process of writing reports for the Atomic Energy Commission on various aspects of ocean disposal. They believed these reports were poisoning the attitudes of oceanographers all over the world; the ones at Monaco seemed to take the academy's studies as gospel. An official from the Foreign Office observed that "the pretensions of the oceanographers have recently been looked upon somewhat bleakly by atomic workers in this field," summing up neatly the attitude of officials and scientists working on waste disposal within the British atomic energy establishment.[11]

Of the American reports, the one on disposal from nuclear ships was the most alarming, because it presumed to speak to the problem of permissible discharges of radioactive waste *anywhere* in the sea, particularly in coastal areas. While appreciating the desire of the AEC and the academy to study the problem in detail, particularly if nuclear ships were to become the wave of the future, British health physicists realized that the universal threshold values for permissible discharges touted by oceanographers—in other words, values that could apply in any waters—would threaten their discharge policies from the pipeline at Windscale. The British policy had long claimed that discharges could not be set broadly, but that they depended on the local environment and economy; Windscale, for example, discharged much more than the study of nuclear ships would deem as permissible. The American report did not provide for discharges into a specific environment that had been studied intensively. As Dunster, the chief of Britain's delegations to Monaco, lamented: "The figures recommended for discharge from nuclear ships into unspecified coastal waters are, as one might expect, substantially lower than those which we find in practice acceptable at Windscale." Thus the British delegates to Monaco geared up to defend this policy against the broadly construed recommendations of American oceanographers.[12]

The Monaco conference turned out to be a failure for atomic energy officials who wanted to nip the oceanographers' revolt in the bud. Sir John Cockcroft, director of the AEA's research establishment at Harwell, openly declared it a victory for the exchange of ideas but privately regretted that the result was much sharper distinctions between positions.[13] Two separate camps had emerged, the oceanographers and the atomic energy scientists, and each had arrived at its own consensus. At the conference, French and Italian oceanographers believed that there was not sufficient knowledge of oceanographic conditions to justify dumping, especially not in nearby shallow seas such as the Mediterranean. They seemed to want to declare the whole Mediterranean as unsafe for dumping. By contrast, the representatives of the atomic energy establishments of three countries (the United States, Britain, and France), found common ground in their view that they already knew enough to make conservative estimates of what could be dumped, and that they could make such estimates about any part of the sea. As Cockcroft acknowledged, few skeptics from other countries were convinced, largely because of the influence of the oceanographers.[14] Frustrated, the leader of the British delegation, Dunster, wrote that the oceanographers' view "was based more on prejudice than knowledge."[15]

The Monaco meeting publicly raised more questions about the validity of radioactive waste disposal than it resolved, and it drew attention to the oceanographers' view that more study was needed. The *Sunday Times* summed it up with a headline: "All at Sea on Atomic Waste."[16] The conference opened British practices to scrutiny and left wide avenues for scientists to ask for more money and to assert roles in decision making. British officials felt they had been outmaneuvered by oceanographers at the international level. In the aftermath of the meeting, one disappointed Minister for Science official predicted that the discharge of waste into the sea was destined to become a "hobby-horse for the mischievous, the ignorant and the timid alike."[17] Another was harsh to Dunster's delegation, writing that, in the future, conferences on technical matters should be recognized as having wide political repercussions. As such, they ought to ensure that the delegates "had a sufficiently high level of political competence."[18] And they ought to ensure that an oceanographer of international standing was included, if only for appearances, to beef up the scientific credibility of the delegation—it was probably worth the cost of an airline ticket and hotel.[19] Like the AEC's experience with Mexico, the AEA had yet to show much skill in the art of diplomacy. The message was clear: the AEA, even with its complement of highly competent health physicists, did not carry sufficient political or scientific authority, certainly not beyond Britain's shores. On the international stage, oceanographers seemed to be winning a battle for scientific authority.

An Unwitting Alliance

The oceanographers' objections were particularly untimely because the Soviet Union had decided to take a hard line at Monaco too. It had given hints of this in 1958, during the second Geneva conference on peaceful uses of atomic energy, quite in contrast to the cooperative spirit of the 1955 conference. The delegates at the meeting repeatedly brought up the question of nuclear weapons testing, and some feared that the Soviets would use waste disposal at sea as a way to criticize American and British nuclear policies. The discussion of Kreps's paper at Geneva, which condemned waste disposal in the capitalist countries, indicated that the thinking on the subject was increasingly falling along cold war lines. Having put so much energy into trying to disprove the circulation of water in the deepest parts of the ocean, during the International Geophysical Year, it seemed that the Soviets might want to turn waste disposal into a new diplomatic tool. As AEA health physicist W. G. Marley put it, "It is quite likely that there is pressure from the Russians to have this particular subject discussed, since denunciation of ocean disposal in any form is one of their main theses at conferences." Soviet delegates routinely criticized waste disposal, and thus the delegates to Monaco needed to be prepared to counter all Soviet objections.[20]

Going into the conference, British officials were afraid it would degenerate into a "ban the bomb" debate if the Soviets pressed matters. Further, they thought that some of the neutral countries in Asia and the Middle East might turn it into a debate for nuclear disarmament or cessation of weapons testing. The Americans surely would resist such tendencies, a Minister for Science official wrote to the British delegation, and "Western European countries and the 'white' Commonwealth can be relied on to take the same line." Non-white Commonwealth countries like India, by contrast, might take the Soviet Union's side, the office of the Minister for Science believed; the United States might have to use its influence to keep unruly Latin American countries from turning the Monaco conference into a disarmament rally.[21]

As expected, the Soviets started a controversy at the start of the Monaco conference. One of the delegates, M. Zemokov, pointed out that "one does not have the right to throw away anything," and that it was necessary to store all wastes. Although perplexed about what precisely Zemokov had in mind as a policy, the other delegates were unable to get him to clarify his position.[22] The next day at a press conference, his compatriot Victor Spitzine made the Soviet views clearer by criticizing the United States and Britain explicitly. He argued that "Soviet scientists had a duty to warn their foreign colleagues" about the these nations' practice of dumping radioactive waste at sea, "which poisons great masses of water for several years."[23] Such declarations against dumping, directed so obviously at political enemies, were well in line with what delegates expected.

Western diplomats did their best to quash the unwitting alliance between oceanographers and the Soviet propaganda machine. Before a general conference of the IAEA in September 1960, diplomats began to panic because the Italian delegation was trying to insert an item into the agenda about the need for studies of waste disposal at sea. This was due, most suspected, to the influence of Italian oceanographers who were using the momentum of the Monaco conference to pressure their government in this direction. British, American, and French diplomats immediately began to put pressure on the Italians to prevent any such item, which they assumed would be used by the Soviet Union as an opportunity to vent some propaganda. Even the secretariat of the IAEA bristled at the idea, informing the Italians that this was a controversial subject that, as a British delegate worded it, "the Soviet Union might be expected to exploit to the embarrassment of some of Italy's friends, such as France and the United Kingdom."[24] The Italians backed off, but the Soviet delegate at the conference brought the subject up anyway, pointing out that ocean disposal was an evil practice and that its main dangers came from reprocessing facilities used to make plutonium for weapons. It soon became clear that the Soviet Union had decided "to flog the subject of waste disposal for propaganda purposes" in its broader political campaign about weapons tests and nuclear disarmament.[25]

The failure of Western atomic energy establishments' views to win the day at Monaco compelled Cockcroft to request a four-power meeting of atomic energy specialists. He wanted it to be a closed conference at the IAEA between the French, British, American, and Soviet members of the IAEA's standing Scientific Advisory Committee. This seemed to be an ideal forum, since Cockcroft already was Britain's representative; he discussed the plan with his counterparts, French physicist Bertrand Goldschmidt and American physicist Isidor Rabi, who agreed. They wanted atomic energy official Vassily Emelyanov to come to Vienna to discuss the Soviet Union's attitude and any genuine problems with waste disposal at sea. Cockcroft hoped that this kind of meeting—that is, with neither reporters nor oceanographers—might be more successful in discussing what he considered fundamental issues in actual operations, rather than general discussions about the behavior of the oceans. The Soviets agreed, adding that a four-power meeting in Vienna (IAEA headquarters) could be followed up by site visits, so that all four could get a real sense of how everyone handled their radioactive waste. They decided to leave it to the United States to make a formal proposal.[26]

Leaving it to the United States proved a bad strategy by Cockcroft. AEC chairman McCone wrote to Emelyanov, including a mention of site visits, and soon the idea got lost in some existing plans for teams of American and Soviet scientists to visit each others' installations. Emelyanov's response indicated that the two superpowers might convene a conference on waste disposal at the IAEA with five to seven persons in each delegation (American and Soviet), and they could also invite one or two specialists each from Britain, Canada, and France.[27] The British were very discouraged to see Cockcroft's initial idea fade. Despite the scale of its discharges, which dwarfed all the rest, Britain was being treated as a junior member of the atomic club. Even worse was the inevitability that a bilateral conference, as one Foreign Office official put it, "leaves the door wide open to hounding the British representatives on Windscale discharges." Since there were no plans for site visits to Windscale, the British would have no opportunity to convince anyone that their operations were done responsibly.[28]

The British government then shifted its focus away from this idea of a four-power conference and concentrated its energy instead upon playing a role in the legal interpretations of previous IAEA activities. The scientific conference at Monaco was not the only meeting designed by the IAEA to help countries conform to the 1958 United Nations Conference on the Law of the Sea. The IAEA also convened a more policy-oriented panel to formulate recommendations to guide national policies and, presumably, any future international agreements about sea disposal. Chaired by Swedish atomic energy official Harry Brynielsson, this ad hoc panel consisted of experts from eight countries and met several times in 1959. After the release of its report in June 1960, the IAEA convened yet another panel, this time of experts to assess the implications of the Brynielsson report for international law.

In the meantime the IAEA convened a brain trust to include key scientists at atomic energy establishments. The British representatives, led by John Dunster, soon realized that the Brynielsson report could vindicate most of their practices. It argued against sea disposal of high-level waste, such as irradiated fuel from reactors, but favored disposal of low and intermediate level wastes under controlled conditions. It also recommended using the guidelines of the International Commission on Radiological Protection to create such controls on a national basis. These were essentially consistent with Britain's practices. The Soviets did not send any participants, and the brain trust interpreted the Brynielsson report to sanction waste disposal. Although the British had been worrying and planning to argue their position against critics at a four-power meeting, which never materialized, they now decided to withdraw the diplomatic pressure for any meeting at all. Although the Soviets would undoubtedly keep up their propaganda attacks, the Brynielsson report should, AEA scientist G. M. P. Meyers observed, shield Britain from any immediate criticism from anyone wanting Britain to follow international consensus.[29]

In fact, British officials used the Brynielsson report as a justification for disregarding its only authoritative competitor, the U.S. National Academy of Sciences' report on effluent from nuclear-powers ships. In December 1960

British scientists and officials met and agreed to dismiss a few inconvenient aspects of the academy's report. They invited John Swallow, of Britain's National Institute of Oceanography, to critique the report; he revealed that the Brynielsson report had thrown out the academy's method of determining if an environment was appropriate for discharges. The academy had identified different zones of the ocean where restrictions should or should not apply. But Swallow noted that these zones often overlapped, and the academy had not taken into account that the build-up of radioactivity from such overlap might exceed the academy's levels. Since the academy had recommended limiting the dumping of some wastes only in zones with deep water, Swallow was suggesting that this ocean zoning might not be an adequate way of determining safe discharges. That was all the AEA needed to hear. The atomic energy scientists, led by John Dunster, took Swallow's point and agreed that the depth requirement was not an adequate way to ensure safe disposal.[30] They did not, however, ask Swallow to devise a better system. Instead, this simply reinforced the notion that the American oceanographers' depth requirement was flawed, did not need to be enforced, and thus they could fall back on their usual ways of determining discharge levels—particularly at Windscale. Moreover, they could claim to have based this policy on the recommendations of the Brynielsson report, which had the political advantage of being based on the work of an international group.

A Living Environment

Shortly after the Monaco conference, France began to make headlines for its activities and for internal disagreements with oceanographers. Like Britain, France had devoted major resources and political commitment to becoming a nuclear nation.[31] The country's first industrial-sized facility for reprocessing spent fuels was built in 1958 at Marcoule, in the south of France. Another would be built in 1966 at La Hague, with a pipeline to the English Channel, and these two facilities soon became France's principal waste-producing sites. Like its American and British counterparts, the French Commissariat à l'Énergie Atomique was aware of the difficulties associated with nuclear safety and saw radioactive waste as a potential political problem.[32] Once operations began at Marcoule, the commissariat envisioned two possible solutions: land burial in "radioactive cemeteries" in the environs of the plant or sea disposal. The main drawback of land disposal was the potential negative reactions of neighboring people. CEA officials hoped—ironically, as it turned out—that sea disposal would help them avoid public outcry. Thus in May 1960 the CEA decided to plan an experimental dump of about two thousand tons of liquid and solid waste—contaminated work clothes, with assortments of plastic, wood, metal, glass, and other materials. These would be packed into 200-liter drums and dumped into the Mediterranean, at a site between the towns of Antibes (near Nice) and Calvi (in Corsica), in water about 2,500 meters deep, fifty miles away from the coast of France and sixty miles from the coast of Italy.[33]

The CEA's official description of the plan observed that there were no currents at the sea's surface in the specified region. "One could hope," it added, "that the currents would be equally nonexistent at the bottom." Some measurements had been taken in 1959 that indicated that such might be the case. Thus the site seemed ideal: it was in relatively deep water, away from the coast, with no discernable current, and not near known fishing waters. There would be no notable risks, the CEA stated; even if all of the drums burst, which was unlikely, the surrounding water would dilute the material so much that the danger to human health would be "completely negligible."[34]

French newspapers picked up the story in October 1960, provoking harsh criticism from oceanographers in France and Monaco. Jean Furnestin, director of the Institut Scientifique et Technique des Pêches Maritimes, pointed out that all of the physical oceanographers and biologists at the recent IAEA meeting in Monaco, without exception, had emphasized the formidable dangers that confronted humanity from ocean dumping of radioactive waste. No one had been able to demonstrate that there were in fact dead zones in the ocean. The Soviet work aboard the *Vityaz* during the IGY, Furnestin pointed out, had proven that even the deepest Pacific waters moved to the surface much faster than previously believed. In the Mediterranean site, he claimed, there were no instruments sufficiently sensitive to tell whether there were currents at that depth or not. Moreover, there was plenty of evidence to suggest that such currents might exist on a seasonal basis, and that the region was unstable. Besides, biologists were unanimous in pointing out the perils of radioactive concentration in deep flora and fauna that could be passed to other creatures at shallower depths. Furnestin argued that approval should not be given without first consulting oceanographers and fisheries specialists.[35]

In Paris, oceanographers at the Centre de Recherches et d'Études Océanographiques also criticized the CEA's action because the commissariat had planned it without any consultation. Vsevolod Romanovsky, the director, learned of the plans from the newspapers and was appalled to find that the CEA suggested to the press that oceanographers—and specifically Romanovsky—approved of the idea. Romanovsky was the one who had conducted the 1959 studies cited by the CEA. But the studies, Romanovsky now stated, had been inconclusive and some were still ongoing. True, he had recommended an experimental dumping, but he had in mind something on the order of ten drums. The CEA was planning to dump 6,500.[36]

Scientists at the Institut Océanographique in Monaco joined the chorus of criticism. In a widely circulated letter, the institute's director Louis Fage wrote that he was stupefied to read the news that the CEA was planning to dump two thousand tons of waste into the Mediterranean. He described the Mediterranean in two words, exclaimed on the page: "mer fermée!" (closed sea!). Aside from directing the institute, Fage also was the president of the Committee for the Exploitation of the Sea, a body consisting of scientists from France and Monaco. In that capacity, he registered strong protest against the CEA's decision, giving several rebuttals to its scientific assumptions.[37]

If these scientists' rage came partly from the scientifically questionable nature of the dumping experiment, it came also from the fact that they had been left out of the decision-making process. They were reading about the experiments from the newspapers like everyone else. Fage wrote that he could not vouch for the veracity of the newspapers' claims, since no specialists in *marine biology* (Fage underlined this also in his letter) seemed to have been consulted. These scientists, not atomic energy officials, had already established the crucial questions on the issue, namely the ecological connections between marine life and human beings. Fage wrote as if a new orthodoxy existed that put biological transfer of radioisotopes at the forefront, at the expense of physical dispersal and chemical dilution. For example, Fage believed that atomic energy officials should pay more attention to plankton, "for they are at the base of the chain in which we occupy the summit." He quoted the findings of American marine scientist Bostwick Ketchum, who had shown that the concentration of radioactive substances in plankton could be up to five hundred times that of seawater. This was a living environment ("milieu vivant") for which the introduction of radioactive waste could be destructive. Fage insisted that it would have been better to hear the marine biologists prior to the decision.[38]

The most formidable opponent was Jacques-Yves Cousteau, a member of Fage's committee and director of the Musée Océanographique de Monaco. According to Fage and Cousteau, undersea photographs had revealed that the deep water in the dumping area did indeed move. Over the years, Cousteau would become a household name because of his books and films about undersea life. With underwater breathing gear, and with manned and unmanned submersibles, Cousteau and his colleagues took photographs of the depths and published them internationally. He already had established his fame with a book, *Le Monde du Silence*, translated into several languages in the 1950s, and a 1955 documentary film of the same name that won the Palm d'Or at the 1956 Cannes Film Festival.[39]

In statements made for French newspapers, Cousteau bitterly criticized the CEA. He proclaimed that scientists had not been consulted. He made it clear that neither the Musée Océanographique nor his ships were involved in any way. He accused the CEA of having acted behind the backs of scientists after the Monaco conference when it had become clear that profound differences of judgment existed between atomic scientists (*atomistes*) and an international group of oceanographers. The latter, he claimed, had categorically condemned sea disposal on the grounds that sufficient studies had not yet been made. He then included a list of the important scientific bodies that had not been consulted, including the International Commission for the Scientific Exploration of the Mediterranean, the (French) Academy of Sciences, the Centre National de la Recherche Scientifique (CNRS), and other institutions. He drew an analogy: it was like announcing that tomorrow morning there would be an experiment to dispose of allegedly inoffensive nuclear waste at the Place de l'Opéra, without first consulting with the mayor of Paris.[40] With such complaints, Cousteau appealed to local officials, who also had not been consulted, and implied that oceanographers, not atomic energy officials, were the true custodians of the seas and the protectors of local interest.

Meanwhile, mayors and city councils of towns all along the French Riviera sent protests to the CEA's high commissioner, Francis Perrin. Of course the CEA sent reassuring replies to them, pointing out that the experiment rested on the firmest scientific grounds. However, the Mayor of Nice, Jean Médecin, sent back a telegram that cut right to the issue, underscoring the power of international scientific consensus. He baldly stated that whatever the CEA's scientific competence might be, it would certainly not prevail over the numerous French and foreign scientists of contrary opinion. Whoever marked this telegram (in the CEA's archives), Perrin or a subordinate, underlined that statement in red and penciled two exclamation marks in the margin. If Nice were any indication, clearly the battle for scientific authority was being lost to the oceanographers. The people of Nice, as Médecin said, stood ready to oppose the CEA and to stop the dumping "by all possible means." Other nearby towns voiced similar sentiments. One anonymous letter suggested to Perrin that if the waste was so inoffensive, perhaps he should put it in his own breakfast.[41]

Due to this intense domestic turmoil, France's plans to dump radioactive waste in the Mediterranean became worldwide news. Prince Rainier of Monaco urged the government of Charles de Gaulle to put off the experiment until scientists knew more about the dangers. The leading French newspaper, *Le Monde*, quoted Cousteau's statements that the CEA did not understand anything about the problems of the sea.[42] Cousteau tried to make it an international issue, saying that it involved all of the countries bordering the Mediterranean, not just France. The *New York Times* called Cousteau the "unofficial leader of the anti-dumping campaign." In the face of the publicity assault, less than a week after making the announcement of the experiment, the CEA backed down and decided to put the project off for a while.[43] In the weeks that followed, Cousteau gave more interviews, stressing how the issue was really an international one and that it could be resolved only by oceanographers. It might end up as a choice for all humanity, he said, between using the sea as a waste dump or preserving the riches within it.[44]

The CEA did not attribute these problems to the press, the local politicians, or the general public's irrationality. Instead, it blamed oceanographers, particularly Cousteau. In an internal note, CEA officials dismissed the idea that the press could have mounted such an offensive or that the population could have spontaneously reacted so negatively. Instead, the escalation of the issue's importance "is a direct function of the declarations, acts, and positions taken by M. J.-Y. Cousteau, director of the Musée Océanographique du Monaco." Quoting American newspapers, they lamented the fact that Cousteau suddenly seemed to be internationally recognized as the leading figure against ocean dumping. The CEA's Department of External Affairs outlined specifically the steps that Cousteau had taken to undermine the commissariat. He began to critique the experiment "violently," sent telegrams to all the mayors in the area, attended all the important local meetings to discuss the issue, attacked Perrin in the press, acted as a "scientific expert" at a major regional meeting, and acted indirectly at other local political meetings to oppose the project. In addition, he had given "innumerable interviews" to reporters for newspapers, radio, and television. Cousteau focused on the CEA's incompetence, calling it childish; its scientists were incapable of understanding the sea, making mathematical calculations that would not even measure up to the standards of first-year oceanography students.[45]

From the CEA's point of view, the task ahead was not to change plans to dump radioactive waste, but rather to repair relationships with oceanographers by ceding to them some scientific authority and possible financial support. Perrin left for a three-week visit to the United States and promised Fage that they would meet upon his return.[46] In the meantime, Henri Baïssas (of the CEA's Department of External Affairs) asked leading French geophysicist Jean Coulomb to help facilitate a rapprochement with oceanographers. Coulomb felt that it would be easy to bring physical oceanographers into the CEA's camp, but more difficult with biological oceanographers. He tried to arrange a meeting with Cousteau but was ignored. A leading CEA physicist, Bertrand Goldschmidt, did finally meet with Cousteau and informally promised to support more scientific work under the IAEA. Baïssas and a colleague met with Furnestin and had a relaxed conversation about the importance of supporting research, and Baïssas followed up with a formal letter stating that the CEA would not proceed with a dump without proper studies by, guidance from, and agreement with oceanographers.[47]

Perrin met with Fage upon his return from the United States, taking other CEA officials with him. According to a CEA internal memorandum, the meeting was very pleasant. In fact, Fage declared himself in support of the commissariat's activities, being convinced that they were harmless. In return, the CEA promised to lower the number of drums dramatically to keep it in line with biologists' views.[48] Baïssas went with a colleague to Monaco to reenlist the scientists there and soon reported his "mission to Monaco" as a success. His strategy was to admit candidly to the oceanographers that the CEA, despite being convinced that its plans were harmless, had committed the error of not sufficiently consulting the scientific community. He promised much closer collaboration in the future, declaring the Mediterranean as a place for experiments, on the order of ten to twenty tons, not for massive dumping. As a result, Furnestin "incontestably" wanted to help them, as did other scientists present. Even Cousteau, cornered by Baïssas during a prelunch cocktail, privately assented to the CEA's plans.[49] In the coming months, Baïssas worked hard to cajole Cousteau and others, careful not to ruffle any scientific feathers, to create an experiment that helped the CEA but also drew on outside scientists' expertise. Such conciliatory maneuvers were necessary, Baïssas wrote, to rupture "the mystical charm that paralyzes us."[50]

Poisoners of Wells and Their Accomplices

The Mediterranean debacle was only the beginning of a series of conflicts of expertise that would keep radioactive waste disposal in the news for years to come. In the case of oceanographers, the solution seemed obvious, if distasteful—cede some authority and promise more funding. This is certainly the strategy followed by the AEC and, after the Mediterranean debacle, the CEA, though the British resisted it tooth and nail. However, there was one critic that could not be satisfied by any strategy—the Soviet Union. While problems with oceanographers divided scientists along institutional and disciplinary lines, the objections of the Soviet scientists reflected the geopolitical tensions of the cold war. The official Soviet line against dumping hardened in 1960, when scientists, lawyers, and politicians began to try working out the details of dumping regulations. The Monaco conference the previous year had resolved little and had simply made differences of opinion clearer. When delegations began to address the legal obligations of the Law of the Sea in connection with radioactive waste, the political rhetoric of the cold war intensified.

By developing recommendations on how to regulate dumping at sea, the IAEA report of the Brynielsson panel implicitly sanctioned the practice. Because the Soviet Union took part in the discussions, Western scientists learned informally that the Soviets released some radioactive wastes into public sewers. One British official concluded that the main difference between Soviet and British practices was one of interpretation about the role of rivers and seas. The Soviets restricted discharges into the environment by ensuring safe levels prior to release, whereas the British released higher levels and depended on the environment to dilute them. Moreover, the Soviets used threshold values as more than ironclad guarantees of safety; their thresholds proscribed levels below which substances did not need to be called radioactive at all. This interpretation probably encouraged the Soviets to claim that they did not dispose of any radioactive waste. Although discharges to the environment supposedly were small compared to those in the United Kingdom, one authority official noted cynically (and correctly) that "it is extremely unlikely that they discharge *nothing* into the environment as they would have had the lawyers believe."[51]

This small difference in policy—and many Westerners doubted that such a difference existed—formed the basis upon which the Soviet Union could take a hard-line stance against dumping. When an IAEA legal panel met in Vienna to discuss the results of the Brynielsson report and to outline a plan of action, it seemed clear that the Soviets intended to widen the apparent cleavage between the American/British and Soviet attitudes. Most of the panel agreed that dumping could be allowed, and that any regulatory criteria should be based upon the Brynielsson report. This is not to suggest that all were pleased with it. If enforced, the depth requirement would compel Britain to abandon its primary dump site in the Hurd Deep. At the same time, the report implicitly condoned its discharges from Windscale. So British representatives were not openly hostile to the Brynielsson recommendations. The same could not be said of the Soviet Union. Its representatives, supported by those of Poland, spoke out against any radioactive waste disposal into the oceans, claiming that scientists had not yet shown that it could be done safely. Indeed, they questioned not merely the Brynielsson Report's legal implications, the topic of discussion, but also the scientific and technical conclusions on which they were based. The Soviets were unwilling to accept the report's validity.

The official Soviet line soon spilled over into the Russian-language media. A *Pravda* article included what one British official described as a "somewhat vitriolic attack on Brynielsson and by inference on Sweden generally."[52] The article was entitled "Poisoners of Wells and Their Accomplices." It accused the American and British "monopolists" of improving upon the "old method" of sinking poison into wells and taking it to a much higher level, by poisoning the world's oceans. Even worse, they were using the IAEA to establish the legality of their "dirty and dangerous business" to lure the international community to approve poisoning the oceans. The article was particularly hard on Brynielsson himself, accusing him of abusing his position and misleading the world: "[United Nations Secretary-General Dag] Hammarskjold's compatriot Brynielsson tries to prove what cannot be proved—that the disposal of radioactive wastes into the sea is not dangerous. In his report, Brynielsson intentionally conceals the fact which leaves nothing of his 'scientific' conclusions . . . that already today as a result of nuclear tests radioactivity in the Pacific Ocean has nearly reached the limit of the permissible level."[53] The article further accused Brynielsson of writing the report "on order" from the atomic monopolies of the United States and the United Kingdom.[54]

The *Pravda* article was a nasty reminder that cooperation with Soviet scientists would have little effect on the Soviet government's propaganda campaign. Britain's Office of the Minister for Science gathered that the attack in *Pravda* had come straight from the State Committee for the Utilization of Atomic Energy, chaired by Vassily Emelyanov, though Emelyanov was unlikely to admit it. The Office of the Minister for Science now oversaw the political aspects of the AEA, as the Office of the Lord President of the Council had in the past. One of its officials, M. I. Michaels, observed:

> In my experience in dealing with Emelyanov, he can be two quite distinct people. In private conversation, he is prepared to talk objectively on technical and scientific matters; in public, he more often acts as a politician willing to mis-state and malign the position of his opponents without any regard to the factual accuracy or the relevance of his arguments. I hope that in the conversations that take place in the near future here that he will act solely as the technical Head of an atomic energy organization; but, he is liable to switch quite suddenly and to behave as a politician.[55]

If scientists working together at the IAEA headquarters in Vienna might occasionally agree that certain practices were reasonable, any consensus soon disintegrated due to the Soviet Union's hostile declarations.

Most Western diplomats interpreted the Soviet complaints as yet another swipe at American nuclear weapons testing; waste disposal was just another way to accuse the West of global contamination. Although the British, American, and French found these criticisms annoying, their attitude toward dumping did not change. But the Brynielsson episode strengthened the resolve of those who wanted to avoid disclosing dumping operations. British officials feared that the Brynielsson Report had stirred up so much international interest that its unilateral dumps into the sea might elicit criticism. For example, Britain's Atomic Energy Authority decided not to make any public statement of its intention to carry out an operation in 1961. To Britain, it was business as usual and should not have been a cause for special comment. Moreover, officials thought it unwise to let on that an operation was happening at all, given that the site for dumping was some 150 miles west of the Madeira Isles, owned by Portugal. "It has been decided here," one authority official wrote, "that this location ought not to be made public in view of the possible reaction of foreign Governments who may feel that their interests are adversely affected."[56]

The AEA drafted a list of possible questions by the press, and the answers that officials should give. The list was a portrait of UK attitudes toward the international implications of waste disposal. Who sanctioned such operations? They were made on the authorizations of British ministries (the Ministry of Housing and Local Government and the Ministry of Agriculture, Fisheries and Food), and because they were done on the high seas, "the question of special rights, therefore, does not arise. . . . In the absence so far of any system of international consultative or regulatory machinery for the sea dumping of radioactive wastes, there has been no formal national or international consultation." Further, British practices had been public knowledge for years. Because they believed ocean disposal was a safe and simple way of getting rid of radioactive waste, the British would continue it indefinitely.[57] The Soviets, according to the list of press questions and answers, were simply ignoring the facts and disregarding the distinction between high and low activity wastes, only the former having been judged by the Brynielsson report as too dangerous to dump. The Soviet attitude, the British wrote, "can only be regarded as part of their political opposition to all operations by Western countries which might conceivably be associated with the military uses of military materials." Besides, there was plenty of land in the Soviet Union in which to dump such materials, compared to the United Kingdom, so naturally, the AEA insisted, the United Kingdom should be obliged to dump more at sea.[58]

The British Office of the Minister for Science cautioned the AEA to tread lightly on these international issues, especially given the Soviet propensity to criticize. "We do not intend to change the practice because of these attacks," one official wrote to the authority, "but it is desirable to minimise the scope we give for further criticism." In particular, it was unwise to refer to "the rather tricky question" of the authorizations made by the Atomic Energy Authority Act of 1954. Other nations might disagree, for instance, that "the 1954

Act, a domestic measure, really empowers the two Ministries concerned to authorise disposal of waste on the high seas." The British attitude about secrecy remained: nothing was officially a secret, but they would not take active measures to discuss their policies. Past disclosures, dating from 1949, and the government legislation founding the AEA in 1954, were sufficient. As a Minister for Science official put it, these policies were in place "before the Russians had begun their campaign; and, although what was said then remains for all to see it seems desirable to avoid repeating it unnecessarily."[59]

British diplomats had a difficult dilemma about how to address Soviet criticism. They wanted somehow to inform the public that establishments in Britain "only release radioactive waste in small and carefully controlled quantities subject to strict rules." It was one thing to convince scientists and government officials of this, but quite another to placate the general public; the latter might even believe that the Soviets, in contrast to Britain, released nothing harmful into the sea. Because Soviet practices were secret, it was difficult to demonstrate with publicly available information that the Soviets, like everyone else, could not possibly be achieving zero release of radioactivity. "It is to my mind essential," an official wrote, "to say that in fact their statement is untrue. We must I believe consider some counter attack which we can make orally and if need be in writing, in public discussion of this subject." They could publicize everything they had learned of the Soviets' secret practices. From various public and secret sources, the British government knew that the Soviets discharged radioactive waste into their sewers, producing exposure levels exceeding by a hundred times the recommendations of the ICRP, and certain rivers contained high levels of waste undoubtedly from nuclear installations. Also, radiation exposure in Czechoslovakian uranium mines was well above acceptable limits, and some rivers in that country were radioactive above commonly accepted norms of safety.[60]

The British were bewildered by people's willingness to believe that the Soviets discharged nothing to the environment. Surely there should be intelligent people who would see through the Soviet claims. As a frustrated official put it, "It is clear that technically speaking an absolutely zero release of radioactivity by a country with a vast weapons programme and a considerable civil programme as well, using quantities of isotopes in hospitals, factories and farms, cannot be achieved, and any claim to this effect ought easily to be demonstrably false."[61] But the secrecy surrounding the Soviet Union made it impossible to point to any data, or any published dissenting statements within the Soviet Union, to make statements more specific than vague counteraccusations. Ultimately, officials within the Foreign Office decided that Britain's hands were tied. They believed that it would be unwise to make allegations based on secret information. Not only could it compromise intelligence sources, but it might also have the opposite effect than that intended, raising suspicions about the government's motivations. "To uncommitted countries," one Foreign Office official wrote, "a violent campaign might seem too obvious a cover for our own activities."[62]

Despite the Foreign Office's reluctance to inflame uncommitted countries, Britain's Office of the Minister for Science launched a fact-gathering exercise for its own information about Soviet waste disposal practices. British intelligence services furnished fragmentary information, explicitly labeled "top secret," gathered in the course of intelligence activities aimed principally at estimating Soviet weapons production capacity. They believed the Soviet Union had two plutonium production sites with chemical processing facilities: one at Kyshtym, in the southern Urals between Chelyabinsk and Sverdlovsk, and the other near Tomsk, in Siberia. No information was published about them except vague or uninformative details about the "Siberian power station." No Westerners had visited the sites, and precious few details were known about them. It was not for want of trying, and the area was a natural target for reconnaissance aircraft: one of these, an American U-2 plane, was shot down in May 1960 as it flew over the Sverdlovsk region, resulting in the capture of its pilot, Francis Gary Powers.[63] The Kyshtym site had operated on a large scale since 1948, and intelligence reports spoke of workmen handling "some kind of mud or sludge" and suffering damage to their lungs and other organs, suggesting (to the British) "a certain casualness in the handling of waste." Beyond that, however, few details were known. At Tomsk they knew only that the Soviets stored some waste, piped some into a nearby marsh, and released the rest into a river, ultimately flowing into the Arctic Ocean—they knew nothing about how much radioactivity was involved.[64]

There was, however, a rare and puzzling public reference dating back to November 1957 to scientific experiments on the effects of radioactivity on aquatic plants and plankton. The "experiment" involved putting concrete drums of radioactive waste into lakes at the Kyshtym site. This seemed to suggest that the Soviets were not totally averse to introducing wastes to the environment. But the British took this to be a cover story for the intensified activities in the region necessitated by a major accident. Between 1957 and 1959, British intelligence services produced several reports about a serious accident at Kyshtym that had led to the Soviet government's evacuation of nearby villages and the confiscation of foodstuffs over a huge area including Chelyabinsk and Sverdlovsk. Presumably, the Soviets now were trying to get rid of the material by dumping it into lakes, calling it an "experiment." Americans at the Central Intelligence Agency suggested to their British colleagues that the Soviet attitude toward the safe containment of wastes had become much more serious after the incident.[65]

At the time, specific details of the accident at Kyshtym were unknown to the West, and the British Joint Intelligence Bureau had no idea if it had occurred in a reactor, in the chemical separation plant, or in some other area. The only certainty was that widespread contamination had occurred. Only later, in 1976, did the Western public get a hint at the truth, largely through the scientific sleuthing of Russian émigré biologist Zhores Medvedev, who ultimately wrote a book, *Nuclear Disaster in the Urals*, about a major explosion of radioactive waste.[66] The Soviet government acknowledged the explosion in 1989, under the influence of President Mikhail Gorbachev's policy of glasnost. The accident, an explosion of a storage tank containing high-level liquid waste from chemical reprocessing, occurred on 29 September 1957. The tank contained seventy to eighty metric tons of waste estimated at about twenty million curies of radioactivity. The explosion itself resulted from chemical interaction and probably did not involve any fission reaction, but it had a force of seventy to one hundred metric tons of TNT. After being ejected into the air, most of the waste fell down in the immediate vicinity, while about two million curies were carried away in a giant cloud and came down as fallout over a large area between Sverdlovsk and Chelyabinsk, leaving a permanent zone of major contamination.[67] In light of the extraordinary environmental catastrophe that followed, the technological failure at Kyshtym stood in stark contrast with the self-congratulation and banner-waving that accompanied the launch of *Sputnik*, just five days later. It also took place eleven days prior to the West's first major nuclear accident, the Windscale fire, which drew widespread criticism but was miniscule in comparison.

Because of the impossibility of using intelligence sources to counterattack the Soviet Union, Michaels and others at the Office of the Minister for Science decided that Britain's only remaining strategy would be to "make great play with the fact that the Western side openly disclose the sites of their reactors, their technical details and the whole arrangements for the disposal of wastes."[68] This kind of transparency on the "Western side" was not altogether true, but compared with the Soviet Union, Western countries appeared relatively forthcoming. "We agree," wrote one British official, "that Mr. Michaels is rather overstating the case when he says that we openly disclose 'the whole arrangements for the disposal of waste.'" Yet "it is the startling contrast with Russian practice which is the main point, and on the whole we do publish freely and frankly."[69]

Even leaving the Kyshtym accident aside, these skeptics probably would have been appalled at the scale of the Soviet Union's hypocrisy, the details of which became known only in the final years of the government's existence. In 1990, the Soviet government declared the entire region around Chelyabinsk an ecological disaster zone—not only because of the explosion's contamination, but also because of the radioactive rivers, lakes, and groundwater. According to a report commissioned in 1991 by President Gorbachev, all of the liquid wastes from chemical reprocessing at Kyshtym had been pumped, without treatment, directly into the Techa River between 1948 and 1951. Until 1956, all low- and intermediate-level liquid wastes were pumped there. They stopped putting high-level waste directly into the river after discovering in 1951 that most of the radioactivity was absorbed by the silt in the riverbed over the first thirty-five kilometers, and they had begun to connect diseases in the local people with overexposure from the river. The Soviet government subsequently banned the public use of the river, without explaining the hazard. In 1951 they began to divert the high-level liquid wastes into nearby Lake Karachai and later into storage tanks, but low-level wastes were put directly into the Techa until 1964, certainly covering the period during which the official Soviet line was that they did not put any radioactive waste into rivers and oceans at all. Before 1952, the average discharge into the Techa was about four thousand curies *per day*, which dropped to about twenty-five per day when they decided to put the high-level waste into Lake Karachai.[70] By way of comparison, the British during this same period were trying to justify increasing their daily discharge from a pipeline into the Irish Sea to several hundred curies per day.

Fearing the public relations and diplomatic backlash of a smear campaign against the Soviets, the Foreign Office struggled to identify a feasible strategy. It chose to take a different course entirely. If Britain truly stood behind its policies, the Foreign Office reasoned, then perhaps a greater degree of openness and candor would help them in their media battle with the Soviet Union. The Soviets certainly appeared to be gearing up for a fight. In late August 1961, the Soviet oceanographic vessel *Mikhail Lomonosov* left Kaliningrad to study the Gulf Stream. According to French newspaper *Le Monde*, its object was to prove that currents were strong enough to bring submerged radioactive waste to the surface, thus endangering human life.[71] To the Foreign Office, this was only further evidence that the Soviet attacks might go on for quite some time. Consequently, it behooved the British to provide the propagandists with as little fuel as possible. A few weeks later, the Foreign Office concluded that it profited the AEA nothing to continue its practice of keeping its disposal sites secret for fear of provoking other countries. Instead, it reasoned that if the AEA believed in its practices, then the sites should be disclosed, not only as a measure of good faith but also to lend credibility to the United Kingdom's practices.[72]

With the Foreign Office putting pressure on the AEA to be more open about its activities to avoid fueling Soviet criticism, British atomic energy officials thought of ways to handle the negative publicity that was to come. Initially its most effective choice proved to be in developing international coordination between American, French, and British atomic energy establishments. Fortunately, the French atomic energy establishment had begun to feel the political strains of their waste disposal policies when the French press began to raise questions about plans to dump radioactive waste into the Mediterranean Sea. During the 1960 debacle, French radio commentators contacted the Atomic Energy Authority, requesting that it provide a spokesperson to answer questions of "public interest" regarding the controversy in France. Officials at the AEA shrewdly demurred, hoping to maintain a working dialogue with the French Commissariat à l'Energie Atomique. As one British scientist put it, "we considered it inappropriate to comment publicly on matters which were currently of embarrassment to them."[73] This was an opportunity, not to comment on others' activities, but rather to coordinate action with another atomic energy establishment facing similar attacks on its waste disposal practices. A week later, the chairman of the AEA, Sir Roger Makins, visited France. There he and his French counterparts decided that already there were adequate means for exchanging views on technical questions through the IAEA. However, they agreed that new links between the two national agencies should be established, particularly for "having discussions on the means for dealing with the public relations aspects."[74]

The AEA thought that the French controversy provided a perfect environment in which to strengthen the ties between the two atomic energy establishments, and it attempted to build a cordial working relationship with the French over the next year. One official noted, "We have every reason to believe that the French will be valuable allies in wider international discussions on these matters rather than the reverse."[75] The French were certainly willing to be such allies, as the CEA was expanding its infrastructure and expected not only to dump wastes at sea but also to discharge wastes into the English Channel. This prospect could not have pleased the AEA more, and the two agencies sought to cooperate in planning for it. To the AEA, there were "political advantages in associating with the French in a field where hitherto we have tended to bear the brunt of Russian attacks on sea disposal of radioactive waste in any form."[76] If closer liaison could be accomplished, they could help each other and aid in the international acquiescence of radioactive waste disposal in the sea.

The AEA and the AEC helped each other as well by sharing copies of their national delegations' secret instruction briefs for meetings at the IAEA.

One of the topics of discussion was the oceanographers' idea to create a new laboratory under the auspices of the IAEA. This laboratory would study the relationship between radioactivity and the sea. Its existence would imply long-range financial support, which was why oceanographers wanted it and atomic energy establishments loathed the idea. The laboratory had the potential to become a font of criticism by oceanographers perpetually casting doubt on waste disposal in order to receive funding. By sharing instruction briefs, the Americans discovered that the British bristled at the thought of more mushrooming scientific projects, but for political reasons they did not want to be seen as the only nation to oppose the new laboratory. Cockcroft's instructions were to "do what he can to curb the [International Atomic Energy] Agency's natural proclivities in this matter" without blatantly taking too strong a stand against international scientific studies. One official urged the British delegation to show a "conspicuous lack of enthusiasm" but admitted that it had become politically impossible not to support such studies. The oceanographers had become too powerful.[77]

The British learned through this cooperation that the United States, by contrast, strongly supported the international laboratory. According to the U.S. delegation briefing (shared with the British by physicist Isidor Rabi), its primary reason for doing so was to counter Soviet propaganda. The Soviet Union opposed dumping in principle, and the United States feared that the environmental effects of waste disposal would form the basis of a major propaganda campaign against the West. The Americans hoped that their support for international scientific work could give more credibility to their policies, particularly if scientists could identify problems and be seen to be researching them. The draft-trading between the AEC and the AEA allowed officials to see that their apparent disagreement was more of a difference in tactics and immediate diplomatic necessity, rather than a genuine divergence of views. Neither side believed that a new international laboratory was necessary to solve real problems of waste disposal.[78]

To consolidate the American-British position, British nuclear physicist Sir William Penney and two colleagues went to the United States in October 1960 and met with General Alvin Luedecke, the general manager of the AEC, along with several of his staff. They appeared to have agreed that an IAEA laboratory in Monaco, with a mandate to study radioactive waste, might become a source of scientific criticisms, not solutions. But despite its own problems with oceanographers, the AEC saw the laboratory as eminently manageable, particularly if what the oceanographers wanted was money and not to protect the oceans from atomic energy. Luedecke pointed out that the laboratory would be staffed with plenty of American scientists whose paychecks would come from the AEC. Luedecke appeared to think that if funding was the oceanographers' goal, they would not bite the hand that fed them. They also agreed that it was a very good idea for the two establishments to continue collaborating in this way, prior to meeting with scientists and certainly prior to any international meeting. They also needed to see if there was any way to do the same with the French, to understand fully their position before coming to the IAEA. Although the French had not been close partners on atomic energy matters in the past, the Mediterranean fiasco gave them every reason to believe that the CEA was confronting similar issues and might want to collaborate.[79]

As noted above, the French were more than willing to compare notes on how to deal with troublesome oceanographers. Francis Perrin met with American atomic energy colleagues in November, after being forced by public opinion to shelve the Mediterranean experiment. By the next month, the French were collaborating directly with the British. The French and British atomic commissions met to deal specifically with their mutual public relations problems. It turned out to be a productive meeting of minds. They felt they had to do something to establish common practices that they could defend against critics, even scientists. The British gave talks on subjects ranging from the technical details of disposal to relations with local authorities. In fact, each side specified two people to liaise directly with their foreign opposite in the future on public relations and technical issues to avoid any semblance of disagreement to be exploited in international meetings.[80]

The existing notes of these meetings between the CEA and AEA, held in the British National Archives, reveal an atmosphere of mutual understanding and solidarity. Their common problem was public relations, and their common headache was the oceanographic community. Geopolitical difficulties seemed nearly trivial by comparison. In the course of the meeting, for example, the British found that their French counterparts felt the same as they did about the Soviets—that they discharged quite a bit into river and oceans, despite their public pronouncements, and were not to be taken seriously. More important were the oceanographers who, they believed, made mountains out of molehills in order to serve themselves. Cousteau, French public relations official Jean Renou told his British counterparts, might be well known for his literary work, but he was not much of an oceanographer. What was needed, he and others agreed, was a push for education about atomic energy so that officials and the general public would not be so easily swayed by prejudiced scientists like Cousteau.[81]

When the AEA's public relations director, Eric Underwood, addressed the joint meeting, he pointed out that frankness, rather than secrecy, usually paid the best dividends for public relations. "Skeletons in cupboards," he observed, "tend to rattle sooner or later." He blamed the scientists for this, for the most part, because they often divulged embarrassing facts even when others in the atomic energy establishment managed to suppress them. It seemed impossible, he pointed out, to repress pride of authorship in any scientist. The best course to avoid misrepresentation was to be frank and to keep a meticulous record of precisely what was said to the public. Yet despite any specific measures that the AEA or the CEA might take, there would always be unknown factors when dealing with public apprehensions. He reminded them of Napoleon's dictum that his first requirement of a general was that he should be lucky.[82]

For the time being, the victory appeared to go to the oceanographers, who won both scientific authority and the promise of patronage. All three establishments publicly claimed that they needed to support more research in oceanography to ensure that their practices were indeed safe. Moreover, they agreed to the creation of a permanent laboratory financed by national governments through the International Atomic Energy Agency. When it was founded in 1961, the Monaco laboratory looked exactly as oceanographers had hoped and as the atomic energy officials cynically had expected. Its first director, oceanographer Ilmo Hela, the former director of Finland's Institute of Marine Research, kept close institutional ties to Cousteau's Musée Océanographique. According to an IAEA press release, the laboratory's first goal was to understand the movement of water and marine organisms and the deposition of organic and inorganic matter, a pretty broad agenda that made no specific reference to radioactive waste. Second was the study of the distribution of radioactive materials in organisms, and last of all was the study of the effects of radioactive materials on marine ecology.[83] If the AEC, AEA, and CEA saw oceanographers as interested in power and patronage rather than solving the waste disposal problem, the establishment of the Monaco laboratory only strengthened this view.

Although the United States was turning away from ocean disposal, it strongly supported its allies' policies. The United States saw the Soviet criticism in the same way as the British did—it was an easy way for the Soviet Union to bring propaganda about weapons testing into the IAEA, where delegates were supposed to limit their discussions to the peaceful uses of atomic energy. Not only did waste disposal provide a common link between civilian nuclear power and weapons development; the scientific controversy over chronic exposure to low-level radiation translated easily from the fallout debates to waste disposal. Because the question of waste disposal was now obviously being used by the Soviet Union as propaganda, the United States, Britain, and their allies met together to coordinate their views on what increasingly had become a cold war issue. The United States suggested pressing forward with an international convention on waste disposal to clarify what was indeed legal. Undoubtedly this was partly a result of its recent quandary about how to deal with Mexico's objection to dumping in the Gulf of Mexico, a difficult decision given the lack of international agreement. But the Soviets did not want to allow any dumping at all, which made agreement about regulations impossible. Britain, for its part, did not want a convention either, because it might compel negotiations about its discharges from Windscale or its dump site in the Hurd Deep.

Leaving aside the possibility of a convention, the United States brought the subject up through the Western military alliance, the North Atlantic Treaty Organization (NATO). The Americans agreed with the British that the Soviet government was trying to prevent everyone else from disposing of their waste at sea while refusing to tell anyone the specific measures they took with their own waste. Delegates to a May 1961 meeting of NATO's Political Committee pointed out that there was no evidence to back up the Soviets' claims of danger, but unfortunately there was not enough scientific knowledge about the seas to declare sea disposal safe, either. The French delegate said that the whole question was being decided by political propaganda, not scientific study. Other representatives—of Belgium, Greece, Canada, the United States, Norway, the Netherlands, Italy, and Turkey—all agreed that supporting scientific research was their only viable option to do something positive, since they had no publicly available information with which to confront the Soviets. This became the official line of the NATO countries—to dismiss the Soviet position as propaganda but to make up for it by publicly supporting scientific research, including oceanography.[84]

The United States was less vulnerable than Britain to diplomatic outcries against ocean disposal. Though the Soviet scientists seemed to imply that the Americans were determined to put high-level wastes into the ocean depths, this was not the case. AEC scientists had decided by 1960 that high-level wastes probably would need to be permanently stored on land, in salt beds. The chief of the AEC's Sanitary and Engineering Branch, Joseph A. Lieberman, pointed out that deep-sea disposal of high-level radioactive waste, mostly from chemical reprocessing plants, seemed pretty unrealistic. This was due in part to the lack of adequate oceanographic information, but, perhaps more important, the logistical problems of handling and transporting the wastes over land and out to sea to proper depths "lead one to rather negative conclusions regarding the disposal of significant quantities of high-level wastes at sea."[85]

In addition, unlike Britain, the United States did not need to discharge wastes from its chemical reprocessing plants directly into the sea. Because most of these facilities were inland, the Americans followed a practice of concentrating their high-level wastes and containing them on site in temporary storage tanks while putting their relatively low-level liquid waste into ponds, lakes, and rivers, all within the United States. The principal burial grounds for the AEC's high-level wastes were at the Oak Ridge, Hanford, Idaho Falls, Savannah River, and Los Alamos sites. However, the AEC acknowledged putting packaged, solid wastes into the sea; compared to the concentrated high-level wastes stored in tanks, these were "of a relatively low or intermediate level." Between 1951 and 1960, the AEC put some 23,000 fifty-five-gallon drums of it into the Atlantic at various locations, comprising less than 8,000 curies at the time of disposal. In the Pacific, disposals had begun in 1946 off of San Francisco, and by 1960 the AEC had put 21,000 drums and 329 concrete boxes there, amounting to some 14,000 curies. Another site, closer to Long Beach, accepted a smaller amount—about 60 curies in 2,950 drums, from 1953 to 1960. Lieberman pointed out that one reason he and other AEC scientists believed these operations were completely safe was "the actual operating experience of the British in disposing of greater quantities of radioactive material in a more mobile (liquid) state."[86]

The Soviet government's decision in August 1961 to resume nuclear weapons testing temporarily took the wind out of the sails of its antidumping campaign. Shortly after the Soviet announcement, British Foreign Office official Anne Stoddart mused: "It will be interesting to see what the Russians say about radio-active waste in the sea, now that they are testing in the atmosphere."[87] And like clockwork, they appeared to relax their attitudes, hinting that they might let up on their condemnations of waste disposal. At the IAEA Annual General Conference in 1962, Vassily Emelyanov pointed out that the Soviet position was never as uncompromising as everyone seemed to think, and that there might be permissible levels of radioactivity to be discharged into the oceans. Emelyanov had written an article for the Soviet journal *International Affairs*, stating that such disposal caused legitimate anxiety because, just as American oceanographers had noted, radioactivity could contaminate marine flora and fauna, reaching humans at the end of the food chain. Thus, Emelyanov asserted, it was "above all necessary to establish scientifically the permissible levels for the content of radioactive substances in the water, the soil, and the air. These levels should be binding on all countries."[88]

By accepting the American views of the dangers of fauna contamination, and by agreeing that such dumping ought to be regulated by the IAEA, the Soviets acquired a renewed moral authority. Rather than denunciating everyone involved in ocean waste disposal, they embraced the concept of international regulation. This apparently conciliatory move nonetheless pitted them against Britain, France, and the United States, all of whom hoped that international standards might leave to individual nations the discretion to formulate policies.

In November 1962, the Soviet periodical *Water Transport* condemned these countries for their systematic disposal of radioactive waste into the world's seas and rivers. *Water Transport* specifically pointed at the United States, for its Pacific and Atlantic dumping, and to the United Kingdom, for discharging wastes from Windscale into the Irish Sea. The newspaper challenged the Western countries to abide by the articles of the 1958 United Nations Convention on the Law of the Sea (UNCLOS), which stated that all nations should take measures to prevent such radioactive pollution. Unlike the West, *Water Transport* stated, the Soviet Union buried most of its waste and reduced liquid waste to safe levels prior to discharge.[89] Unlike its previous denunciations of the dumping practices, this one claimed its own adherence to an international agreement that was being systematically violated by the West.

Both the American AEC and the British AEA scrambled to respond to the accusation. The *New York Times* quoted an AEC spokesman saying that the agency required such dumping to be made at a minimum depth of 6,000 feet (a thousand fathoms), to ensure dispersal and safe dilution. It also claimed that "extensive environmental surveys" showed that there was no radioactivity above natural background radiation in the water, sediments, or in any marine life. Both statements, any reader of reports by the National Academy of Sciences should have realized, were not altogether true of past actions. But in any case, the AEC told the *New York Times*, "principally because of economic considerations," most radioactive waste in the future would be buried on land.[90] The editor of *The Times* of London must have alerted the AEA to the fact that it would announce the Soviet accusation, because the AEA's rebuttal appeared just underneath the report on the Soviets. The article reflected the AEA's position that discharges into the sea "cannot reasonably be considered an automatic infringement" because the IAEA recommendations, not as yet defined, certainly would allow for some dumping. At present, the levels of discharge were drawn from the recommendations of the International Commission on Radiological Protection, "by considering the possible routes, including fish and edible seaweed, by which members of the public might be irradiated."[91]

These bitter salvos across cold war lines coincided with extraordinary geopolitical tensions. The possibility of genuine cooperation or even mutual understanding seemed remote—and certainly no one expected an abatement of Soviet propaganda. Nuclear weapons testing was a constant and bitter dispute, and the scientific controversy about fallout heated up in the late 1950s. In addition, the United States had been caught spying on the Soviet Union from a U-2 spy plane in 1960, and the early 1960s saw President Kennedy facing off with Soviet leader Nikita Khrushchev over Berlin, where a new wall was being built around the western sector. Worst of all was the Cuban Missile Crisis in 1962, probably the closest the two superpowers ever came to an actual war. The silver lining of that crisis was that it convinced world leaders to get serious about banning nuclear weapons tests. They did this in 1963 with the Limited Test Ban Treaty, which banned testing in the atmosphere, oceans, and space.[92]

After the negotiation of the Limited Test Ban Treaty, Western atomic energy officials had high hopes for the international endorsement of radioactive waste disposal on the high seas. After all, they believed that the Soviet attitude was tied primarily to its government's criticism of nuclear tests. In April, several Soviet atomic energy officials visited the United Kingdom to observe methods of waste disposal. Health physicist John Dunster felt that the visit helped to penetrate the wall of inflexible opposition by the Soviet Union. He also felt that the Soviets had been impressed both by the standards of the plants and equipment and by the thoroughness of Britain's ocean dumping work. The British made it clear that the authority conducted scientific work and monitoring programs in areas of discharge and dumping, and the Soviets appeared suitably impressed.[93]

Much of the discussion during the visit turned on the question of radioactive effluent discharged into rivers by nuclear installations. From Soviet institutions comparable to Harwell, the delegation claimed, the liquid wastes were kept to the local standards for public drinking water before they were discharged. In the absence of treatment plants to accomplish this low level of contamination, higher-level discharges were permitted to sewers, but only if they were suitably diluted by inactive wastes in the same effluent discharge. The highest activity wastes were collected and stored in vaults on land. As they already suspected, the British learned that the critical (apparent) differences between UK and USSR policies were twofold. First, in the USSR, effluent had to conform to sanitary norms for drinking water at the point of discharge, while effluent from UK sites depended on environmental dilution. Second, packaged wastes were stored on land in the USSR, whereas the United Kingdom put wastes in drums and dumped them in the ocean.[94]

Despite the continued difference about the role of rivers and seas—as part of the process to achieve safety or as the receptacle of "safe" wastes—Dunster felt that the visit had been a success, in that the Soviets would be more sympathetic to ocean disposal in the future. This was particularly so with regard to effluent, rather than deep-sea dumping. AEA officials explained that, in a small but densely populated country such as the United Kingdom, the discharge of low-level effluent to sea was far safer than on land. The leader of the Soviet delegation, B. S. Kolychev, appeared to accept that for certain countries perhaps it was indeed the best option.[95] The Soviets were impressed by the measures taken to provide the safest conditions for disposal. The presentations on British monitoring work particularly "took the Russians by storm," wrote Dunster. Although he expected that the Soviets would not relent on the political front, Dunster believed that "at least a number of their influential technical people now recognise our point of view and some of them at least sympathize with it."[96] Kolychev even pointed out that there was a minority of officials in the Soviet atomic energy establishment that favored ocean dumping. Dunster observed in a report of the visit: "Once it was clear that we were not committed to sea disposal for doctrinaire reasons and compared it with other methods of disposal with the aim of achieving the lowest practicable dose to man, he seemed to be in general sympathy with our views." Further, the Soviet scientist told Dunster that his government's position probably was unreasonably inflexible.[97]

These remarks by Soviet atomic scientists seemed out of step with the official Soviet line, but most Western scientists interpreted Soviet criticisms as politically motivated. Few, if any, took the public Soviet line as a serious scientific argument against existing policies, because Soviet scientific work always led to the same politically generated result. In 1964, the head of Environmental Research at Atomic Energy Research of Canada, Colin Mawson, complained about Russian scientists' odd conclusions. For example, one of them, L. I. Gedeonov, had claimed during a recent conference that the sea was a better mixing medium than previously believed. This might have been interpreted as a vindication of the Western view that the ocean could be part of the disposal process. But that was not Gedeonov's conclusion. Baffled, Mawson exclaimed in a letter to a Harwell health physicist: "His conclusion was that wastes must *not* be put into the sea!" It turned out that the Soviet scientist was trying to demonstrate deep downward penetration of fallout from nuclear tests. Greater diffusion indicated that the fallout was spreading; but from a waste disposal point of view, such spreading was the key to dilution. Mawson took the Russian's opposition to sea dumping as a mindless conclusion dictated by his country's official line. Others likewise found it equally ludicrous when Mawson calculated the intake of fish based on Gedeonov's figures for the concentration of strontium-90. Mawson described the scene as he put a question to Gedeonov at the conference: "I assumed 250 g per day intake and a concentration factor of 1000 (!) into fish flesh, and came up with about 1/100 of the MPI [maximum permissible intake]. I then pointed out that this was the result of dumping some 5 x 10^6 curies of Sr-90 into the sea from the air. His answer to this was that fallout concentrations were high enough to damage production of caviar—roars of laughter from the audience!"[98] Soviet criticisms did little to persuade Western scientists, largely because they seemed too obvious a product of political necessity.

For the time being, the British and their allies could laugh. But the continued opposition to waste disposal by the Soviet Union could still hurt Britain, which relied so heavily on ocean disposal. Its principal ally, the United States, had stood up to the Soviet Union, even marshalling allies through NATO. But the American decisions to bow to diplomatic pressure closer to home, from Mexico, and the shocking effect of the political ruckus from Atlantic states—leading the United States to stop dumping at sea—painted a grim picture for the British. New strategies would be needed in the future if they were to continue to rebuff political opponents and Soviet propaganda, and doing this alone seemed nearly impossible.

Chapter 7 Purely for Political Reasons

When british health physicist H. J. (John) Dunster visited Lisbon in April 1967, the city recently had constructed a new suspension bridge that boasted the longest suspended span in Europe. Overlooking the wide Tagus River, the bridge was named after Portugal's longtime prime minister, António de Oliveira Salazar. Dunster had come to the city to participate in what promised to be a contentious meeting of the European Nuclear Energy Agency (ENEA), which was planning to dump thousands of drums of radioactive waste into the ocean far off Portugal's coast. Dunster had a lot at stake; he had been an influential health physicist advising on waste disposal policy in Britain for many years, and everyone knew that Britain was behind the ENEA dumps. He also was the author of the specific hazard assessment that had to be "sold" to the Portuguese. He was not surprised to find some resistance to his ideas among the Portuguese delegates, and he tried to warm up to them by putting his safety assessment into a hypothetically local context. He said that these radioactive wastes were so safe that, were it not for the solid components in them, they could just as easily be taken out onto the Salazar bridge and put directly into the Tagus River.[1]

Dunster's attitude's outraged the Portuguese. Even if they could agree with the safety assessment, there was more at stake than the safety of the Portuguese people. The ENEA operation undoubtedly would make world headlines, and everyone would know that the first Europe-wide dumping operation had taken place not far from Portuguese beaches. What would become of the tourist industry that drove an important segment of the country's economy? With the contentious nature of the effects of radiation exposure, and the political controversy that surrounded French efforts to dump in the Mediterranean in 1960, surely such an operation would scare away tourists and beachgoers. Besides, there were some oceanographers in Portugal and Spain who claimed that the currents off the coast would bring the radioactivity close to the shore. Dunster had tried to brush aside these objections with his flip remark about the Salazar Bridge, adding that he had come to give a scientific assessment to a scientific audience, and that all nonscientific objections to his assessment were fundamentally illogical.[2]

Dunster had been criticized nearly a decade earlier for his lack of political aptitude in contentious international meetings, yet here he was again, lecturing oceanographers and political representatives on the rectitude of British policies. But by this time Dunster was well aware, as were many of his compatriots, that political and diplomatic concessions needed to be made from time to time. Throughout the 1960s, Britain's Atomic Energy Authority developed strategies to protect itself against international criticism by doing just that. It acquiesced in what its health physicists saw as senseless monitoring programs, and it developed a keen sense of public relations. Most important, Britain embarked on an effort to cooperate with other nations in dumping radioactive waste at sea. By doing so, British officials at the AEA and Foreign Office hoped to share the burden of Soviet criticism among the several small European states that had begun to accumulate bothersome amounts of radioactive waste. After the United States abandoned the sea as a major waste disposal option, creating new allies became a central concern. In the process, however, Britain created new opponents, particularly the politicians in those countries closest to the dump sites, such as Portugal and, later, Ireland. British efforts to normalize radioactive waste disposal by widening participation also widened the diplomatic problem beyond the Soviet Union. Worse still was the growing general awareness of the practice, which threatened to conflate radioactive waste disposal with the other forms of marine pollution that grabbed headlines in the late 1960s.

Much Too Dull for Drama?

In 1959, the American magazine *National Geographic* planned a story about French bathyscaphe dives in the Pacific and the studies of deep-sea currents. As in the past when the IGY scientists had used deep-sea studies to search for stagnant water, the issue of radioactive waste arose. The magazine wanted to include an editorial note explaining the situation thus: "The British, in fact, are at present pumping liquid low-level waste without any containers into the Irish Sea, to the amount of 10,000 curies a month. The water dilutes these materials so much that there is no appreciable increase in radioactivity in the Irish Sea." Checking its facts, *National Geographic* asked the British Embassy in Washington, D.C., if this was an accurate appraisal of the situation.[3] There was, in fact, nothing inaccurate about it. But the Atomic Energy Authority was acutely aware that, expressed in such terms, it seemed like a huge amount of radioactive waste. Responding to the British Embassy, the AEA observed that it was "about right for actual discharge," but that it was very dilute and constantly checked by scientists. The authority advised the embassy to play up the fact that the area was constantly monitored; such an emphasis might "make it appear less unfavourable."[4]

National Geographic was not interested in antagonizing the AEA or the British government and made an editorial change that was even more favorable, changing the wording slightly: "The British, in fact, believe that they could safely dump liquid low-level waste without any containers into the Irish Sea, to the amount of 10,000 curies a month. . . . Smaller quantities are already being discarded in this way. The water so dilutes and disperses these wastes that there is no appreciable increase in radioactivity in the Irish Sea." The embassy was pleased. It implied only the belief that the large amounts were safe but stated that in practice only smaller amounts were dumped. This was not true, of course, but British officials did not feel obligated to correct *National Geographic*'s very accommodating factual error. As the British Embassy wrote to the AEA, "This seems to me to be a less unfavourable statement than the factual one originally made and I informed the editor that so far as I was concerned it was quite acceptable."[5]

The brief exchange between *National Geographic* and the British Embassy (and the internal exchange between the embassy and the Atomic Energy Authority) was a symptom of the British government's views toward public information at this time. Certainly government officials did not want to draw attention to the practice of waste disposal, as it was an increasingly sensitive issue. In addition, the attitude toward environmental monitoring as a defensive tactic is especially illuminating. Although *National Geographic* did not mention monitoring (given its level of accommodation, it did not need to), AEA officials clearly believed that doing so would help dissipate criticism. Constant monitoring of the Irish Sea was becoming a crucial component of the AEA's claim of safety, despite the fact that many in the AEA considered monitoring entirely superfluous from a policy point of view and irrelevant from a scientific point of view. Its only utility was in the realm of public relations.

The term "monitoring" itself held some power; it suggested a certain level of vigilance in scientific testing, thus ensuring safety. But to those in the atomic energy profession, the scientific merits of monitoring were hard to defend. Even a large-scale monitoring program was unlikely to provide conclusive data in favor of, or against, dumping. At worst it was just another project to pay for. Although some saw environmental monitoring as an important aspect of any waste disposal operation, conducting such monitoring for ocean disposal presented serious challenges. As one scientist wrote in *Health Physics*, "the very features which make the ocean an attractive disposal area, also account for several monitoring perplexities." The excellent dilution of the ocean easily masked the migration of radioactive materials; the vast depths provided ideal isolation but also hindered sampling due to inaccessibility; the deep sea, chosen for its distance from man's food chain, yielded few specimens for adequate ecological studies; and the composition of seawater created complicated problems for radiochemists.[6]

Some recognized, however, that monitoring was an important component of public relations. For example, when discussing the work of the IAEA's international laboratory at Monaco, British atomic energy scientists belittled it as useless except for its role in appearing to monitor the environment. Although publicly Sir John Cockcroft stated to the IAEA that Britain enthusiastically supported future scientific work at Monaco, really he did so only grudgingly, seeing it as the brainchild of opportunistic oceanographers. But he and others were finding a silver lining in the laboratory, as well as in any kind of monitoring work in home waters. Leading health physicist John Dunster acknowledged that more research could be done, "if only to demonstrate that current practice by Great Britain in this connection is safe." The value of international research, he wrote to a colleague, was principally to allay the fears of people who are swayed more by political arguments than scientific ones. At the national level, he suggested, they should start to monitor their principal dumping areas, such as the Hurd Deep in the English Channel, not because they expected to find significant levels of radioactivity but because they did not wish to face the criticism that they had no data. From the AEA's point of view, such environmental surveys were superfluous, costly, and purely political.[7]

But political expediency proved necessary in the early 1960s, particularly after the United States Atomic Energy Commission began to direct more and more of its waste products toward land disposal, apparently abandoning the sea—and casting doubt on the reliability of the United States as Britain's partner in international negotiations about waste disposal at sea. At the end of 1962, the United States had dumped some 86,000 containers of radioactive waste into the oceans since commencing such operations in 1946. But between 1959 and 1962, the AEC had dealt with at least eight instances of waste materials being recovered by ships or being washed ashore to be discovered on American beaches. In some cases, evidence that the containers belonged to the AEC was inconclusive, or they were false alarms, or they were hoaxes. But at least one was a genuine container of radioactive waste. A 30-gallon drum was netted in 1960, in only 275 feet of water, twelve miles off the coast of Massachusetts. Combined with the National Academy of Sciences' controversial report identifying several sites off the Atlantic coast for dumping, including a shallow area in Massachusetts Bay, the AEC decided that the political fallout of waste disposal at sea was too high. One AEC spokesman said that "it was in the furor over that report that people thought we were dumping in those close-in areas, in those quantities, that we in effect stopped all ocean dumping for all practical purposes."[8]

The United States, with its huge land area, could afford to divert its attention to land-based disposal and thus attempt to avoid such problems. By contrast, Britain depended upon the sea for its waste disposal far more than any other nation. It had long been committed to it, and the visibility that came with the Soviet criticism only increased with international developments in the 1960s. One of these came from the IAEA itself, which began to track waste disposal with its Register of Marine Disposal of Radioactive Materials. This was designed to provide an internationally accessible account of how much radioactive waste each country dumped into the oceans. When the figures for 1959–1960 became available to the AEA in early 1963, the Office of the Minister for Science did not like what it saw. The United Kingdom appeared to be making by far the most use of the oceans for radioactive waste. One official, B. C. Peatey, wrote testily to the AEA, "As you will see, the U.K. takes up a great deal of the draft, India is silent, and the whole thing is pretty unbalanced. The U.S. contribution is in a completely different form from anyone else's."[9]

The Office of the Minister for Science dreaded the Register, which promised to make Britain look out of step with other countries. Certainly there were operations by other countries that had not been reported. The British knew, for example, that France dumped package wastes into the Atlantic Ocean, in addition to the discharges to the Seine and Rhone described in the Register, but that French officials were unlikely to admit these officially. Britain had dumped Belgium's waste, and the American figures did not lend themselves to easy comparison. There was precious little information about the practices of some other countries. There was, however, little to be done about their lack of candor. Although Britain would suffer from being so forthcoming while other countries were not, health physicist R. J. Garner insisted cheerfully that "we do have to recognise, of course, that the U.K. does discharge quite a lot!"[10]

In addition to discharging "quite a lot," Britain had to contend with growing domestic and diplomatic difficulties related to its packaged wastes dumped at sea. In the summer of 1961, English writer Hugh Popham wrote to the British Admiralty asking if he could take part in some of the operations of the Royal Fleet Auxiliaries to help in the background research for a novel. He hoped to acquaint himself with the RFA's activities, being particularly interested in "such jobs as the dumping of radioactive waste from nuclear power stations." Although the Admiralty never considered letting Popham come along on one of the cruises, it did ask the Atomic Energy Authority its opinion about how much Popham ought to be told, if anything, about what they did. "Would there be any objection to the publicity which a novel of this sort might arouse?" one Admiralty official wondered. "Would this publicity cause undesirable political complications?"[11]

Although Popham had not intimated that he intended to incorporate any kind of negative attitudes toward the Atomic Energy Authority in his novel, the question itself proved problematic. A novel was not the appropriate medium to discuss the subject, most atomic energy officials felt. "One would hope," one health and safety official wrote, "to persuade him that the work is not of a sufficiently spectacular nature to be of any special interest for his purpose." They wanted to discourage him from writing any novel that involved radioactive waste. Yet, if they refused to offer any information, it might appear that they were not willing to justify their practices.[12] Still, the AEA replied to the Admiralty that it should not be difficult to get Popham to understand their position: "In our experience novelists and playwrights are only interested in atomic energy themes if they offer the possibility of something going seriously wrong and thus providing dramatic situations in which the personalities of the work can be appropriately highlighted. If Mr. Popham is given a broad picture of what actually happens during a sea dumping operation he will probably draw the conclusion that the subject is far too dull for his purpose."[13] These attitudes were consistent with the Atomic Energy Authority's efforts to reinforce the banality of dumping radioactive waste at sea. Because the Soviet Union "constantly plays on the alleged iniquity of discharging such wastes to the environment," it was wise to provide Popham with little ammunition. As one official put it: "The duller the information given to Mr. Popham, and the less opportunity for sensationalisation, the better."[14]

The situation had changed dramatically from 1957, when AEA officials could not conceive of anyone trying to take measurements from a Windscale pipeline. Now news articles routinely looked for inconsistencies, problems, or anything that could be sensationalized. In 1960 even U.S. Atomic Energy Commission sanitary engineer Arnold Joseph was perturbed by something he had read in a magazine about British practices. He contacted the British Embassy to ask for clarification about an article in the magazine *Reporter* that said that if the United States had dumped as much radioactive waste into the sea as Britain had done from Windscale since 1952, American beaches would be the leading source of radiation exposure for millions of Americans. The article quoted John Dunster as having said that their safety calculations assumed that no one spent more than a hundred hours per year at the beach. The *Reporter* pointed out that many Americans in places like Cape Cod and Florida spent hundreds of hours at the beach annually, and that children usually played at water's edge, where most of the radioactivity concentrated in the sand. The British Embassy had to reassure Joseph that this was not a legitimate criticism. In response to his phone call, they sent a transcript of the official response from the Atomic Energy Authority, which said that 100 hours was simply an arbitrary figure used to make calculations. The AEA stated: "It would be safe, in fact, to camp out permanently on these sands."[15]

On another occasion, not publicized, engineers were trying to decide whether to build one of the Windscale pipelines underneath a roadway. They believed that as long as double piping was installed, the risk would be negligible. One official pointed out that, compared to the risks the AEA ran at other locations, such as an experimental reactor at Winfrith Heath, this seemed trivial. "I understand that the effluent pipeline at Winfrith Heath passes under ground owned by the War Office," he said. "This ground is used for firing practice and it would seem that the proposal to run under a road at Windscale would be considerably less hazardous than is the case at Winfrith."[16]

The AEA's efforts to insist on the nonsensational nature of dumping radioactive waste were frustrated by a series of incidents in the early 1960s. In June 1962, as British AEA and French CEA officials met to discuss prospects for further collaboration, the French announced that fishing trawlers in the Bay of Biscay twice had discovered drums of waste in their nets. On the first occasion, the fishing vessel had been trawling at relatively shallow depth, about one hundred miles from the coast; on the second occasion, the vessel had been about sixty miles from the coast. At least one of the three drums recovered on each occasion was marked as the property of the UK Atomic Energy Authority. Here the international implications of radioactive waste disposal crystallized, as the waste from one nation's nuclear establishment ended up in nets filled with fish.[17]

Over the next few months, the AEA conducted an inquiry into the matter and drafted an internal report summarizing the event and identifying problems with selecting the disposal area, packaging the wastes, and navigating on the high seas. The members of the board of inquiry determined that weather conditions had forced the ship's master to rely on dead reckoning for navigation, and thus the drums had been dumped in the wrong place. For them, this meant two things: the dumping area was too small (and thus did not allow for enough margin of error) and the navigational equipment was not accurate enough.[18] Having chalked up the 1962 problem to navigational errors, the

AEA conducted twelve more dumps in 1963 and 1964, taking care to dump some fifty miles beyond the Continental Shelf.[19]

Outside Britain, observers were sharper in their judgments. Some French officials attributed the incident to British incompetence and disregard for the safety of human life. At an international scientific colloquium held at the French CEA research center at Saclay in October 1962, British health physicist R. H. Burns overheard the director of Saclay harshly criticizing the AEA. In Burns's view, the director was giving an exaggerated version of the incident to an international group of delegates. Burns reported, "The group of people to whom he was talking contained several of the American delegates and he asked M. DeJonghe (Belgium) to interpret for him. M. DeJonghe in his translation, endeavoured to 'play down' the incident but there was no doubt that the version, as given, gave rise to considerable interest and criticism."[20] The Belgian, Paul DeJonghe, probably wanted to "play down" the incident because Belgium recently had turned over some of its own waste to be dumped by the AEA. When the conversation ended, DeJonghe made his way toward the British delegation, which tried to reassure him that no Belgian waste had been involved. The incident had caused, in one AEA official's words, some "rather unfortunate publicity," and it convinced the British to divulge their conclusions about the navigation problems "to let the Americans and if necessary the Belgians know the true facts without letting them have a copy of the [board of inquiry's] report."[21] Although it was reluctant to divulge such information, the AEA found that disclosures to a limited international group could be quite useful. The British needed to reassure dumping-friendly nations that their methods were safe and that this problem had occurred only because the carefully planned methods had not been followed.

On 25 May 1964, French officials informed the British scientific attaché in Paris of another troubling incident. Once again, fishing boats on the high seas, this time trawling off the coast of Brittany, had discovered drums of waste in their nets. All of them were of British origin, and some, they said, contained radioactive material.[22] The incident prompted a brief reassessment of a dumping operation planned again for the Bay of Biscay. The AEA acknowledged that, in light of the recent embarrassment, it might choose a different site, perhaps one farther from the coast and away from fishing areas. But ultimately the health physicists insisted that the present location, despite the possible public sensitivity, was the most practical.[23]

Nevertheless, the AEA had to be cautious in proceeding with its dumping plans because of internal divisions. The two international incidents were having an adverse effect on the willingness of British ministries (particularly the Ministry of Agriculture, Fisheries and Food, or MAFF) to approve further dumping operations. Anticipating this reticence, the AEA quickly identified a sacrificial lamb: the Hurd Deep in the English Channel. It had planned to abandon it as a dumping ground anyway, because of the recommendations of the IAEA's Brynielsson committee, and here was an opportunity to do so with the appearance of having conceded something.[24] However, the immediacy of the situation evaporated when British investigators determined that the recovered drums had not contained radioactive materials at all. Questions of accountability and safety seemed to fade into irrelevance, and the authority looked forward with renewed confidence that government ministries would not interfere with its future dumping operations. To the AEA officials, all was well. As soon as they were satisfied that an incident had not occurred, all their reservations vanished.[25]

Still another incident was successfully covered up and never reached the news media, and it involved a ship, with Spanish fishing boats in view, trying collect floating contaminated debris. On 13 July 1964, the M/V *Halcience*, docked at the Royal Naval Armament Depot in Gosport, was loaded with nearly a thousand tons of radioactive waste. Most of the waste (778 tons) came from the Atomic Energy Research Establishment (AERE) at Harwell, and the rest came from the Atomic Weapons Research Establishment (AWRE) at Aldermaston. The concrete case on one of the Aldermaston containers, which were of new design, had broken up while being loaded. Inside, a drum of oil "had 'blown up' due to gassing and had apparently burst." As the loading crew explained to the AEA's escorting officer, William H. King, the container had been dropped a few feet during the loading operations. The lids of most of the Aldermaston containers, they said, had come off at some point during the loading procedure. King made a note of it, judged that the containers were shoddily constructed, and did nothing more.[26] The *Halcience* set sail on 17 July, bound for the Bay of Biscay. The AEA had not taken long to decide that this location, a large box on the map in the Bay of Biscay, was best. It was suitably deep—more than 2,500 fathoms. Any drums dumped in the area would "almost certainly" sink to the bottom of this box and stay there, far beyond the Continental Shelf. The site was about sixty miles from the 1,500 fathom line. Any location farther away would, unfortunately, take the vessels beyond the range of navigational accuracy.[27]

In the early afternoon of the first day of dumping, the crew spotted two white packages floating about 150 yards from the vessel. Through binoculars, they appeared to be polythene (a plastic). Over the next two hours, seven packages appeared, all over an area of about one-quarter square mile. While continuing to dump, the crew retrieved the floating objects: "The first item retrieved was a polythene beaker with a pair of surgical gloves marked 'Made in England.' Next came 6 polythene canisters sealed in a P.V.C. bag and neither of these items gave a Beta/Gamma reading when checked. The last item retrieved, however, was a small bottle (1 pint approx.) labeled 'Milk Sample Bottle. Property of A.W.R.E., Aldermaston' with an extension number which was not easily recognizable." Alarmed by the bottle of milk from Aldermaston, King requested that no further Aldermaston drums should be dumped.[28] Over the next several hours, more items were observed and retrieved. Some items, such as a surgical glove, appeared innocuous. Others were more intimidating, such as a parcel of small bottles marked "U-235." One package, dripping its contents on the deck of the ship, caused alpha contamination up to 600 curies per second. King concluded that at least two or three Aldermaston containers had broken up prior to achieving sufficient depth to prevent items from floating to the surface.

In his report, King noted some of the basic problems in dealing with such accidents. One was the difficulty in retrieving packages from the sea. They accomplished this only through skillful maneuvering by the ship's master and the adroit handling by crewmen of a boat hook or bucket on the end of a rope. He wrote that it "cannot be emphasised too strongly that had the conditions not been perfect, many packages could have been missed." As they recovered the items, they became acutely aware of the many Spanish boats who fished in the dumping area throughout the day. It was essential that they not see any of the waste or somehow retrieve the packages. Given this risk, the *Halcience* searched the area for more packages on the following day. Finding a floating drum, the crew unsuccessfully tried to remove it with the boat hook. Instead, the hook punctured the drum, and it sank. The *Halcience* returned to Gosport without dumping the rest of the Aldermaston containers, which meant that many of the Harwell containers underneath were also returned. Examining some of the remaining drums, he found that many of them appeared to have the same poor quality concrete cap that he had witnessed at port. That, combined with "remembering the recent deliberations regarding dumping in the Bay of Biscay," convinced the escorting officer of the *Halcience* to return with a large amount of waste materials still onboard.[29]

The *Halcience*'s recovery operation prevented an international incident. Yet this close call compelled Britain to reevaluate dumping operations. At an AEA executive meeting, the immediate question was not one of safety but one of publicity. So far there had been none at all, but some wondered if it might be appropriate to issue a warning to vessels and coastal inhabitants. In the authority's view, however, there was "so far no evidence that this was necessary."[30] Despite such confidence, MAFF official A.J.D. Winnifrith was skeptical. After the mistake of 1962, he wrote, "this incident, combined with the accounts in the report about the containers used for the waste at Aldermaston, is still most disturbing." Winnifrith felt that the time was right to appoint a committee to "examine the whole question of the dumping of radioactive solid waste at sea."[31]

The AEA reacted defensively to Winnifrith's suggestion, claiming there was no basis to reevaluate the "whole question." No wholesale reevaluation of sea disposal was in order, as "it is surely accepted that some material must be dumped in the sea, under, of course, adequate safeguards."[32] After all, there had been only two instances of failure to comply with the parameters of authorization over the span of several years. Still, Winnifrith pointed out, the two instances raised the question as to whether any margin of error at all could be accepted. If there could be no acceptable margin of error, the practice itself might not be feasible.[33] "You may be quite right when you say that some material must be dumped in the sea," he wrote, "but I am not sure that this is self-evident." He continued: "There really is a problem about dumping at sea. We have had too many uncomfortable moments lately when containers have been trawled up off the coasts of France. The sea is still probably the right place but I suspect we need more stringent rules about the safeguards. And some things they dump are probably too risky for the sea, whatever the safeguards."[34] Probably in an effort to diminish the defensiveness of the authority, Winnifrith noted that he did not have in mind a formal committee to make policy changes, but rather a series of discussions between interested parties.

In the end, no major reevaluation of radioactive waste disposal at sea occurred. Looking into the question of safety, the AEA eventually determined that the only real problem was that the appropriate precautions, in terms of canister design and handling procedures, had not been taken.[35] In the policy review that ensued, the AEA defined the issues in terms of technology and responsibility, not in terms of whether to dump or not to dump. The most obvious technological challenge was in the design of the waste receptacle, and the problem of polythene, which often floated even when canisters imploded at the expected depth and could be seen rather easily. Francis Arthur Vick, who had taken over Cockcroft's job as director of the Atomic Energy Research Establishment in 1960, noted that no matter how successful an operation was, polythene items most likely would be picked up in nets or on beaches eventually. "By then," he added confidently, "it is hoped that the items will have been well-washed."[36] But they could not count on fishermen and beachcombers to understand that. Thus, they needed to devise a way to keep polythene from floating, by weighing it down, incinerating it, melting it around metallic waste, or shredding it to prevent people from ever recognizing it.[37] The reevaluation of policy envisioned by Winnifrith ended up having little to do with policy and much to do with identifying technical changes and improvements to prevent further unfortunate publicity. The AEA conducted fewer dumps than usual in 1965 (two dumps) and 1966 (one dump), staying away from the Bay of Biscay and dumping about thirty miles off the Continental Shelf southwest of Ireland.[38]

After this undisclosed Bay of Biscay incident, the AEA recognized that *all* atomic energy establishments in Britain and elsewhere should have to conform to the same standards, lest bad publicity for one of them should damn the practice for all. Harwell, for example, could not simply blame Aldermaston for having faulty containers. The effects of bad publicity would have fallen upon them both equally. The Harwell group had to shoulder some of the blame for the incident, because it had agreed to dump materials for another establishment "without fully realising that they must in consequence accept the full responsibility for the material dumped." Presiding over the dumping should have forced Harwell to make certain that all packages, not merely its own, met with the proper standards. As Authority official R. F. Jackson wrote, "This seems a parallel with the kind of responsibility you accept in taking somebody else's parcel through customs!"[39] This attitude regarding publicity marked a contrast to the UK's earlier satisfaction that accidents in international waters should have no legal ramifications, and the new attitude of establishing common practices would have wider ramifications for other nations' nuclear establishments.

New International Partners

Happily for Britain, France had begun the construction in 1962 of a nuclear reprocessing facility similar to the British facilities at Windscale and Dounreay (Scotland). Thus it soon would face similar choices about waste disposal. The location of France's discharge pipeline would be La Hague, on the Cherbourg peninsula, which meant that the wastes would be put into the English Channel. The Commissariat à l'Énergie Atomique conducted a number of experiments to plan the precise location of the pipeline, in order to match the tides and currents and ensure maximum dispersal of the wastes. From the point of view of Britain's Atomic Energy Authority, the construction of this facility was a positive step in consolidating its own policies, because they were based not only on French studies but also on advice by scientists from Britain.[40]

As the CEA scientists tried to prepare for the discharges, British scientists monitored their activities carefully. The AEA invited some of the French scientists to Windscale in order to review British practices and compare them with those planned for La Hague. John Dunster was one of the

British scientists who reported on the meeting. He and his colleagues learned about the scale of hydrographic surveys conducted by the CEA, which used dyes and drift cards to understand the local currents and the level of dilution to be expected. The CEA would also make studies of plankton to understand the biological transfer of radioisotopes. Although impressed by their preparations, Dunster probably saw in them the hands of marine scientists. The French were doing a great deal of scientific work, Dunster's report stated, but they did not appear any closer to an estimate of permissible discharges than they had been a year before. From Dunster's point of view, the hydrographic studies alone, which showed a great deal of dilution, should have allowed them to set a policy. "Overall they seem confused as to the immediate objectives," the report stated, "and in their own minds have not separated waste management operations from scientific studies on the physical, biological and chemical partition of radionuclides in the environment." Since they were planning to have the facility running by 1966, Dunster's report expressed surprise that the French spent so much time on scientific projects and not much on setting policies.[41]

The British AEA scientists grew increasingly disdainful of the French sluggishness, especially when a team led by Henri Jammet revealed in early 1966 that waste disposal decisions still had not been made. This was probably unfair, because the AEA scientists did not face the level of hostility from oceanographers at home that the French faced from Cousteau, Fage, and others. From the British perspective, however, the "preliminary" studies for waste disposal had no logical end; the French needed to develop real plans, not just fund scientific work. Radioactive waste already had been produced within several facilities and it was just being stored in tanks until decisions could be made. As British health physicist I.G.K. Williams put it, "French practice has thus involved postponement of a final solution and there is growing recognition of the need for a *disposal* policy rather than a storage one." Thousands of drums were sitting at different sites, particularly in the south at Marcoule, waiting for a policy. Williams judged that the French had "a general tendency to regard all solid waste as dangerous and to assume that disposal practices need to be accordingly elaborate."[42] He noticed that economics seemed only recently to be regarded as relevant. He believed that the French reluctance to make a decision or to economize had to do with the likelihood of public outcry, fueled by oceanographers' objections, as had occurred in 1960. He likened it to the 1954 public outcry over Britain's desire to put radioactive waste in the Forest of Dean. In the British case, public outcry hastened the decision to dump at sea; in France, the opposite had occurred, with the people wanting to protect the Mediterranean. He wrote: "That experience—which might be regarded as the French Forest of Dean—had made it clear that disposal in the Mediterranean was a non-starter. Dr. Jammet indicated, however, that the French were not expecting significant objections to dumping in the Atlantic."[43] The dump sites in the Atlantic (including the English Channel) had the advantage of not being within a closed area like the Mediterranean, and the nearest country to the action (Britain) had a strong interest in continuing sea disposal.

Given the prolonged French deliberations, Britain had every reason to suspect that national atomic energy establishments might approach radioactive waste disposal at sea with trepidation. After all, the Americans had publicly stated that they would concentrate disposal on land. But British officials, both inside and outside the Atomic Energy Authority, wanted to fight these tendencies. Some suggested that it was preferable to be politely paternal than to cave in to sensational reactions to radioactive waste based on ignorance. Health physicist I.G.K. Williams told the AEA's public relations director that "public reconciliation with radiation will be encouraged more by being prepared to handle such a reaction with urbanity than by pandering to it."[44]

Prior to the Mediterranean controversy, France had tried to work through EURATOM, the European nuclear organization that it had helped to create in 1957. That organization's purpose was to combine the efforts of Europeans not yet in the nuclear club to promote nuclear power. Certainly the EURATOM nations, including Italy, had been consulted about France's plans to dump in the Mediterranean, but the British government had not been (it was not a member). By the end of 1960, however, the French and British atomic energy establishments were more eager to work together to protect their waste disposal operations. It would have been foolish for the British to antagonize the French, since Britain's principal dumping ground was not far from the coast of France. And for their part, the French were in need of allies to help justify their own policies. As Harwell scientist D. E. H. Peirson pointed out to a colleague in 1961, the French were still sensitive about waste disposal, which was understandable since "they took rather a mauling over their proposal last year to use the Mediterranean as a dumping ground."[45] The British Atomic Energy Authority met with the French Commissariat à l'Energie Atomique and EURATOM officials on several occasions after 1960, discussing not only the technical aspects of waste disposal but also public relations strategies.[46] But there was no convenient forum, outside the IAEA, in which the French and the British could build a larger international coalition of dumping nations.

Conveniently, a new body came into existence at the end of 1960: the Organization for Economic Cooperation and Development, based in Paris. Eighteen original member countries ratified the convention in 1961, including

France and Britain, and many more would follow. Though the organization was not limited to Europe (the United States also was a member), one of its agencies was the European Nuclear Energy Agency (ENEA). This body would soon become the favored vehicle for cooperation among European states, particularly in the area of radioactive waste disposal at sea. It had the advantage of bringing European countries devoted to nuclear power together with Britain, which thus far had pursued nuclear issues with an orientation more toward the United States than continental Europe. And unlike the IAEA in Vienna, it also had the advantage of not including the Soviet Union.

Discussions at the IAEA implied a possible ban on sea disposal at some point in the future, and the British maneuvered to prevent this from happening. Given the inclusive nature of the IAEA, particularly with the influence of the Soviet Union, active measures to preserve British policies seemed to warrant new strategies. One of these was to carve out a separate international community of European states that, despite the politicization of the IAEA, could act unanimously because of members' shared concerns. Britain's involvement in the ENEA made this possible. In the mid-1960s, Britain's main fear was a total ban on waste disposal, which would limit even small amounts being dumped in the Hurd Deep, the Atlantic, or the North Sea. It stood ready to adopt an array of strategies to either prevent this from happening or, failing that, to circumvent any rules that might come about. Writing to a colleague, Ministry of Housing and Local Government official B. Taylor judged: "We shall doubtless be able to discuss the tactics on this later. One way round the problem might be to introduce suitable definitions for such terms as 'radioactive waste' and 'radioactive material.' Playing with definitions would allow for greater flexibility in interpreting future regulations.[47]

Britain tried to assemble diplomatic allies besides France who could help legitimize using the oceans for waste disposal. Through the ENEA, Britain attempted to negotiate with Norway and Germany, the Federal Republic, about the use of the North Sea. They all agreed that it would be a good idea to control discharges of waste (from Britain's vantage point, well-defined control was superior to a ban). For Britain, creating any such international agreement could be useful in justifying its policies, particularly "in view of the somewhat uncertain state of play with regard to the proposed IAEA Legal Convention." At the time, the Germans were planning an experimental dump in the Atlantic. Although they realized that any effective monitoring in the deep Atlantic would be unrealistic, they hoped to photograph drums, utilize dye to measure the diffusion of water from the dump site, and conduct other tests to experiment with a dumping operation. The Norwegians seemed interested, because Norway had hundreds of drums of its own radioactive waste to dump, as did Sweden. What attracted them all to the idea was the prospect of basing any future dumping procedures on preliminary scientific research. It seemed quite possible, the British representative to ENEA observed, that Germany and Norway might decide to join forces.[48] As one British representative reported at the meeting, the Norwegians wanted to protect their fishing interests, the Germans were interested in marine research, and the United Kingdom needed to be there "because of its interest in keeping the whole matter in perspective!"[49]

Keeping matters in their proper perspective—this phrasing was common in correspondence between British officials during the mid-1960s as they confronted growing opposition to waste disposal practices. It was a way of thinking about the problem that crossed borders as well, and the ENEA was a perfect strategy to encourage allies to keep radioactive waste in its proper perspective. This meant encouraging candor with local authorities, as in the case of France's discharges into the English Channel (La Manche) near the British-owned islands of Jersey and Guerney. But it also meant that it was more important to tell foreign atomic energy establishments about dumping operations than it was to tell the general public about them; the French, for example, received coordinates of specific dumping sites, but British newspapers sometimes remained totally ignorant. The CEA, but not the press, could be counted upon to keep the proper perspective.[50]

For the British, there were a few negative outcomes of Europe-wide cooperation. The whole notion of conducting scientific studies beforehand was a slap in their face, because they had been dumping radioactive waste at sea for nearly two decades, under the guidance of health physicists. It implied that the British practices were not based on science. At the same time, this might be outweighed by the diplomatic dividends; if such European cooperation could help ocean dumping survive, then the British might be better off ignoring the insult. The political advantage of sandwiching waste disposal with science—that is, with preliminary investigations beforehand and monitoring afterward—might help satisfy critics at home and abroad. Indeed, the latter attitude won the day among British atomic energy officials. In preparing for a study of Atlantic waters, they recognized that such scientific studies "will appeal to those countries such as Norway and Federal Germany which are very conscious of the public relations aspects of waste disposal (whether their own wastes or other people's)." In the case of Norway, the time was ripe to get them to agree to some kind of controlled use of the North Sea. Because its internal politics favored fishing interests, a diplomatic opportunity to get the Norwegians to sign onto radioactive waste disposal might not come again. The ENEA, then, increasingly appealed to Britain, which could "undoubtedly find in this forum friends and allies for sea disposal who might be valuable in wider international discussions."[51]

Other countries saw opportunities in purely European-wide cooperation, too. French representatives proposed in 1961, and again in 1965, that the OECD be used to carry out a program of research on marine pollution. The European group would be "strictly limited to research," without policy recommendations, to study the effects of pollution, including that caused by the disposal of radioactive wastes. Although some countries (like the United States and the Federal Republic of Germany) favored the idea, the British had taken a quite negative attitude toward it, claiming that other bodies were handling the problem.[52] They undoubtedly felt that framing research questions on radioactivity as "effects of pollution" was prejudicial from the start. They did not want to associate radioactivity with marine pollution—they did not consider it in the same class as oil and chemical pollution. Reacting to the proposal, John Dunster pointed out that radioactivity seemed to be "dragged in by the heels as an afterthought," and that whoever designed the proposal was more interested in pollution in general. He suggested that the whole question of radioactivity be "handed over bodily to ENEA" to put it into nuclear-oriented hands, rather than pollution-oriented ones.[53]

In fact, this attitude was mirrored by officials, British and non-British, within the ENEA itself. They believed that those concerned with marine pollution were worried about industrial pollution and oil spills, not radioactive waste, and associating the two would be extremely unhelpful for the development of nuclear energy. The ENEA thus carved this element of marine pollution out, keeping the study of it well within an organization populated primarily by scientists working for atomic energy establishments.[54]

Why was the ENEA becoming such an attractive choice for atomic energy officials? The British continued to see the ENEA's work—even what might have been considered strictly scientific—as complementary to their own policies. In fact, doing experiments under the ENEA could placate foes of dumping who complained about the lack of adequate environmental monitoring. Perhaps, they reasoned, the German plans for preliminary research might lead to a relaxation of demands for expensive and useless monitoring. AEA health physicist M. Phillips pointed out to a colleague in the Ministry of Agriculture, Fisheries and Food, "the whole exercise is primarily a public relations one—to reassure those who need reassurance, and to substitute a paper exercise in place of the more grandiose schemes for large scale monitoring and experiments that have been mooted."[55] Indeed, by the end of the summer of 1964, plans were under way to conduct an experiment, and other countries in the ENEA—France, Denmark, and Belgium, in particular—gave their consent.

Moreover, the British attitude prevailed: rather than see it as a preliminary step in regulating dumping in the North Sea, they regarded it as a way to study how much the sea could safely absorb: "The original suggestion that methods should be sought of applying restrictions on the use of the North Sea as a depository for solid wastes was abandoned and replaced by a proposed study on the possibility of using the North Sea as a disposal site for such wastes." This subtle distinction had the merit of implicitly encouraging the ocean's use rather than discouraging it.[56]

It was not long before Europeans got more ambitious. In addition to studying the sea together, why not dump together, too? Several countries had significant amounts to dump by the mid-1960s, and a joint operation might preempt any diplomatic criticism being directed at any particular country. The Swedish government authorized the disposal of some of its radioactive waste products in international waters. It scheduled 830 concrete containers to be dumped in the Atlantic.[57] In the Federal Republic of Germany, waste disposal on land had become an acutely difficult problem. By mid-July 1964, all of the storage depots had nearly reached their capacities, due to the immense amount of radioactive waste produced by nuclear research centers. The government's plan for the immediate future was to use an abandoned salt mine for the wastes, but the storage problem, hitherto most sharply felt by the United Kingdom, would soon lead the Germans to look to the sea.[58] They proposed to combine operations with Sweden to dump into the Atlantic, at a different location than the usual ones utilized by the British. Soon other countries wanted to take part, either with wastes of their own or with national observers. The French, for example, saw the operation as a great opportunity "from a psychological point of view," because it would show the general public how careful and cooperative the atomic energy establishment was being.[59]

These countries were not keen, however, to simply copy British practices. The Federal Republic of Germany's Ministry of Scientific Affairs had determined that the use of salt mines for disposal would be rather expensive, and that the sea might prove more economical. But it wanted to avoid committing to any method prior to weighing both the economic and the scientific arguments. It was ready to initiate an experimental dumping operation of about one hundred drums of radioactive waste, the location of which would be based upon the advice of oceanographers at the Hydrographic Institute in Hamburg.[60] The Germans hoped to proceed with as little international controversy as possible, so they eschewed any unilateral action that could be construed as being out of step with oceanographers' recommendations. They wanted to include a number of postdumping checks, whether they were photographs of the drums on the seabed, the use of dyes to track dispersion, or some other kind of environmental monitoring program. The British regarded these ideas as public relations maneuvers, but the Germans were adamant. To consolidate the plan's credibility further, the Germans asked for the advice and approval of the ENEA. They also invited other countries to participate in the experiment or to be associated with it somehow.[61]

The British atomic energy scientists felt that they should be involved enough in the German enterprise to ensure that the practice of sea disposal was not ruined. R. H. Burns explained that if, for example, the German containers were to burst open because of a design problem, "then there is a real danger that sea dumping would be damned. At the very least it could give our critics further ammunition."[62] He and others wanted to avoid further allegations that a few nations were poisoning the sea with their waste disposal practices. Here was an opportunity to have a voice in how other Europeans went about waste disposal, and to promote British procedures.

The Germans had tried to ensure that its experiment would be perceived in a positive, scientifically oriented light (indeed, going so far as to call their dumping exercise an "experiment"). Thus, they chartered the German research vessel *Meteor* to conduct a preliminary scientific voyage to help select the site for the large-scale operation. The Germans felt that it was absolutely necessary to have an intensive program of research associated with the dumping, and that this was the best way to achieve it. The Germans took the research agenda, dismissed by the British as a publicity gesture, very seriously.[63]

At a 1965 meeting, representatives from the United Kingdom, West Germany, France, the Netherlands, Belgium, and Italy (as an observer) concluded that the Atlantic Ocean was an entirely acceptable dumping ground for radioactive waste, and that they should move forward with an international dumping operation. This was a delightful turn of events for Britain, which had shouldered most of Europe's public relations burden on sea disposal. The British Foreign Office rejoiced at the plans, saying that "it is very much in the UK's interest to obtain international recognition of the safety of dumping solid radio-active wastes at sea."[64] In early 1966, a working party between the AEA and Britain's Ministry of Agriculture, Fisheries and Food affirmed that ocean dumping was the best disposal method for certain kinds of radioactive waste. Thus, even if a suitable land disposal site were found, the United Kingdom would look forward to an extended period of ocean dumping. It was in its best interests to gain some international acceptance of the practice.[65]

The advantages of the ENEA over the IAEA seemed enormous. In planning for a 1966 IAEA symposium on radioactive waste disposal, British officials expressed little hope that anything of substance would be accomplished, particularly because of the Soviets. Although the United Kingdom, France, and the United States intended to discuss effluent from their major installations, the Soviets would not be so forthcoming. They planned to limit discussions to disposal in the Black Sea, "a sea into which they discharge very little, if any, radio-activity and where most of the countries bounding the sea are not in a position to be critical of the Soviet material." As usual, the Soviets were expected to be critical of the West for poisoning the oceans.[66] The ENEA, by contrast, was limited to a group of Western nations with whom Britain could deal more easily and avoid the kinds of barbs it routinely received from the Soviet Union in IAEA meetings. Up until this point, the British had been "liable to be pilloried internationally for a practice condemned by many, largely as a result of Russian initiatives." Thus the AEA made the success of the ENEA experiment a major policy objective.[67]

The British health physicists who were pushing ENEA cooperation readily agreed to act primarily behind the scenes. As the *Meteor* made its way to the waters over the Iberia Abyssal Plain in early 1966, the Germans acknowledged that the success of the future dumping "experiment" depended largely on British participation. Yet one German official "hinted delicately that too much U.K. dominance might not be universally acceptable."[68] The British agreed. Their internal correspondence reflected an attitude that if other countries chafed at UK dominance, so be it. "For the reasons you appreciate," AERE director Francis Arthur Vick wrote to health physicist I.G.K. Williams, now working for the ENEA, "the U.K. must not be suspected of making the running in this exercise," but would be satisfied simply by sufficient involvement to ensure that certain specifications and procedures were followed.[69] In this, Britain was hugely successful. Obviously they were the only ones with long experience in the subject, and the other countries' representatives treated them with deference. "The UK experience is respected and, at times, a little embarrassing," Britain's representative at the ENEA George W. Clare confessed. Clearly other countries were looking to Britain for a major role in planning, even if the public face of the ENEA suggested German dominance.[70]

Clare seemed somewhat gleeful about Britain's ability to bring the others to heel. For example, he pointed out that he had expected some resistance to Britain's container designs, but no complaints materialized. He probably should not have expected many objections, because these atomic energy officials saw the same opportunities in ENEA that the British did. International arrangements provided ammunition to be used in domestic atomic energy struggles. Clare pointed out that the French clearly wanted international action; "I am sure the French CEA have only won this first stage of the sea disposal battle in France on the grounds of joint participation." Moreover, most of the representatives thought that the British were too restrictive, particularly with their self-imposed embargo on polythene (plastic) in the drums. The reason for the embargo was that such materials, which could float, were the principal cause for anxiety during the 1964 incident in the Bay of Biscay. Surprised that the others should oppose banning the use of polythene, Clare declared: "I am sure this attitude is mainly due to the 'niggers in their own woodpiles'!" This crass comment alluded to the fact that the other countries had already planned to dump polythene and were too far along in their planning to reconsider. Almost jovial to see so many countries wanting to dump, Clare thought the ENEA was a great coup for Britain; it would mean the Atomic Energy Authority could use this international blessing against the British government ministries who had insisted upon eliminating polythene.[71]

Not everyone was so self-congratulatory about the liberal attitudes of the ENEA countries. Some chastised Clare for forgetting that although broadening the practice could strengthen Britain's policies, lax standards could lead to major incidents by other countries; this, too, could doom radioactive waste disposal at sea. For example, Clare had rejoiced that Britain's two-stage concrete casting method of pressure equalization (in the drums) had been accepted. This method presumed some permeability of the concrete. But Aldermaston engineer R. Pilgrim indignantly predicted that "this unreliable method of achieving pressure equalization will let the side down—as indeed we were let down in 1964." By "side" Pilgrim meant "team," using English sports imagery to illustrate that they were all in it together, and clearly the ill-conceived drums had put them all at risk before. If one day someone were to develop better concrete—"and we know from war-time experience that the Germans in particular do produce very good concrete"—the design would fail. Although some came to Clare's defense, saying that international deference to British experience was a good thing, Pilgrim reminded them that the 1964 blunder also was part of that experience.[72] Many British officials, like Pilgrim, were concerned with protecting the integrity of Britain's existing policies, not to mention its reputation.

The downside of cooperation was that the public statements of participants continued to imply negligence on Britain's part. In October 1966, the ENEA drafted a press statement to explain the purpose of the international dumping exercise. The AEA, feeling somewhat abused by the draft, demanded that it be rewritten. The original version noted that "many European countries are experiencing difficulty in dealing with the disposal of their solid radioactive waste," and that "traditional methods of burial or storage . . . present difficulties in densely populated regions." Thus, the draft stated that ENEA was now "undertaking a study to establish a safe and economic means of disposing of such waste into the Atlantic Ocean." To the British, this read like an indictment. It was bad psychology for a press release, the AEA declared, to open with the statement that some countries were having difficulties. The term "traditional methods," attached some novelty to ocean dumping, despite the fact that Britain long had considered it the preferred, and thus the traditional, method. The AEA was dismayed by the implication that, despite Britain's long history of dumping and its professed commitment to safety, only now was a responsible body coming forward to establish safe means for radioactive waste disposal in the ocean.[73] AEA officials griped about the implications, but on the whole they suffered the implied insults in the interest of promoting international acceptance.

Although the planned ENEA operation was called an experiment, none of the British scientists or officials involved viewed it that way. Writing to Colin A. Mawson, a colleague in Canada, John Dunster observed that it was called an experiment "for a complex series of reasons, one at least of which is concerned with allaying possible alarm amongst member states on the Atlantic coast." Some states, he reasoned, would be less concerned with an experiment than with any hint at continued, routine operations. The scientific studies aboard the *Meteor* were genuine, to be sure, but these were unlikely to have any real bearing on anyone's actual practices:

> The German hydrographers are doing some studies in the area, and while this will be interesting as oceanography, there seems little doubt that they will contribute nothing to the assessment of the problems of waste disposal in this particular area. I am sure that no one concerned with the operation seriously regards it as an experimental release of radioactivity to be followed up by scientific studies of the movement of the activity through the environment . . . [A]ny monitoring which may ultimately be done in this area will be purely for public presentation and not for science.[74]

For scientists like Dunster, the *Meteor*'s cruise was something of a farce; there was no imaginable set of results that an oceanographic expedition could achieve that would alter Britain's dumping practices.

Fulfilling this prophesy after the cruise a few weeks later, Dunster wrote to the ENEA Secretariat that the expedition itself had provided some interesting results, but "they do not make necessary any reconsideration of the safety of the operation or of the choice of disposal area." In fact, the Germans' finding appeared to agree perfectly with health physicists' views, according to Dunster. These results had shown a great deal of mixing in the water column and also some vertical biological transfer of radioactive elements through the food chain. Still, these issues had all been considered in the ENEA's hazard assessment (created by a panel chaired by Dunster), and the health physicists did not need to hear about them from German oceanographers. There was no surprising information in the German study, the British seemed to claim, and there certainly was no reason to treat the study as a revelation.[75]

What Dunster and others emphasized repeatedly in internal correspondence was the unscientific nature of monitoring for radioactivity, particularly in the open sea. They treated monitoring as a political activity, useless from a scientific point of view, but quite useful from a public relations point of view. Still, many health physicists balked at monitoring for precisely these reasons; they did not relish the idea of creating ostensibly scientific projects for purely political reasons, because they eventually would become so obvious a ruse as to damage their reputations. The more intelligent critics surely would realize that the monitoring programs encouraged by the AEA and other establishments were misleading and built on faulty logic. "If these critics are able to show that the [monitoring] survey has been designed in such a way that it cannot provide a logical reassurance," he wrote in 1966, "then they have legitimate grounds for suggesting that the monitoring operation has been conceived cynically. We must not put ourselves in the position where this point could be made with justification." Those who distrusted the atomic energy officials, Dunster seemed to suggest, would have their antinuclear feelings vindicated by such ideas because they would suggest that the British were hiding something.[76]

Despite Dunster's objection, he and others found themselves in the position of having to pretend they cared about the monitoring results and about the work of oceanographers in general. It was somewhat awkward since no British oceanographers participated in the high-level international scientific discussions at the ENEA. This contrasted sharply with the participation of leading French and German oceanographers such as Vsevolod Romanovsky and Hans Kautsky. After the *Meteor* cruise, Denmark's representative at the ENEA approached the British to say that Danish oceanographers aboard the vessel *Dana* had been studying the vertical diffusion rates of radioactivity in the sea, and this data might be pertinent. The Danish representative thought he was being useful by keeping everyone informed. Dunster and others were counseled by their compatriot working at the ENEA, I.G.K. Williams, not to brush off the Dane: "you will, I know, agree that it would be politically prudent to accept his offer to try to keep us posted on the results obtained."[77] There was no question, however, whether there were any possible results of the Danish expedition that would change British attitudes. Similarly, when the British received preliminary copies of a paper by German oceanographer Günther Weichart on the limitations of chemical mixing in the ocean, Dunster concluded that there was nothing in it to alter their plans, though there was a great deal of interesting data for oceanographers. From his point of view, health physicists had already assessed the hazards, and the oceanographers had no special knowledge to offer that would change Britain's long-standing policies. Before flying to an ENEA meeting in Paris to discuss the scientific merits of the dumping site, he observed to Williams, "As I see it, our main task will be to relate the two fields of work without making it too obvious that the exercise is more diplomatic than scientific."[78]

British health physicists and other atomic energy officials saw in German oceanographers' work the same sort of opportunism for which they had criticized American and French oceanographers in the past. After reading the results of the *Meteor* expedition, Dunster commented that Hans Kautsky not only played up the possible biological transfer of radioactivity but also claimed that that future plankton studies might be required.[79] German newspapers quoted oceanographers at the Deutsches Hydrographisches Institut saying that there was still no satisfactory answer to the problem of waste disposal. The *Meteor* work was only the beginning. Solutions to the problem would probably require further investigations of mixing, the behavior of various chemical elements, and the absorption of radioactivity by flora and fauna. This would have to be done in all the different areas where sea disposal was contemplated, because the variables would change in each environment. Reading press translations from the British Embassy in Bonn, health physicists cringed at such reasoning. One of them, David Richings, wrote to Dunster that the German oceanographers were just trying to make a case for increased funds, and they did so at Britain's expense; by claiming that the problem had not yet been solved, they simply helped "fringe critics to continue their sniping at the British."[80]

British health physicists certainly did not believe oceanographic observations should be a factor; indeed, the specific current patterns mattered little to them. But for political reasons, they attempted to speak oceanographers' language, albeit without much conviction. Although he did not ask them to join the British delegation to ENEA, Dunster asked British oceanographers to submit any relevant data they might have. The reason was that, because of weather problems during the *Meteor* cruise, the Germans had to curtail some of their observations; this would make it less likely that the *Meteor*'s work would seem complete enough to satisfy critics. But French oceanographer Vsevolod Romanovsky volunteered to supplement the *Meteor* work with a bibliographic study, analyzing any relevant past work. John Swallow of the (British) National Institute of Oceanography complied, writing a letter to Romanovsky in Paris about British oceanographers' understanding of the currents off of Spain and Portugal. He pointed out that there was evidence of significant variable currents in deep water, which suggested opportunities for horizontal mixing even at great depths. This was not specific to the region in question, of course. But they did have "some observations tending to support the idea of a strong flow of Mediterranean water along the southern continental slope of Spain and Portugal and thence presumably northwards along the western continental slope off Portugal."[81] There were, however, no direct observations of this. Swallow's obvious lack of significant previous attention to the area in question and his noncommittal language—"tending to support" and "presumably"—revealed more about the chasm between oceanographers and atomic energy affairs in Britain than about the current patterns off Portugal.

Environmental Diplomacy

Not all countries were pleased with the idea of dumping radioactive waste at the German oceanographers' site over the Iberia Abyssal Plain. This was especially so for Portugal, and to a lesser extent Spain, the closest countries to the proposed action. The Germans, who had gone to great lengths to provide the ENEA plan with sound scientific justification and political endorsement, had chosen the site based on the recommendation of German oceanographers. They had not, however, given much consideration to the psychological effects on nearby inhabitants. These concerns came to a head in April 1967 when ENEA delegates met in Lisbon to finalize preparations for the dumping operation. Forty people from ten countries attended, as did representatives of EURATOM, the ENEA, and the IAEA.

As expected, the fragmentary nature of the *Meteor* observations became a point of contention with the Portuguese and Spanish representatives. Romanovsky's efforts to analyze the bibliographic evidence did little to assuage them, and the delegates got bogged down in making inferences, based on past studies, about what actually occurred in the sea. The Spanish and Portuguese claimed that ENEA's promise to precede the operation with valuable scientific studies had not been met. In addition to disputing the oceanographic claims, they took issue with Dunster's methods for evaluating health risks because they were entirely theoretical and not based on studies of the specific site. Dunster's efforts to emphasize "the importance of providing safety without either restriction or waste of money and scientific effort" did little to impress the Portuguese who would have liked to see more scientific effort before giving approval. Dunster pointed out during the meeting that he recognized that there were genuine "emotional" issues that needed to be overcome, one of which was the natural inclination to dislike having radioactive waste dumped off one's coast. A second emotional issue, he added in a swipe at oceanographers, was "the preference of the marine scientists to have disposals carried out in an area where investigations had been performed." He regarded both views as "illogical," then made his comment that were it not for the solid components, the Portuguese could safely dump all the material over their own Salazar Bridge into the Tagus River.[82] Unimpressed, a Portuguese representative, Júlio Galvão, implored them to delay the plan or move the site farther away from the coast of Portugal. According to the French delegates' report to colleagues back home, it seemed as if he had been ordered to make the request.[83]

The request, and the responses to it, revealed the divergent priorities of the different member states. Portugal was primarily interested in protecting its tourist industry. France, as a major contributor of wastes, was eager to hasten the project, so it wanted to stand firm on the original plan. But if moving the site would lead to swifter progress in getting the operation under way, they would support it. The Germans, however, had staked the whole operation on the idea that it had the firmest scientific foundations. Thus they voiced the loudest objection to Portugal's request to move the site, because it would undermine the credibility that the oceanographers lent to the operation. They threatened to withdraw completely if the site was moved. Of the dumping states, Britain appeared the most conciliatory. Given how little weight British health physicists gave to oceanographic arguments, this should have come as no surprise. Their hazard assessments could apply broadly to a number of sites all over the Atlantic. A slight move away from the coast was probably worth placating the Portuguese. The dire threat of Germany's withdrawal, however, was a serious one; it might send the message that German oceanographers had determined that the operation was unsafe. The British representatives determined that "to stop the operation or even to move the dumping area further north would undoubtedly cause much more public comment and trouble than merely absorbing any Portuguese protest and continuing the operation." If they were to stop the operation, British officials would have to make public statements of explanation, and Europe's press would create havoc for them. If they were to move the dumping location, the Portuguese undoubtedly would make some statement that tourists were no longer in danger, which would imply that indeed there had been a danger with the initial site.[84]

After some discussion, the ENEA participants joined to oppose Portugal's objection. The Germans were angry at the suggestion that "a manifestly irrational attitude by a relatively small member state should hold up what most other countries deemed a desirable as well as a safe operation." Concurring, the French representative felt that the Portuguese objections were based purely upon emotion; the French had been monitoring dumping operations for years and they felt it was harmless. The Spanish, who also had objected, now admitted that this operation should probably go forward in the interest of promoting internationally agreed-upon parameters and control for dumping operations.

The British representatives obviously wished to see the operation go forward and appreciated the Spanish support. However, they disagreed with the Spanish reasoning, as the whole point of working with ENEA was to avoid deferring to an even more politicized body like the IAEA. British representative David Richings noted that "my view, privately expressed to some representatives, was the British may not go along with this sentiment where I.A.E.A. was concerned, but may well be happy to co-operate under E.N.E.A. arrangements."[85] Together, the larger ENEA countries were prepared to override the protests of Portugal, in the interest of setting European standards for waste disposal at sea.

The Portuguese tried to present some economic and scientific arguments, to no avail. First, they felt that the operation would have a negative impact upon tourism. But the French and the Dutch representative assured the Portuguese that, if their experiences were any indication, no such problems would arise. Tourism, they said, had not declined in areas of radioactive waste discharge in their countries. In fact, the Dutch even reported (informally) an increase in tourists to the area of their effluent pipeline. Although a couple of short articles about radioactive waste disposal had appeared in French newspapers, the French reported that there was no decline in people visiting the coast. Thus, having gathered such ad hoc views and anecdotal evidence, the participant countries told the Portuguese that their economic point had no substance.[86]

The Portuguese then turned to scientific arguments. German and French oceanographers had determined that only very slow currents, moving northward away from land, were present in the area. This had convinced the Germans that the site was safe and had provided the operation a clear scientific blessing. But Portuguese and Spanish oceanographers claimed that currents actually moved eastward, toward the coast of Portugal. Trying to settle the matter, British representatives brushed aside the oceanographers' concerns; if the Germans were correct and the area was stable, with slight northward movement, the operation could be justified. If the currents were variable, as the conflict seemed to suggest, then the conditions were even better, because the radioactive materials would be well dispersed. In other words, they did not really require accurate oceanographic knowledge; they could still dump. Despite some "long and somewhat emotional" interactions with the Portuguese, the other delegates all agreed that the dumping would be safe. Although the German and French delegations saw this as a victory over the unscientific objections of the Portuguese, it also amounted to a tacit capitulation by their oceanographers to the health physicists whose hazard assessments did not depend on specific regional studies of oceanic conditions.[87]

Curious about the outcome of the Lisbon meeting, the United States Atomic Energy Commission's Chief of Environmental and Sanitary Engineering, Walter

Belter, wrote to the AEA for details. Dunster confided that "the proceedings were a little unreal," because his own hazard assessment, despite its occasional weaknesses and oversimplifications, left a huge amount of room for error, putting the radiation levels below international recommendations (from the ICRP) by a factor of about a million. His impression, he noted, was that he had convinced those who were not determined to remain unconvinced. There would always be some people, he complained, who think that more research ought to be done before anything is put in the sea. Repeating his skepticism about marine scientists' motivations, he added, "Presumably if these same people had geological interests they would feel the same about land disposal."[88] Certainly this was a timely comment, as perhaps Dunster was aware, because the AEC and the National Academy of Sciences recently had been embroiled in a debate about an academy report that sharply criticized geologic disposal.

Gaining multinational acquiescence in sea dumping required a trade-off. The ENEA operation brought waste disposal at sea increasingly into the public consciousness, especially in light of Portugal's intense protest. No longer was the Soviet Union the only source of diplomatic protest. More than ever before, the AEA was concerned with matters of publicity. By early 1967, it had decided to use the port of Newhaven as a collection and shipping port for its radioactive wastes for the ENEA operation. In March 1967, the AEA received a letter of complaint from representatives of the Newhaven (Sussex) Fishermen's Association, who expressed their deep concern about having such wastes brought into their port. In addition, "the secrecy with which this matter appears to have been negotiated, has in itself added considerably to our concern." The association feared an accident, such as a collision at sea, which would have a catastrophic effect on the fishing industry.[89] The AEA replied that it had been dumping radioactive waste successfully and safely for years and that the ENEA operation was only a novelty in that it was an international operation.

Despite these words of reassurance, even some within the AEA believed that the stakes had changed with the ENEA operation. The AEA's director of public relations, Eric Underwood, argued that it should not be treated as a routine procedure. Based on the Newhaven letter and the Portuguese anxiety about the effects of the dumping on its tourist industry, Underwood anticipated an assault by British journalists seeking "to gain further publicity of a dramatic sort." One official noted, "In Underwood's view we cannot rule out the possibility of some journalists wilfully seeking an opportunity to, for example, photograph a child or children on or near a train with a background of drums and/or radioactive signs." Any incident, however small, had the potential of being blown out of proportion by the press.[90]

Members of the press, Underwood believed, were consciously awaiting the start of the operation, just to find something sensational to report. Journalists were on alert, "in a way that we have not had to face before." Underwood envisioned reporters sneaking around to get close enough to dangerous materials merely to demonstrate that a mischievous child could do it. He feared drums rolling off of trucks and the railway officials being unwilling to handle the materials. Although such incidents were unlikely, "we shall be laying up a lot of future trouble for the Authority if there *is* an incident and we have demonstrably *not* prepared ourselves for it."[91] Less than a decade earlier, the AEA would have found these scenarios ludicrously improbable.

Despite these worries, the operation commenced in late May 1967 with no discernable protest and certainly no fanfare. Over several days the ship, M/V *Topaz*, stopped to collect radioactive waste drums in Newhaven (United Kingdom), Emden (Germany), Ijmuiden (The Netherlands), Zeebrugge (Belgium), and Cherbourg (France). The *Topaz* dumped it all at sea off Portugal in June.[92] The lack of protest undoubtedly was due to the ENEA's public relations strategy, which favored retrospective openness. It did not want to provide information that might prevent an operation from occurring; however, it felt that the public should be made aware, through official publications, of all the details of the operations. "Experience has shown," an ENEA steering committee concluded, "that it is not sufficient . . . in a matter of this nature to show by scientific arguments alone that an operation is safe. References to atomic energy still too often breed public apprehension." The only solution, it concluded, was to publicize details of the operation as openly as possible—after the fact—to make the public aware of what had transpired. In due course, perhaps over a great deal of time, the public would recognize that the atomic energy establishments were acting in good faith and according to high safety standards.[93] Most of the members accepted these conclusions. The French, in particular, given the debacle in the Mediterranean in 1960, were eager to adopt some intelligent public relations strategies. In its internal report about the ENEA operation, the French CEA observed that the 1960 controversy occurred because it had prematurely divulged information about the dump without first preparing public opinion for it. Reflecting on the psychological aspect of the ENEA operation, in which discretion about details was maintained until the last possible moment, the CEA congratulated itself that the operation was completed without incident.[94] For all these European countries, disclosure after the fact appeared to be the ideal middle ground between openness and expediency.

Chapter 8 Confronting Environmentalism

BY 1970, BRITISH RADIOBIOLOGIST Alan Preston had been working for the Ministry of Agriculture, Fisheries and Food for nearly twenty years, at its Fisheries Radiobiological Laboratory at Lowestoft. During that whole period he and his colleagues had been studying organisms in the Irish Sea, into which the pipelines at Windscale discharged their effluent. Like other Lowestoft scientists, he based his work on the assumptions set down by Harwell scientists in the early 1950s that permissible levels of discharge should vary depending on the environment. The amount of discharge into the Irish Sea should, Preston and others believed, depend upon the most likely pathway to humans. If the edible seaweed *Porphyra* absorbed a great deal of radioactive ruthenium, for example, and this was harvested and eaten in great quantities as "laver bread" by the residents of South Wales, this represented a critical pathway by which dangerous materials could get to humans. In fact, this was indeed the dominant critical pathway that governed disposals into the Irish Sea. To Preston and other British professionals, there was no logical permissible amount of waste or discharge that an international body could set. The past two decades of thinking on the subject held such universal thresholds of environmental discharge to be fallacious; the only real thresholds were those set by health physicists in bodies like the (British) Medical Research Council or the Radiological Protection Board, the (American) National Committee on Radiation Protection or the National Academy of Sciences, and the International Committee on Radiological Protection. These levels defined the dangers of exposure to humans; thus, Preston and others reasoned, discharges ought to be limited only by the pathways by which those levels could reach humans in that specific environment.[1]

Preston was one of several British radiobiologists advising on waste disposal policies who began to worry in the late 1960s and early 1970s because of the rise of the environmental movement—or, as he might have put it, the environmental lobby. An increasing focus on the sea, largely due to some disastrous oil spills in the late 1960s, was accompanied by repeated references to the dying ocean due to the effects of pollution. British scientists saw a troubling development among environmentalists: they wanted to protect the environment itself, not just humans. This echoed the recent criticisms made by Soviet scientists that certain radionuclides might concentrate in organisms or their habitats and endanger them, regardless of the permissible levels based on exposure to man.[2] Lowestoft scientists like Preston and D. S. Woodhead were particularly vocal in defending the critical pathways approach, which they did in public addresses and articles in the early 1970s, saying that the protection of public health would ensure more than adequate protection of the organisms in the environment, since they assumed that if levels were safe for humans, they were safe for marine organisms—even if the Russians disagreed.[3] But these scientists knew that environmentalists wanted to protect the sea itself, a consideration that never had been a criterion of waste disposal assessments. They also knew that in the future, radioactive waste disposal might be limited not just by critical pathways to humans; it might be extended to the protection of marine biota as well. Such an approach seemed in keeping with the spirit of the time, but if accepted by scientists and politicians it could damn waste disposal at sea forever.

At the turn of the decade, atomic energy establishments and the scientists who advised them confronted their most serious opponent yet: the environmental movement. The political tide of environmentalism stripped Britain of its longtime ally, the United States, which began to oppose ocean dumping in the early 1970s. The United States had abandoned the practice for economic, political, and diplomatic reasons years before, without condemning it, but now it stood against it explicitly, putting Britain in an awkward position. The Europeans who dumped together under the ENEA in the late 1960s also became sheepish, with only a few die-hards wanting to conduct any dumping operations while major environmental conferences were discussing the effects of marine pollution. British efforts to dissociate radioactive waste from general marine pollution failed, and the 1972 Convention on the Prevention of Marine Pollution by Dumping of Wastes and Other Matter (the London Convention) included radioactive waste in its list of regulated materials.

Yet in all of this the health physicists and fisheries radiobiologists in Britain were not idle; they had battled for years to protect their practices—against the Soviets, against interloping oceanographers, and other apparent foes—and they were not content to abandon them to what they saw as an irrational bandwagon movement by politicians. Seeing the inevitability of the London

Convention, they accepted it and worked hard behind the scenes to protect the legitimacy of their own assumptions. Ultimately they succeeded, despite the appearance of a major environmental policy change. They ensured that only dumping waste from ships was regulated, not pipeline discharges, thus preserving the main avenue by which radioactive materials reached the sea. They continued to set thresholds by influencing the discussions that hammered out definitions of high and low levels of radioactive waste. And they never budged on the critical pathways approach. In the end, they could congratulate themselves on surviving what they viewed as a temporary surge of interest in environmental affairs, and to their surprise they had an international convention that did more to bless existing practices than restrict them.

American Dilemmas

British health physicists were chagrined to see how fast the political winds were changing in the United States. When in June 1969 David Richings and John Dunster met with Martin B. Biles of the AEC's Division of Operational Safety, it was clear that the United States could no longer be counted upon to back British or European policies. The Americans had initially asked to send an escorting officer on ENEA dumps, but eventually they withdrew. The Americans were worried about legal liability, should anyone interpret the American role as a policy change; the United States had, according to the British, abandoned the practice of dumping solid waste at sea, "on the grounds that they could easily bury it or store it ashore and the objections were too troublesome to deal with." Richings wrote in a note for the record that the agitation in the United States seemed much more serious than that in Europe. And worse, American policy seemed to give in to it. He was astonished at the attitude of Biles, who "said that he thought the A.E.C. were in no position ever again to 'get away with' a Hanford situation where millions of curies were discharged into the river."[4]

Although Biles claimed that the United States would not commit itself to a position in which it would condemn others for dumping waste at sea, the Americans were committed to land disposal. From Richings's note for the record, he and Dunster were worried that the Americans were moving away from the concept of basing risk assessment on dangers to humans. "If they did move away from the yardstick of mankind," Richings observed, "there was no rational basis for the determination of the acceptability of discharges." In addition, the British scientists hoped the Americans would agree with the British position that safe levels should not be required at the point of discharge, but that the surrounding water should be considered part of the process of dilution. But Richings was gloomy: "Dr. Biles seemed to me to believe that public opinion, informed or uninformed, would carry the day and they may well get into an irrational situation."[5]

The sense of American abandonment continued in the next few months as the International Atomic Energy Agency convened scientists to discuss the possibilities of accidental releases of radioactivity at sea. In the course of the meetings, the delegates realized that it had been about a decade since the Brynielsson report had discussed the legal implications of dumping at sea. In light of scientific progress and the heightened concerns about marine pollution, some reasoned, it might be time to rethink the Brynielsson panel's recommendations. Most of the countries in the IAEA were thinking of establishing some kind of threshold figure for maximum permissible concentrations of radioactivity in seawater, and the Americans appeared willing to consider it. Alan Preston lamented that this could drive the Americans away from the British camp: "if the Americans adopt this sort of approach, they will be closer to some of the views held by the Agency and the USSR than they have been in the past, and the UK might find itself to some extent isolated from its previously staunchest supporter." Certainly there was evidence that some countries would ally themselves with Britain—Italy was one possibility, because it planned to discharge wastes into the Adriatic, and it would want some diplomatic support if Yugoslavia made a fuss over it. But losing the United States to public pressure might offset the gain of any European state.[6]

The United States' previous decision to concentrate its dumping activities on land gave it far more flexibility than Britain about sea disposal. Although the 1960s saw renewed faith in the power of science and technology to solve problems, and some extraordinary technological feats, this did little to alleviate concerns about the atomic age. Largely because of corporations' willingness to assume the costs of building reactors (instead of relying on U.S. government subsidies), the 1960s was a boom decade in the ordering and construction of nuclear power plants. But the end of the decade was a high point of concern for environmental degradation, and nuclear power was not spared criticism. A popular book by Sheldon Novick, an administrator at Washington University in St. Louis, criticized the spate of reactor constructions as hurried and unsafe. In *The Careless Atom*, Novick warned of future reactor accidents and argued that not enough care was taken in site selections, especially in earthquake-prone regions such as California.[7] The world seemed to be rushing forth unwisely amidst mixed signs of technological progress. During the summer of 1969, the world watched a man walk on the moon. The same summer, the *New York Times* published letters urging the use of space for radioactive waste. One proposed an artificial junk pile, "a constantly growing moon as our use of the atom pile increases." It would be cost-prohibitive, the writer warned darkly, "that is, unless we measure it against the cost of transporting the last sickly remnants of humanity away from a no longer habitable earth."[8]

A couple of major oil spills in Europe and the United States focused attention on the pollution in the marine environment and the disastrous effects on sea life. In March 1967, the supertanker *Torrey Canyon* crashed into some rocks off of Land's End, England, slowly releasing its nearly 120,000 tons of oil into the surrounding area. British airplanes scrambled to the site and bombed the ship, hoping to sink it and burn up its oil. Although they succeeded in creating an inferno that could be seen a hundred miles away, oil slicks spread out across the whole sea between England and France. The financial toll amounted to some $14,850,000, the largest single ship loss in history. An exposé by Richard Petrow in 1968 observed that the environmental damage was compounded by the use of chemical detergents in the sea and beaches; designed to clean up the oil, the detergents also killed wildlife.[9] In the United States, in January 1969, an offshore drilling platform operated by Union Oil had a major blowout, and for eleven days the rupture released some 200,000 gallons of crude oil into the ocean. The beaches near Santa Barbara, California, became covered in oil, killing wildlife and drawing national attention to an ecological catastrophe. The event is often credited as the spark for Earth Day, an annual day of environmental awareness, and it spurred the creation of several groups devoted to political activism for environmental causes, such as Get Oil Out (GOO). President Nixon said that the Santa Barbara oil spill touched the conscience of the American people.[10] Together, the *Torrey Canyon* and Santa Barbara oil spills catalyzed environmental consciousness about marine pollution, providing ominous signs that even the vast ocean was vulnerable to human interference.

When President Nixon addressed the UN General Assembly in September 1969, he favored international action on environmental issues, and he pledged his strongest support for the United Nations Conference on the Human Environment, scheduled to take place in Stockholm, Sweden, in 1972. In the months that followed the pledge, Nixon's acting secretary of the interior, Russell Train, suspected that this was just talk; when he heard rumors that the administration had balked at the cost of the conference, he urged government officials to seize the moment and to assert a positive, leading role in the international environmental movement. "Quality of the environment is attracting world-wide interest and attention," he wrote to the Department of State. "The opportunity is ripe for U.S. initiatives."[11]

Train need not have bothered, because Nixon already had several initiatives in play. He was quite serious about the Stockholm conference, despite

Train's apprehensions, and he was quite serious about the environment. He appointed the cabinet-level Committee on Environment and set up a standing committee within it, led by Christian Herter, to deal with international issues. In addition, he created a new environmental group within NATO, the Committee on Challenges of Modern Society. And in general, the Department of State used its Office of Environmental Affairs to forge cooperative relationships with other countries, to keep in line with Nixon's stated commitment to addressing international environmental challenges.[12] In October 1970 the Department of State made this commitment an official policy, cabling all American embassies in Europe and directing them to "exercise affirmative leadership" in the major organizations devoted to the environment—the United Nations, the Organization for Economic Cooperation and Development, and NATO.[13] The administration listed forty-six subjects needing to be tackled first, most of them categorized according to location (i.e., atmosphere, oceans, coasts, cities), but others were more specific: pesticides, toxic metals, noise, and man-made radioactivity.[14]

Nuclear power and radioactive waste soon were drawn into more visible environmental controversies, particularly those related to marine pollution. One critic of Novick's book, *The Careless Atom*, pointed out that its crusading tone was better suited to describing oil spills off the beautiful California coast. Nuclear power was a solution, not a contributor, to pollution. Radioactivity was easily the best understood and most easily measured of environmental hazards, and the record of safety in nuclear power far outclassed other power-generating industries, such as coal mining.[15]

The argument that nuclear power was a solution to pollution had only limited appeal. Skepticism of the AEC's attitudes toward protecting the environment continued in 1970, particularly as rumors surfaced that the AEC had quashed a damning report about geological disposal by the National Academy of Sciences. Idaho senator Frank Church accused the AEC of suppressing the four-year-old report that had called into question AEC's environmental policies toward areas nearby the nation's major reprocessing facilities. The AEC could not deny that the report had been written, but they refused to acknowledge that it had been suppressed. It claimed that there were many reports written by the academy that had never been published. All the senator had to do was ask and he could have a copy.[16]

There was more substance to Church's claim than the AEC admitted. The report in question had been a culmination of several years of academy criticism of the AEC's practices. By the late 1960s, the AEC's attitude toward geologists mirrored the British AEA's attitude toward oceanographers—they were opportunists who expected more scientific work to be done and unrealistically expected a few committee reports to force a restructuring of the entire atomic energy establishment. The geologists, in turn, increasingly saw the AEC as self-serving and complacent about dangers and unwilling to bear the necessary financial burdens to protect future generations. This conflict began in the late 1950s, when the Committee on Geologic Aspects of Waste Disposal was formed by Princeton geologist Harry Hess as part of the academy's Division of Earth Sciences. Hess's group presumed that its goal was to avoid radioactive waste coming into contact with any living thing, and it questioned the wisdom of the AEC's practices at Hanford and at Oak Ridge. In 1960 Hess wrote to AEC Chairmen John McCone stating that none of the installations generating high- or intermediate-level wastes were located in areas with geological conditions suitable for disposal. Further, none of its waste disposal practices could be considered safe. The AEC did nothing to satisfy the committee, and Hess resigned in 1963. His replacement was John Frye, who did nothing. The reason he did nothing was that the committee was only authorized to make studies at the AEC's request—and the AEC never requested anything.[17]

John Galley was the third chairman of this committee, taking over in 1965. Largely through the efforts of U.S. Geological Survey scientist M. King Hubbert, who had been part of the committee from the beginning, Galley's group received a request from the AEC to study the four major chemical processing plants at Hanford, Oak Ridge, Savannah River (South Carolina), and Arco (Idaho). Hubbert had, in fact, threatened to embarrass the AEC by terminating the committee due to lack of activity. Ultimately, the AEC might have preferred the embarrassment, because the Galley report would prove so inflammatory that the AEC would attempt to shut down the advisory capacity of the academy committee altogether. In the meantime, frustrated at lack of liaison through the official committee, a new committee was formed at the academy that had no terms of reference set by the AEC. This ad hoc group, chaired by Abel Wolman, also took on the subject of geologic disposal of radioactive waste. Thus there were two separate committees on the geological disposal of radioactive waste, one permanent and the other ad hoc; both were critical of the AEC's practices.

Wolman tried to get AEC chairman Glenn T. Seaborg to admit to some of the AEC's failures. He felt that the AEC's overriding problem was that it was primarily an operating agency that did not plan for the long term effectively. Wolman's views, conveyed to Seaborg in a personal letter from academy president Frederick Seitz (who had replaced Detlev Bronk in 1962), challenged the AEC to spell out it practices, goals, as well as the reasoning for them, projected over the next twenty to thirty years. "There is a need," Seitz wrote, "for a long-range, comprehensive plan that will elucidate the principles and practices needed to solve not only present problems but those of the future." Moreover, the plan needed to account for the effects of accidents, natural disasters, and any foreseeable man-made changes to the environment.[18] Were the AEC to take the time to make such a long-term study, the National Academy would be in a better position to provide as useful a service to the country as possible.

Seaborg reacted defensively to Wolman's and Seitz's recommendations, claiming that the AEC "has always considered the satisfactory treatment and permanent disposal of all waste materials as a prime requisite for all program activities." In order to achieve this—satisfactory treatment and permanent disposal—the AEC constantly was reevaluating its processes in light of new technology. Seaborg was satisfied that existing relationships between the AEC and the academy were working and had no intention of launching a two- or three-decade projection into the future.[19]

Enclosed with Seaborg's reply was an AEC report of its own policies, covering the years since the congressional hearings about waste disposal in 1959. The basic message of the eleven-page report was that the JCAE had determined there were no harmful effects and that the problem of radioactive wastes "need not retard the future development of the nuclear energy industry" as long as public health and safety continued to be safeguarded. Now it was 1965, and the AEC concluded: "We believe these conclusions are still valid." In fact, the basic assumptions of the AEC had never changed, though the scale of production and disposal had increased. Billions of gallons of low-level wastes were produced each year, but the AEC insisted that it should not concern anyone. Not all radioactive wastes were the same, the report argued, despite the tendency of most people to ignore the differences and focus purely on the word "radioactive." Only a small amount of the waste was high-level, needing stringent controls. There were two basic rules of thumb—high-level wastes ought to be concentrated and contained indefinitely, away from humans, whereas low-level wastes could be dumped and dispersed, mixing them with the air or seawater, thus rendering them harmless. There were two fundamental problems—the long-term storage problem of wastes requiring isolation and the scientific problem of judging the capacity of the sea or air to sustain low-level wastes safely.[20]

The members of Wolman's committee were disappointed with Seaborg's lack of enthusiasm. The AEC seemed to be distancing itself from the academic scientists, claiming that congress had vindicated its practices in 1959. Oceanographer Milner Schaefer observed that Seaborg's response seemed adequate, "at least for the time being," but the fact remained that the AEC did not appear to have definite long-range plans for waste disposal. Although Seaborg mentioned that the recent report explained such things, Schaefer could not find them there. The whole report tended to be rather self-congratulatory, with too little attention paid to the future. The AEC seemed to think, Schaefer wrote, that the future problems of disposal, including accidents such as those from nuclear-powered ships, could be handled well enough with present information. That was just not the case, Schaefer argued. "I may, however, be unduly suspicious," he noted, but he did not believe that the AEC studies were sufficient. Another committee member, Gordon Fair, added his own reservations, noting that AEC studies on waste disposal "are no longer being reported as fully to the interested professions as they might be." If the AEC felt satisfied it was making progress, it ought to publish quarterly or semiannual reports, or at least send representatives to conferences more often.[21]

Others used stronger words. One of them was M. King Hubbert, who questioned the AEC's handling of low- and intermediate-level wastes. The AEC was only now changing some of its geologic disposal practices by injecting low-level wastes into shale 700 to 1,200 feet below ground. Before that, "these same liquid wastes were discharged into bulldozed earth tanks where by leakage they were contaminating a considerable area of countryside and also the local streams." As for the solid wastes, such as rags and boxes, they were buried without any barrier between them and the groundwater table. This meant that rainwater passed through the trash on its way to the ground water reservoirs. Hubbert's inspections of AEC sites had revealed trenches of radioactive trash that had become filled with such water. The AEC hoped it returned to the surface, though it might just as easily percolate downward.[22]

Even worse, according to Hubbert, normal practices at Arco and Hanford were to discharge wastes into seepage ponds or directly into wells. "At Arco," he wrote to Wolman, "this body of groundwater is flowing southward where it becomes the source of municipal and irrigation water supplies a few tens of miles to the south." Hubbert complained that he and his group had repeatedly protested the handling of low-level wastes to the AEC. "Invariably the defense offered was that any other method would be more costly." Hubbert stood firm in his belief that the budgets for waste disposal did not reflect the magnitude of the problem, and that several times the present amounts were needed to address the problem adequately. He felt that Seaborg had been vague and evasive, with "what seems to me to be a persistent tendency to minimize their potential danger, and to over-emphasize their 'harmlessness.'"[23]

In the meantime, John Galley's committee offered a few specific reasons why he thought the AEC had run afoul of its mission and failed to acknowledge the future risks of waste disposal. First was its overconfidence. The AEC believed, without substantial evidence, that contaminated water never reached the water table at some of its major sites. The most troubling side-effect of this was that three of the AEC's largest plant sites, where the largest amounts of wastes were produced, "are located over some of the largest fresh-water aquifers in the United States." Over the next century, America's booming population would be drawing heavily from them. Second was its failure to allow for changes in conditions in the coming centuries. Third, and perhaps most important, was its own perceived need to dispose of wastes cheaply, by doing so in areas that were not necessarily well suited to it. Seaborg's claim that the AEC showed due attention to long-term problems just did not match with the facts.[24]

As it turned out, Galley's committee was so inflammatory that the AEC turned against the NAS and put enormous pressure on it to quash the report. After its initial release in May 1966, the AEC reminded NAS geologists that the report was privileged information. Others in Washington wanted a copy of it, and the report itself did not explicitly refer to the security restrictions which, in the minds of AEC officials, obviously ought to apply to it. The Public Health Service, for example, knew of its release and requested a copy from the academy. The academy did not send one. Instead, its officials assured the AEC that it would respect its wishes to keep the report private until the AEC wished it to be released. To all those who had received a copy, the academy sent stickers clearly stating its privileged status, and it instructed them to affix it to their copies.[25]

After long deliberation, the AEC acknowledged that a conflict in philosophy had emerged. It leaned heavily on the statements made by the ICRP in 1964 that the limitation on exposure to whole populations "necessarily involves a compromise between deleterious effects and social benefits." A balance must be struck, the ICRP had stated, that reflected all of the factors deemed important by society. Thus it was natural that there could be no universal level of exposure, because "the factors influencing the balance of risks and benefits will vary from country to country and . . . the final decision rests with each country." The Galley committee, AEC scientists argued, had taken a completely different approach that was out of step with the norms of atomic energy establishments and radiation protection organizations everywhere. It began by asking, What is safe? This approach contrasted with the AEC's, which asked, At what point do benefits outweigh risks?[26] In his letter accompanying the AEC's comments, Shaw added that the report already had been seen by all the relevant people. Thus, "we do not believe that additional distribution or publication of the report is warranted."[27]

Galley's group denied that this was its philosophy at all. It too subscribed to the rule of balancing risks and benefits. The difference was that Galley's group wanted that balance to "include a long-range relationship involving risk to distant generations against cost to the present generation." That meant, at bottom, that any calculus of cost and benefits should naturally and heavily emphasize considerations of safety rather than cost. As for abandoning the atomic energy industry altogether, Galley's group did not shy from this possibility. "In the end," they concluded, "it will be the taxpayer and the consumer who will determine this alternative."[28]

A few days after receiving the responses made by the Galley group, the AEC decided to terminate the committee altogether. The AEC pretended that this action was part of its routine annual procedure of reviewing the need for its various advisory committees. Although it had existed as part of the academy since 1955, the committee had been created to advise the AEC's Division of Reactor Development and Technology on ground disposal. Now, "because a major part of our ground disposal R&D program is reaching a successful conclusion, it would appear that there is not sufficient justification for continuing the advisory services of the group at this time." Milton Shaw, the director of the division, politely asked Seitz to convey to Galley the AEC's "sincere appreciation" for the "valuable guidance and assistance" provided over the years.[29]

The academy's president was surprised that the AEC would take such a step. After all, the "advisory" role of the academy had been established a decade earlier largely to lend some credibility to the AEC's practices, by providing an impartial scientific assessment of waste disposal on land and at sea. The earlier decision to turn away from the sea had allowed them to sever some links with oceanographers. Now the AEC was trying to do the same with the geologists, but without giving up land disposal. Was the AEC truly confident enough to ignore the National Academy altogether? Despite Shaw's observation that the studies were reaching a successful conclusion, Seitz reminded him that "everything I have seen during recent years suggests to me that the geological problems associated with radioactive waste disposal will grow rather than diminish in the period ahead." Besides, the AEC was not in the position to hide what the academy already knew, that its scientists had grave concerns about the future implications of present practices. He did not want to terminate the committee unless he could be assured that the gap left would be filled by some other body, preferably outside the AEC. Putting it delicately, he wrote that such a body would be needed "to cover the not negligible possibility that the topic of geological aspects of disposal will continue to be of great public concern."[30]

Although this was not the first time the AEC's actions had left a bad taste in the academy's mouth, Shaw's move to terminate the committee soured the relations. The chairman of the academy's Earth Sciences Division, J. Hoover Mackin, wrote that the events fell into a "familiar pattern" that had been repeated with three successive academy advisory committees on the subject.

"The chairmen of those committees look back on much of their work as an exercise in frustration and futility." To boil the problem down to one—which Galley had been trying to do already—Mackin observed that the AEC was not paying close enough attention to the possibility that, in the long-term, present disposal practices put harmful quantities of radioactive materials into public water supplies, which "is clearly a matter of grave national concern."[31]

Mackin looked back at the previous work, all of which had pointed to the same conclusions and all of which seemed to be rejected by the AEC. He believed that the AEC hid behind criticisms of scientists' overstatements and minor errors. Its commentary was "mostly a legalistic play on words." Mackin wanted the AEC to stop beating around the bush about costs and benefits. This was the language of engineers and contractors familiar with mundane business enterprises and the imperative of achieving maximum efficiency. The fact was that they were all talking about endangering either the future health of the population or the future of atomic energy. "The real question," he wrote, "is whether the nuclear power industry can now, or in the near future, stand the cost of containing the wastes that would otherwise dangerously pollute the environment." If it could not, there were lots of alternatives, but turning away from the safest methods of waste disposal was not an acceptable one. In his view, the AEC could talk all it wanted about how its current studies showed how harmful levels of radioactive materials were not reaching the water table, but that was beside the point. All of the groups who had studied the problem under the academy, "including the best men in the country, were decidedly skeptical after having viewed the evidence in the field that the long-lived radionuclides are certain to *remain* where they are."[32] Mackin noted that he shared the skepticism of the committees. Shaw's termination of the committee only showed that "we have reached an impasse."[33]

The Attack on Ocean Dumping

These revelations in 1970 of heretofore unaired complaints in the academy report followed even more damning criticism of the AEC by two of its own scientists, John Gofman and Arthur Tamplin, who said that the AEC risked genocide with the levels of radiation exposure it permitted from the nuclear power industry. They pointed out that there were no threshold levels of safety for radiation exposure and that the links between radiation exposure and all forms of cancer were very strong. Beginning in the fall of 1969, they called for a tenfold decrease in the permissible doses recommended by the National Committee on Radiation Protection. Both the AEC and the NCRP, backed by key congressional leaders, denounced the findings. But these scientists created the most serious cleavage of scientific opinion about radiation effects since the fallout debates of the early 1950s. And unlike the earlier controversy, this one addressed the causal effects of radiation in inducing disease in one's lifetime, rather than in creating mutations in the generations to come. Making such links made them few friends in the AEC, which already struggled with scientists over interpretations of existing permissible doses. More important, as historian J. Samuel Walker has noted, their scientific findings severely damaged public confidence in the AEC's commitment to safety.[34] Opening the AEC to public scrutiny like never before, the statements of Gofman and Tamplin were explosive. The *Wall Street Journal* likened them to the Trojan Horse, an "enemy within" that brought about the fall of Troy. Such imagery hinted that the AEC as an institution, and perhaps nuclear power as an industry, were under immediate threat.[35]

Renewed questions about dangers from radiation, combined with an expanding consciousness about marine pollution, drew further attention to radioactive waste disposal at sea. A Santa Barbara–based group, the Center for the Study of Democratic Institutions, sponsored a six-day conference in Malta in June 1970 to address the international implications of sea pollution. About 260 scientists from sixty countries attended, and one of the conference's emphases was radioactive waste. They discussed Gofman and Tamplin, wondering if their estimates of radiation exposure should have an impact on waste disposal at sea. They discussed the inaction of the IAEA in bringing about legally binding regulations. They also considered the threats posed by nuclear ships, which could contaminate the sea when sunk. Soviet scientist G. G. Polikarpov arrived with new evidence for the old criticism that the British were contaminating the Irish Sea. He stated that the levels of radioactivity in the sea were high enough to cause deformity in the backbones of some fish embryos. Others praised the British and the Americans, saying that the real problem remained ahead of them, as less responsible nations would build installations without rigorous safety standards. Oceanographer Roger Revelle, then at Harvard's Center for Population Studies, was one of those praising the existing controls, which he called "one of the bright spots in the postwar world." Reporting on the conference, *New York Times* journalist Eric Pace surmised that the general feeling of the conference was that international regulations were greatly needed under the auspices of the United Nations to protect the world from future abuses.[36]

The political winds in the United States increasingly favored regulations on ocean dumping. This was partly due to a controversy in 1970 over nerve gas. Environmentalists and many politicians bitterly opposed the U.S. Army's plans to dump thousands of unwanted M-55 rockets, filled with deadly Sarin nerve gas, into the Atlantic Ocean. One organization, the Environmental Defense Fund, fought nerve gas disposal in the courts but ultimately lost the battle. In August 1970, the canisters of nerve gas were dumped about 280 miles off the coast of Florida, in the Atlantic Ocean. This occurred despite widespread protests, and condemnations by the secretary general of the United Nations, U Thant, who said that it was a clear violation of the 1958 Convention on the Law of the Sea. The State Department denied that it was a violation and pointed out that the most competent scientists had been reviewing the problem for over a year to determine that it would be absolutely safe.[37]

Although opponents failed to prevent the nerve gas dumping, they got a promise from Secretary of Defense Melvin R. Laird that there would be no further dumping of nerve gas at sea. And as journalist Richard Halloran wrote in the *New York Times*, they succeeded in drawing further public attention to the question of how to dispose of dangerous wastes without harming humans or the environment. Certainly the Army had more nerve gas to get rid of, as well as other materials from the manufacture of chemical weapons. And the Navy had been dumping conventional munitions into the sea for many years. It acknowledged that since 1964 sixteen shiploads of them were dumped at sea, a great deal off New Jersey. One of New Jersey's senators, Harrison A. Williams, was shocked to hear it, and he claimed that the Navy might have violated environmental laws.[38] The nerve gas controversy catalyzed demands for more controls, or even a ban, on dumping hazardous material at sea. A congressman from Florida, Dante Fascell, called for an international ban just as the Army's waste ship was cruising off the coast of his state.[39]

The United States' Council on Environmental Quality, created by President Nixon in the wake of the Santa Barbara oil spill, issued a report in October 1970 that pushed American policy even further against polluting the ocean. The council, headed by Russell E. Train, called for new laws to require dumping permits from the newly created Environmental Protection Agency. The report was not a scathing criticism, but it did urge that no further chemical munitions be dumped at sea. In general, it argued that there was no serious problem at present, but that there ought to be laws in place to prevent abuses in the future. If any agency felt the brunt of the report, it was the Army Corps of Engineers, which dredged a lot of sand and silt ("dredge spoils") during its lake and coastal construction projects and then dumped them back into the sea at its own discretion. Requiring approval from the EPA would, Train's group argued, prevent any conflicts of interest. As for radioactive waste, the council recognized that the United States had shifted its focus to land. In 1960, the United States dumped some 76,000 containers of radioactive waste into the ocean, they reported, whereas in 1969 it dumped just 26 containers. Thus the recommendations really targeted other forms of industrial and military wastes.[40]

President Nixon welcomed Train's report and promised to promote the proposals within it. Reasserting that it was not a reflection of any current dangers, he praised it as "acting rather than reacting to prevent pollution."[41] It was Nixon, in fact, who had requested the report. Ever since the Santa Barbara oil spill, he repeatedly had shown himself to be against ocean and lake pollution. His strong environmental agenda likely was designed in part to deflect criticism of the ongoing American war in Vietnam.[42] Earlier in the year, he sent a message to Congress proposing a ban on dumping wastes, like dredge spoils, into the Great Lakes. He pointed out that seventeenth-century French explorers had called Lake Huron "the sweet sea," a name rendered ironic by the pollution of modern industry, technology, and population growth. He expressed his hope that a study of ocean dumping might prevent "the same ecological damages that we have inflicted on our lands and inland waters."[43]

The president and his environmental council did not imply that current measures had imperiled the seas, but others made a stronger case. Waste disposal policies were killing the oceans, environmentalists claimed. An article in the *New York Times* dubbed the sea off of New York Harbor one of "the seven wonders of the polluted world," the vast amount of toxic garbage threatening to make the metropolitan area "the first excrement-locked region in the world."[44] In October 1971 the *New York Times* published a long article entitled "We Are Killing the Sea around Us," in which freelance writer Michael Harwood laid out the environmental argument that the ocean was dying. He laid most of the blame on sewage, toxic industrial wastes, oil spills, and other pollutants, not radioactive waste, but the imagery of the dying ocean was strong. He drew attention to biologists' claims about the synergistic effects of all pollutants in the ocean, and radioactive waste certainly was a contributor. Harwood mentioned that the IAEA had standards for radioactive waste disposal at sea, "but no nation is forced to meet those standards. Some do; some don't." He hoped that when delegates met in Stockholm in 1972 for the United Nations Conference on the Human Environment, they would translate words into action—to plan a real treaty regulating waste disposal at sea. He cited Max Blumer of the Woods Hole Oceanographic Institution, who said that a polluted ocean would remain irreversibly damaged for generations. Harwood posed the question: "And who would care to argue that a dead ocean would not mean a dead planet?"[45]

In the United States, politicians raced to become guardians of the environment. In 1971, the Senate voted 73 to 0 to restrict the dumping of wastes, including high-level radioactive waste, beyond the territorial jurisdiction of the United States. Its Commerce Committee argued that the oceans were not indestructible and could no longer be considered the "universal sewer of mankind." The House of Representatives voted on similar legislation 304 to 3. Very few wanted to stand against the tide of environmental protection of the oceans.[46] During the 1972 presidential campaign, President Nixon highlighted his own environmental agenda, criticizing Congress for inaction on over thirty proposals he had sent them since 1969. In October 1972, shortly before the election, Nixon signed into law a prohibition on ocean dumping of high-level radioactive wastes and materials for chemical, biological, and radiological weapons. In the United States, politicians across the political spectrum had begun to characterize themselves as champions of the environment; consequently, pro-dumping advocates diminished dramatically.[47] Of course, in the United States, this could be done without much sacrifice. The AEC already had decided not to dump high-level radioactive waste at sea, and government officials already had promised not to dump nerve gas. The newly baptized environmentalists in Congress and the White House thus were making a largely symbolic gesture. The only ones seriously distraught by the law were the British, who felt they had been hung out to dry.

Fighting the Tide

Fortunately for Britain, it had seen this defection by the United States coming, and it had devoted substantial energy toward cultivating more reliable European allies. But even the international dumping operations sponsored by ENEA were not immune to pressure from what British health physicists derided as the "environmental lobby." The countries nearest to the disposal operations felt this pressure most acutely, and they routinely demanded guarantees of safety from the ENEA. The ENEA could offer none except the reassurances by health physicists, so it fell back once again on promises of monitoring. This they did reluctantly at times; it was the politicians, rather than the atomic energy establishments, who wanted monitoring to be done. As the ENEA planned dumps in the last years of the decade, requests for monitoring became routine. The Spanish, Portuguese, and Irish all sent representatives to insist upon it. Although the other countries' health physicists repeated that there were no scientific reasons for expecting water samples to yield any conclusive results, they were undeterred. Malachy Powell, an official from Ireland's Department of Health, observed that scientific futility was beside the point. Negative results would satisfy the Irish people, while a lack of monitoring might alarm them. Powell privately admitted that he believed such monitoring was the only way to satisfy critics, if their experiences with the Windscale discharges into the Irish Sea were any indication. Samples of water and seaweed in the Irish Sea was their present strategy of trying to convince the Irish people that British discharges were not affecting them. Although they increasingly saw monitoring the Irish Sea as politically necessary, British health physicists could not help but balk at the idea of "useless, fruitless sampling" of dumping areas farther out in the Atlantic.[48] By contrast, German oceanographers were quite eager to do sampling, and they took up the Irish Sea project with enthusiasm. Led by Hans Kautsky of the Deutsches Hydrographisches Institut, German oceanographers aboard the *Meteor* did more than just take some samples. Their aim was to select suitable long-term monitoring stations in the Irish Sea to keep track of the concentration of radioactivity there.[49]

Ireland in 1969 took on the role Portugal had played two years before, objecting to the cavalier attitudes of its neighbors who dumped radioactive waste so close to its shores. Industrial chemist George Clare referred to it as the "Irish Affair," and it began as ENEA representatives met in March to plan the next dump. From the first day, the various representatives received word from their embassies to expect a distinctly "Irish attitude," that every precaution must be taken to eliminate all risk—not only of health dangers but also of unfavorable public reactions. The delegates were each called away from the meeting, country by country, as their embassies called to notify them. The Irish asked for assurances from the ENEA that no marine organisms would be harmed, and asked that an Irish official join the ship's crew as an escorting officer, to reassure the Irish people that the materials were dumped in the right place and under the safest circumstances.[50]

The Irish representatives to the ENEA raised a number of concerns about the safety of dumping radioactive wastes at sea. Yet few appeared to take these objections seriously, including the Irish themselves. As one French representative surmised, the intervention made by the Irish was "not really made in the goal of suppressing the operation but to cover themselves vis-à-vis Irish public opinion." The Irish seemed more concerned with keeping the ship itself far from the sight of the Irish coast than with the effects of the radioactive materials.[51]

The Irish put radioactive waste squarely into the same category of other ocean pollutants currently absorbing international attention. With the outcries about the *Torrey Canyon* oil spill, they felt there was every reason to expect radioactive waste to meet with fierce resistance. Tadhg O'Sullivan from the Irish embassy in France addressed the group, pointing out that Ireland's intervention did not reflect suspicion of the ENEA's actions or the safety of the operation. As Clare reported it, "he was a persuasive speaker—and certainly would not have disgraced anyone from the old island, gifted with the 'blarney.'" Clare observed that this smooth talker accepted that the scientific justification for dumping was sound, but that the main concern was the psychological effects on the public. Recalling the dramatic decline in fish prices after the *Torrey Canyon* oil spill, he pointed out that news of the dumping operations could affect Ireland's fishing and tourist industries.[52]

O'Sullivan acknowledged that the United Kingdom had dumped unilaterally for years, but now that it was a joint operation the ENEA should take care to go about it responsibly. Reading a meeting report, a British official underlined that particular remark and put two exclamation marks on it, a scribble that reflected deep frustrations at the implication that Britain's unilateral actions automatically amounted to irresponsible behavior. But in the course of the meeting, the other representatives pointed out that the ocean was not a cheap expedient for wastes. In fact, it usually was much more expensive to do than to dump them into the ground nearby the nuclear facilities—as the Americans had learned—or even to spend the money to take the wastes out to abandoned mines. But because such ground disposal raised numerous problems with groundwater, and often much greater problems with public relations, the sea was an attractive choice despite the cost.[53]

The ENEA failed to persuade the Irish to abandon thoughts of moving the site, conducting monitoring operations, and other activities for publicity's sake. O'Sullivan continued to push for them—health physicist David Richings observed that the Irishman "launched a full-scale attack" on the ENEA group planning the operation, using "a whole range of social/political/economic/technical arguments."[54] But the old arguments used against Portugal prevailed. Although ENEA was willing to allow an Irish escorting officer, it would not budge on monitoring or a site change, because both decisions would undermine the scientific credibility upon which the sites had been chosen. To yield any meaningful results, monitoring would have to be done on a huge scale, and it would end up being more expensive than the whole operation. If such funds were to be spent, they might as well be given to oceanographers and marine biologists to do more general studies rather than try to sample the dumping area. Disdainful of the Irish attitude, radiobiologist Alan Preston noted that they clearly were trying to make up for not having attended previous ENEA meetings, and they wanted "a minor movement of the disposal area, say ten to fifteen miles further offshore, which would demonstrate to the Irish public that their government had some influence in keeping the operation as far from the Irish shore as possible." Other countries were in unanimous agreement that this would be a foolish gesture, and in fact a second choice of sites would be closer, not farther from the coast. Ironically, Preston mused, this would probably be preferable to the Irish, because it would put the site about equidistant to Irish, British, and French shores, and that might seem more equitable to them.[55]

Although the ENEA overpowered the Irish and Portuguese objections to their site choices, clearly these dumping operations had become politically vulnerable. Trying to press forward with a 1970 European dump, British officials met separately from their European counterparts to discuss the political ramifications. Given the public outcries against marine pollution, and the government support of such outcries even in the United States, they considered waiting until the political situation cooled a bit. They knew that the stakes were high, given the "risks of ignoring possible reactions in the current climate." The old argument, that caving in to political pressure would undermine their scientific credibility, proved extraordinarily difficult to maintain. But they did maintain it. Experienced hands such as Fred Morley and John Dunster refused to make an exception in 1970. After all, they had been doing this routinely since 1949. Giving ground to the environmentalists would not mark them as conciliatory; it would brand them as having been mistaken all along.[56]

But the pressures to relent on the issue of dumping at sea did not just come from other countries and environmentalists. They also came from within the British government. One official in 1970 pointed out to the AEA that there was growing international opposition not only to unilateral national dumping but also to any dumping; if true, that might spell the end of the ENEA dumping, upon which Britain had staked so much.[57] This was a precarious time for Britain's atomic energy establishment, which was completely reorganized in 1971. The health physicists who had dominated its Health and Safety Branch now moved to the independent, newly created National Radiological Protection Board. Many of the key figures in radiation protection lost their close identification with the Atomic Energy Authority, and the AEA became a much weaker body (and, indeed, this would continue in the 1970s as its responsibility for weapons also disappeared).[58]

Intense pressures now fell on the British health physicists in the National Radiological Protection Board. They did not want to budge, and for good reason. Although they now had no reason to toe the AEA's line, they had to protect their own scientific credibility. After all, they had authored the AEA's policies on radiation protection. But further, they saw opposition to sea dumping as political, biased, and fundamentally illogical. Looking back on the Portuguese and Irish opposition to dumping at sea in 1967 and 1969, they were dismayed that the image of dumping drums should have such a hold over the imagination. After all, the amount of radioactivity dumped from ships far from shore was vastly less serious than the routine pipeline discharges right offshore. As health physicist David Richings put it in 1970, "I do not know why such excitement is caused by the dumping of material in view of quantities of waste discharged through the Windscale pipeline—but it is one of the facts of life."[59] Apart from the public relations and diplomatic problems created by a few dumping incidents, the only serious debates within the British government had arisen over the permissible levels of effluent discharge. Yet the environmentalists seemed relatively silent on that, as did the American legislation, which did nothing to limit discharges into the sea. Strangely, the international opposition to such practices was comparatively small.

The Human Environment

In 1970, national representatives at the IAEA discussed the possibility of establishing a central repository for data about releases of radioactivity into the environment. The United States proposed an ambitious network of monitoring stations, though the idea of simply getting data from member states was probably ambitious enough. The British had given such information before, only to see it become fodder for Soviet propaganda. Deciding how to respond, some British officials saw an opportunity to lay the groundwork for defending their policies at the upcoming conference in Stockholm, Sweden: the United Nations Conference on the Human Environment. As one official put it to British health physicists, "It might well be desirable for the UK to demonstrate that on the nuclear front we have nothing to hide, and this might most satisfactorily be achieved by pointing up the flow of 'environmental' information to the IAEA."[60] As for the network of monitoring stations, this seemed a great example of the growing disparity between American and British views. British radiochemist A. W. Kenny dismissed the idea, pointing out that the British approach was to measure how individual humans or populations were irradiated—in other words, to trace the pathway of dangerous radioactive materials to man—not simply to take measurements uniformly around the world, which could lead to irrelevant but highly politicized conclusions.[61] Others, such as Alan Preston, argued that it might be better to go along with the American ideas to some extent, "in the interests of not widening the differences in approach to environmental problems which already seem to be arising between the US and the UK in a radioactivity context."[62] But this view did not prevail; it was more important to the British to protect the scientific credibility of their policies than to pander to public opinion. The AEA had spent most of the previous decade trying to persuade British ministries and fellow dumping countries in ENEA that monitoring in the open ocean not only was useless from a scientific point of view but also wasteful of time and money. Health physicists were not going to back away from that position now, just because the Americans were promoting monitoring.[63]

The British struck a middle ground here by declaring openly that monitoring was sometimes important for public relations. In 1971, Alan Preston and his colleague at the Ministry of Agriculture, Fisheries and Food, P. C. Wood, tried to put monitoring programs into perspective in an article in the *Proceedings of the Royal Society of London.* They noted that although most people understood monitoring to be the studies designed to ensure that pollutants behaved as expected, after the fact, "in practice monitoring cannot really be separated from the assessment of potential damage which should precede it." What Preston and Wood described as "in practice" was really the attitude of most British health physicists about monitoring, namely that preliminary studies counted as monitoring just as much as studies done after the fact. Indeed, the studies that went into the planning of dumping and effluent discharge were more important to the AEA and its authorizing ministries (like MAFF) than the commonly held understanding of monitoring as something that occurred after the fact. A good monitoring program for the protection of humans and their resources, Preston and Wood declared, had three elements: (1) an assessment of the degree of exposure to humans and the environment, (2) some kind of scientific investigation related to this assessment or to the effects of exposure, and (3) maintaining good public relations. None of these explicitly required scientists to survey the environment after disposal occurred.[64]

In some cases, they believed, routine after-the-fact monitoring was illogical because contamination would be impossible to detect. Even such cases, however, "may still merit a limited monitoring programme for public relations reasons, but careful appreciation is required in deciding the scale of such effort." If this were done, Preston and Wood argued, the surveys ought to be very limited in scope, taking just a few critical measurements. They pointed out that a larger effort could imply, in the eyes of the public, a lack of confidence in the operation as a whole. After a few years, even this cursory monitoring could be abandoned "and the situation can be monitored through control of the discharge alone."[65] Although they were speaking of marine pollution in general, Preston and Wood undoubtedly had radioactive waste discharges in mind when making these statements, as they were a perfect reflection of the British government's actions, and Preston knew the issues intimately. By "monitoring," they meant that they could measure routinely the levels of radioactivity they put in the sea at the point of discharge, not necessarily analyzing the behavior of radioactive materials in the sea. The only requirement of monitoring, the authors pointed out, was to show that established levels were maintained at the point of discharge.[66]

Even when analyzing the actual environment seemed warranted, according to this view, taking samples of seawater was rarely justified. Instead, a more convenient and cost-effective method was to identify the most likely route by which pollutants would reach humans. They could then analyze the concentrations of radioisotopes in particular marine animals or sediments, depending on what had been identified as the most likely, or most critical, pathway. As mentioned previously, this "critical pathways" approach formed the backbone of Britain's practice of monitoring marine pollution. Preston and Wood acknowledged that the synergistic effects of various radioisotopes, and perhaps other pollutants such as pesticides and heavy metals, might complicate this method considerably. But they also pointed out that the effects of radioactivity were probably among the best understood of all marine pollutants, especially compared to heavy metals, and that the monitoring programs currently in place were adequate to the task.[67]

Other European countries did not prove as resolute in defending waste disposal practices at sea as Britain. Although Britain had hoped to make international dumps in 1971 and 1972, it realized it would be difficult to "rustle up enough contributors," as one official put it, due to increased sensitivity about the environment and awareness of all kinds of pollution. This was one reason that the ENEA eventually dropped the "European" from its name in 1972, trying to widen participation by including Japan. Although in 1970 the British reserved the possibility that they "might have to go it alone," the prospect of returning to unilateral dumps was a grim one from a political and diplomatic point of view.[68] To lessen the impact, they hoped to use the ENEA flag even in a unilateral dump. But the other countries would have none of it, particularly the French, who made it clear that the ENEA flag could only be used if at least three nations took part. The issue never came to a head; as it turned out, Belgium, the Netherlands, and Switzerland all hoped to dump some wastes despite the potential political cost. Discussing some of the inconvenient logistical demands made by these countries, AEA chemical engineer J. B. Lewis let slip his country's opportunistic outlook toward international cooperation: "It was not prudent to protest too much . . . as otherwise suspicions would be aroused that the UK did not really wish to cooperate with ENEA but merely wanted an excuse to use the ENEA flag."[69] In the course of planning the operation it turned out that the other countries thought it best to have two separate ships conduct the dumping, one entirely British and the other carrying the rest. Internal memoranda reporting these meetings reveal a great deal of dissimulation on the part of the British representative; despite wanting a unilateral operation, he offered to carry Belgium's waste in a "meaningless gesture" that would never come to fruition because of the cost, and he had other improbable but credible suggestions, as he put it, "up my sleeve." In the end, he said that "I believe that I created the correct impression that the UK were genuinely interested in a joint operation."[70]

Despite the success in using the ENEA flag in what amounted to a unilateral operation, Britain's Atomic Energy Authority was losing control over the ENEA; its scientists learned with disgust that the site for the 1971 and 1972 dumps had been chosen without their input, and the choice reflected the politics of the day. The new site would be 750 kilometers from both Portugal and Ireland, reflecting neither health physicists' nor oceanographers' priorities. Lewis complained about the fait accompli: "This was the first intimation that Mr. Burns and I had of the new site. We pointed out that it was another day's steaming and also out of effective Decca navigation range. However we were informed that nothing could be done about it."[71] The decision incurred not only more cost but also more possibility of a navigational blunder; nevertheless, it made politicians happier. This must have reinforced the existing views among atomic energy officials of the irresponsibility of such political expediencies.

Irresponsible or not, political expediencies became increasingly attractive as the Stockholm Conference approached. Officials scrambled to satisfy environment-related concerns, not wanting their countries to be targets of criticism. The prospects seemed quite real that environmentalists would successfully associate radioactive waste disposal with the other pollutants that made headlines for killing the ocean. The IAEA feared this and tried to prevent it by starting an international register to keep tabs on each nation's releases of radioactivity. According to British observers, it reflected the IAEA secretariat's belief "that it would be wise to demonstrate that the IAEA is doing something in the matter so as to pre-empt criticism, or even action, by other bodies not directly concerned with atomic issues."[72]

Meanwhile the French CEA felt the sting of the ENEA dump in August 1971, despite the fact that no CEA wastes were in it. Environmental groups immediately lodged protests with the government, and the newspaper *Paris-Jour* called it a grave affair of pollution and "an incredible story." It stated, rather misleadingly, that the operations had been conducted in the greatest secrecy and that the French authorities had agreed to it. Another newspaper, *L'Aurore*, provided conciliatory but equally misleading information that the dump was an experiment to be studied by experts in oceanography and radiation protection.[73]

As the Stockholm conference drew near, ENEA countries were afraid to go forward with their planned dump in 1972. Although they realized that backing off of their plans would give the appearance of giving in to public outcries, they did not want to fuel any controversy that might come up during the conference. Their main priority was to get through the Stockholm conference unscathed, without drawing too much attention. Unfortunately, the ENEA had scheduled its dumping operation to take place simultaneously with the conference. As British health physicist David Richings put it, "this was courting trouble and enabling certain people to make a series of hysterical statements which might get widespread publicity." He and others successfully lobbied ENEA chief E. Wallauschek to postpone the operation until after the Stockholm conference, along with the political environment it surely would create, had passed.[74]

The Stockholm conference, held 5–16 June 1972, consolidated the environmental movement considerably, giving rise to international agreement on an array of environmental proposals. But in some ways, the conference just reflected cold war politics. For example, representatives from the People's Republic of China politicized the event by criticizing the United States at length. In addition, the Soviet Union continued to condemn the West; it boycotted the conference because of the exclusion of the German Democratic Republic, pulling out all of its allies except Romania with them. From the sidelines the Soviet media emphasized the destruction of the environment by the American military in Vietnam. American intelligence services concluded that, despite the Soviets' growing awareness of domestic ecological problems in regions such as Lake Baikal, on the international scale "they still seem to view environmental protection primarily as an issue for diplomatic exploitation."[75]

These roadblocks did not amount to much, however, and the conference seemed to be a success. The resistance to environmental issues by developing countries did not materialize as strongly as some had expected. As an American embassy official in Brazil put it prior to the conference, "there is going to be a continuation of the feeling among many of the underdeveloped countries that being concerned about the environment is, in the final analysis, a rich man's game."[76] But at the conference, only China, Tanzania, and Algeria took strong hostile positions, and most developing countries proved cooperative. As for the absence of the Soviet bloc, American representative Christian Herter observed: "No one seemed to care."[77]

The delegates at Stockholm voted to recommend over a hundred specific proposals to protect the environment. For the United States, it solidified the American position as a supporter of environmental issues. The United States proposed a United Nations environmental fund of $100 million and pledged to make up $40 million of it, while other countries immediately pledged about $25 million. In fact, the conference recommended most of the American environmental goals, including a whaling moratorium, a world heritage trust, a global monitoring program—and a global ocean dumping convention. The conference's support for this last item, already part of Nixon's legislative agenda, was immediately brought to the attention of Congress for action.[78]

Meanwhile, ENEA waited through the Stockholm proceedings with trepidation. The conference ended on 16 June 1972; the same day, with the conference safely behind them, ENEA got back to business and started a dumping operation. The cargo included 7,600 drums of waste from Belgium, the Netherlands, Switzerland, and the United Kingdom. To manage the political situation, the ship (*Topaz*, operated by a private firm) included observers from Ireland, Japan, and Belgium. The site for disposal was the same as the 1971 dump, chosen for political reasons roughly equidistant from the shores of Ireland and Portugal, far from the coast of France and well outside the Bay of Biscay, in water roughly 4,500 meters deep. The ENEA secretariat issued a statement a couple of months later observing that all of its operations were experimental in nature, which was somewhat misleading because it was really an organizational and logistical experiment, not a scientific one. It pointed out that the object was to devise, at the international level, "with the help of specialists in marine biology, oceanography, marine radioecology and radiation protection, disposal methods . . . which will provide the maximum guarantee for the protection of man and the marine environment." Though packed with references to scientists, the sentence truly stated that the experiment was designed to smooth out the method—that is, the logistical problems associated with getting several countries to dump together. Two other features of the statement were misleading. One was the explicit mention of marine scientists as crucial to policymaking, despite the fact that they played almost no role at this point in ENEA operations. The other was the mention of the "marine environment" as worthy of protection. This undoubtedly was a nod to the Stockholm conference and its goals, but protecting the environment, including marine biota, never had been a concern in waste disposal policy unless it implied danger to humans in particular.[79]

The political pressures generated by the Stockholm conference had led several countries to limit the amount of waste that they intended to dump at sea in 1972. Though perhaps prudent in the short term, it proved very inconvenient the following year. It meant that in 1973 there was an overabundance of material to dump, which proved embarrassing to European countries not wanting to give the impression of escalating their activities after the Stockholm conference. Not surprisingly, Britain felt this inconvenience most acutely. Not only did it have its usual amount of material and what it had held back in 1972, but it also hoped to get rid of about 16,000 tons of plutonium wastes that had built up over the years and were being stored on land (at

Drigg). Thus Britain faced a conundrum, knowing that it had a comparatively large amount of waste to dump in an international climate that seemed to stand firmly against the practice. Much of the material was tritium, which did not concern health physicists as a serious danger, but as one of them put it, "to the public, all curies are the same."[80]

The rise of environmentalism played havoc on waste disposal attitudes during this period. Although none of the ENEA countries opposed dumping on scientific or technical grounds, they feared the uproar in the press. AEA chemical engineer J. B. Lewis, who represented Britain in some of these Europe-wide meetings, put it thus: "Current opinion, even among those who feel that sea dumping must continue, is that there should be a pause for 1 to 2 years during which time all the technical, legal, and above all, political issues are disposed of *to the satisfaction of the environmental lobby*. It is not enough to convince technologists, the public at large must be reassured."[81] The ENEA countries were afraid to dump. Moreover, they did not want Britain to conduct any national dumps either. The uproar could jeopardize the future of ocean disposal. They already had received an enormous amount of bad press, including a false report in French newspapers that the 1972 dump had occurred in the Bay of Biscay. One tactic might have been to organize a sanctioned dumping operation under the banner of the International Atomic Energy Agency, but the resolute opposition to ocean dumping by the Soviet Union probably would have rendered this impossible. Indeed, routinely negative views (and a repetition of the false French report) expressed in *Pravda* did not bode well for any international consensus. Thus the countries wanting to dump despite prevailing environmental politics—Britain, the Netherlands, and Belgium—faced an array of European nations standing in their way.[82]

The Dutch were particularly eager to press on despite objections from environmental groups. B. Verkerk, a scientist from Reactor Centrum Nederland, proposed that the NEA (recently the organization had dropped "European" from its name) be sidelined for the time being. More dedicated countries could continue dumping according to the principles that had been hammered out in the ENEA over the years. If the NEA was not willing to take on a major dumping operation in 1973, then it could be done on a more limited, yet still multinational, scale, between the United Kingdom, the Netherlands, Belgium, and whoever else wanted to join in.[83] This provided an attractive middle ground for Britain, which was internally facing a major crisis about where to put its radioactive wastes if the sea no longer could be considered an option. Without this offer from the Netherlands, the British likely would have made a unilateral sea dump and would have had to face the unpleasant political and diplomatic consequences by themselves.[84] In the end, most of the NEA countries did not want to participate in a dump in 1973, largely because of the uncertainty surrounding international negotiations for a global dumping convention. But none explicitly opposed a dump. The chairman of an NEA meeting in May 1973, Italian physicist Carlo Salvetti, pointed out that there seemed to be little enthusiasm for "bringing about a marriage of the atom and the sea," but that there were no grounds for stopping the operation.[85]

The Meaning of the Global Dumping Convention

The basic strategy of atomic energy establishments thus far had been to avoid the environmental trends by disassociating themselves from marine pollution. Although there had been oil spills, nerve gas controversies, and industrial waste in lakes and rivers, they pointed out, with some justification, that radioactive waste had never experienced significant periods of laissez-faire neglect. They argued that scientists, particularly health physicists and to some extent marine scientists, had studied the effects of radioactive waste at sea and atomic energy establishments adhered vigilantly to national and, more recently, international standards. British radiobiologist Alan Preston boasted to colleagues that Britain's "exemplary control and management" had allowed it to keep radioactive waste out of the Oslo Convention, a 1972 agreement to regulate marine pollution in the North Sea.[86] The Stockholm conference had not dealt specifically with radioactive waste, but clearly it had not been similarly excluded. And now it seemed clear that radioactive waste would be addressed in a new global dumping convention being negotiated by governments. Although the strategy of complete disassociation had failed, there was still hope that scientists could persuade governments not to ban dumping completely.

Many scientists in European atomic energy establishments instinctively feared the consequences of a global dumping convention. However, some began to see it as a way to save, rather than condemn, radioactive waste disposal at sea. If properly handled, such a convention could extend the life of radioactive waste disposal at sea while satisfying the environmental lobby at the same time. As Alan Preston observed to David Richings, now at Britain's National Radiological Protection Board, they could support a ban on dumping highly radioactive waste at sea, leaving it to the IAEA to define the precise meaning of "highly radioactive." All other radioactive wastes would be permitted. If such wording were preserved, it would give dumping nations a wide latitude for interpretation and might even make it possible to dump most of their wastes without getting approvals from radiation safety bodies. As Preston put it, "Neither of these provisions will be in any way embarrassing to the UK, and

I would have thought could well give us endorsement at [the] international level in non radiological protection fora for the continuation of disposals." Preston was worried that dumping countries might not see the opportunity and would fight any kind of ban by the convention. But he argued persuasively, "We (MAFF) have fought hard for this and would not take kindly to having it all brushed aside by ill-founded nervous apprehension in the [Nuclear Energy Agency] directorate. We believe in the correctness of what we are doing and have clearly succeeded in convincing others of it—how ironic if NEA should let it all go now!"[87] He and others believed that it would be possible to work within the convention and use it to continue their existing policies while seeming to concede a victory to the environmentalists.

During the negotiations for the London Convention, Britain's Atomic Energy Authority finally lost the control over waste disposal operations that it had built up over the years. Preston, for example, was not part of the AEA, nor were health physicists David Richings and John Dunster, who had moved on to work for the National Radiological Protection Board. They were old hands from the Authority's Health and Safety Branch, to be sure, and they protected the policies they had built; but they did so from outside the authority. The AEA's newer scientific leadership did not shape policy as successfully. This caused no small amount of bitterness from chemical engineers such as J. B. Lewis, who now were responsible for waste disposal in the AEA. Instead of helping to formulate policy, he was left only with the task of executing it. Lewis complained that Richings kept him out of the loop in his negotiations with the IAEA and the NEA: "Richings told me that it was not my place, nor that of the Authority's, to get involved in the 'politics' of disposal." As with the American AEC, the AEA was vulnerable to charges that regulating its own safety was a conflict of interest. The international negotiations were happening above the AEA's head, and it was very conscious of its lack of leadership role, particularly given the history of dominating the early waste disposal activities in the ENEA. With a global dumping convention in the making, the stakes were now even higher. Lewis grumbled, "Who, may I ask, is really speaking for the UK? Everybody but the Authority apparently."[88]

This internal squabble did not reflect a change in policy; instead, it underlined how invested health physicists like Richings and Dunster were in the United Kingdom's policies. Richings wanted to interact with his international counterparts directly, and perhaps he suspected that scientists in the AEA would not see the opportunities inherent in the global convention. He knew, for example, that the IAEA was keeping in close contact with the NEA secretariat, comprised primarily of E. Wallauschek and J. P. Olivier, and listening to its advice. As international debates now would turn on what "high level" wastes were (because they would be banned), it would be crucial to concentrate not on resisting the ban but rather on influencing these definitions. For example, the British wanted to keep plutonium out of the "high level" category, because they made up a large part of what the British put in the sea. As Richings explained to Lewis in the AEA: "I detected a doubt in some minds on the categorisation of alpha emitters, and in particular, plutonium. In my view we should take this opportunity of pre-conditioning the minds of Wallauschek and Olivier on the desirability of not including alpha emitters as such in the definition of 'high level wastes,' thus leaving national authorities free to determine on merits whether any particular waste of this category should or should not be dumped in the sea."[89]

Now was the time to act behind the scenes, arranging informal meetings only, rather than making formal representations by the Atomic Energy Authority. The task of exercising indirect influence over the IAEA by "pre-conditioning the minds" of experts in other bodies naturally should fall to British experts—the health physicists—acting in advisory, unofficial capacities. Richings added, "we should be experienced enough to avoid washing any dirty linen in front of them."[90]

The situation was precarious because the American delegates to the IAEA already had devised a troubling definition of "high level waste." The United States defined it as "aqueous wastes resulting from the operation of the first cycle solvent extraction system, or equivalent, and the concentrated wastes from subsequent extraction cycles, or equivalent, in a facility for reprocessing irradiated fuels." This definition targeted wastes generated at reprocessing facilities such as Hanford in the United States, Windscale in the United Kingdom, and La Hague in France. As worded, the definition had the potential to have enormous ramifications because it might include both liquid and solid wastes. From Britain's point of view, the global dumping convention was going to apply to canisters of waste being dumped at sea only. One AEA chemist acknowledged that this definition "would be embarrassing to Windscale as it might rebound and affect their pipeline disposal. (Their medium level streams could legally be called high unless the wording of the USA definition was not amended.)" Given the huge amount of waste water discharged into the Irish Sea from Windscale, the American view was rather threatening. They hoped that they could negotiate other delegates away from this position, particularly by convincing them that neither tritium nor plutonium were high level enough to be included, despite both of them fitting the American definition.[91] All these discussions indicated that the London Convention itself need not be earth shattering. The post-convention negotiations at the IAEA about definitions would, by contrast, be crucial.

A major effect of the London Convention was the reaffirmation of threshold levels in environmental policymaking. Specific benchmark levels by which outsiders could discern "safe" from "polluted" had been criticized in the past by those who believed, as American and British scientists had feared during the major radiation studies of 1956, that politicians and laypersons would manipulate or simply misunderstand their implications. The London Convention identified such thresholds by establishing black and gray lists; the black list designated banned materials, whereas the gray list designated materials requiring only special permits from national authorities. High-level radioactive waste was put on the black list, but its meaning would have to be negotiated across national lines at the IAEA. Those most dependent on disposing of waste at sea knew quite well how flexible such definitions could be. In addition, while high-level solid wastes would be banned, equivalent effluent wastes would be forgotten; the much larger quantities of liquids could not compete with the political firestorms produced by a few drums of solids. The London Convention applied only to dumping, not the general introduction of radioactivity into the sea.

When the IAEA convened a panel in June 1973 to define high-level radioactive waste, the British were surprised at how yielding their counterparts were. For one, the British wanted to exempt their large volume of tritium from being regulated, and the other countries did not seem to mind. The panelists watched a film of the first ENEA dumping; although it was produced by France's Commissariat à l'Énergie Atomique, Britain's practices were on display more than anyone's. After some preliminary reviews of the convention and the technical background, the panelists started the serious business of hammering out a definition of "high-level waste." Ultimately this definition conformed to the expectations of major dumping nations: "For the purposes of Annex I [the black list] to the London Convention of 1972, high-level radioactive wastes or other high-level radioactive matter unsuitable for dumping at sea means any waste or other matter with a concentration of beta activity (excluding tritium) exceeding 1000 Ci per tonne gross weight, or a concentration of alpha activity of half life greater than 100 years exceeding 10 Ci per tonne gross weight. Both of these activity concentrations are to be taken as averaged over 100 tonnes."[92]

Any amounts of solid radioactive waste less hot than the above definition, with fewer curies (Ci) per ton of gross weight, could be handled according to each nation's own authorization procedure. Moreover, the definition did not mention the source of the waste (the United States' proposed definition had identified wastes generated by chemical reprocessing facilities), nor did it mention any particular element except to exempt one, tritium (the British hoped to avoid any direct reference to plutonium).[93] Although this definition changed in subsequent years—identifying different levels for strontium-90, for example, and changing the alpha activity half-life to fifty years—it still left a great deal of flexibility to national atomic energy establishments, who now had internationally accepted threshold levels to legitimize dumping operations.[94]

The soft line taken on the definition of high-level waste should not have been such a surprise. After all, from the politician's point of view, the important part was over: the convention itself was signed and the dauntingly labeled (if as-yet undefined) "high-level waste" was banned. These two facts would satisfy most domestic critics. Still, the participants clearly wanted to use the meeting as a vehicle for making more general statements about the future of the oceans. Because the IAEA was going to leave the granting of special permits to the discretion of individual nations, some of these nations decided to point out that they would not be exercising their rights very often. French and German delegates drafted a note for inclusion in their agreement, stating that the coming decades would see further development of the use of the sea's resources, and that the countries of the world ought to take care not to squander these with excess pollution. Although they did not condemn sea dumping, they wanted to state plainly that the convention should not encourage it, either. In fact, France, the United States, and Canada all gave formal declarations saying that they did not intend to use the sea to dump radioactive wastes.[95]

As the negotiations were under way for a global dumping convention, the world received ambiguous signs from the Soviet Union. Although historically opposed to dumping radioactive waste at sea, they seemed willing to let the convention happen despite its reinforcement of Western European policies. There was some mystery to this. British embassy official R. A. Vining, in her letter from Vienna (the headquarters of the IAEA), observed that the Soviet expert in the IAEA's Waste Disposals Section had returned to the USSR, ostensibly because of his mother's illness, during the crucial period of deciding the procedures for dealing with wastes not explicitly banned.[96] But in subsequent IAEA meetings, the Soviets alienated their colleagues, threatening to derail consensus by reserving the rights of nations to make changes after the delegates' agreements. In doing so they succeeded in negotiating a new perspective (and for dumping nations, ostensibly a new limitation) into the convention. G. G. Polikarpov, from the Institute of Biology of the South Seas (in the Ukraine), persuaded the other delegates to agree that elements of the biosphere needed to be protected, and not necessarily because of a connection to humans. He also introduced a sentence to the convention that required concentrations of radioactivity to present no unacceptable risk to man or to marine ecosystems. Adding "or to marine ecosystems" violated, as the British internally lamented, the long-standing critical pathways approach to determining risk; but it was a change to which they consented in this diplomatic context. Because the convention did not apply to liquid discharges anyway, British scientists let it go. "This was an undesirable addition," British scientists John Dunster and Alan Preston pointed out in a note for the record, "but was the price paid for a unanimous agreement on the final text."[97]

The restrictions on radioactive waste in the London Convention were much lighter than those on other materials. An unexpected result of this was a willingness to emphasize the radioactive nature of wastes rather than to de-emphasize it. That some saw this as an opportunity to dump more toxic wastes under the banner of "radioactive" was revealed during the 1973 voyage of the *Topaz* under the auspices of the NEA. The only countries involved were the United Kingdom, Belgium, and the Netherlands. On the Belgian portion of the dumping operations, the crew at Zeebrugge loaded drums contaminated with cyanide onto the vessel along with radioactive materials, under the weak premise that the waste had been produced at the Belgian Nuclear Centre, in Mol, and had become slightly radioactive. This was certainly true, but the amount of radioactivity was small; the Belgians were only too happy to classify the waste as low-level radioactive waste, with the minimal restrictions under the London Convention, rather than as high-level toxic waste, with much tighter controls. But other government officials in Belgium protested this and alerted the press. Many NEA members soon protested, believing that it would undermine the credibility of radioactive waste disposal. The captain of the ship was confused about what to do and used the emergency communication channels to contact the British Foreign Office and the *Topaz* owners, William Robertson Ltd., in the middle of the night. The ship turned around after dumping some of the cyanide, returning the bulk of it to Belgium.[98]

The main concern about the *Topaz* was the need to distinguish between radioactive wastes and toxic wastes, to avoid any associations between one and the other. Many NEA representatives resented Belgium's move, particularly because the Belgian representative had been instructed to keep quiet about it to the other NEA representatives. Making matters worse, the Belgians spilled considerable amounts of radioactive cesium inside one of the holds before finally handing the ship over to the British for their leg of the dumping operation. Hoping to avoid a reprise of the cyanide incident in the public press, Belgium avoided decontaminating the vessel and simply paid the expenses for the British government to do it in a port of its own, but not before traveling to Ijmuiden to pick up some Dutch drums.[99] The incident did nothing to decrease apprehensions about radioactive waste, which now would be associated with deadly cyanide.

The *Topaz* saga, as some British authorities called it amongst themselves, enabled the ship's owners to wield some power against the British government, which had negotiated the use of the ship and, more important, paid the ship's insurance indemnity. William Robertson Ltd. complained that the ship had been left contaminated for some time and, besides, the Belgians had not paid the proper insurance indemnity for chemically contaminated waste like cyanide. Obviously, things had not gone according to plan, and the insurance indemnities ought to reflect such uncertainties, they argued. After the incident, Harwell chemist J. B. Lewis had a bitter encounter with representatives of Robertson, who threatened to refuse contracts with the AEA unless they agreed to pay higher insurance indemnities; in addition, they insisted that the AEA contract with them on all dumps, radioactive or not, turning its back on competitors. Lewis accused them of trying to blackmail the Atomic Energy Authority. But to colleagues, he conceded that if the threat were carried out, it might be impossible to find another ship to dump radioactive waste. "The situation will probably get better," he wrote optimistically, "if and when IAEA confirm[s] the respectability of sea dumping." In the meantime, they would do well to try to accommodate the ship owners. They did precisely that, and the *Topaz* continued to be used in subsequent years.[100]

In truth the IAEA could never confirm the respectability of dumping, despite ultimately declaring very loose restrictions on the dumping of radioactive waste at sea. Its implicit sanction of the practice, vindicating years of actions by Britain, did not detract from politicians' sensitivity to an environmental movement that would not go away. In December 1974, IAEA Director General Sigvard Eklund transmitted the agency's negotiated interpretation of high-level radioactive waste to member countries. Reflecting the continued stigma associated with waste disposal, he reminded them that the definitions and recommendations "should not be construed as encouraging in any way the dumping at sea of radioactive wastes and other radioactive matter," and that the subject would continue to be reviewed periodically. He mentioned that such a review already was under way, in which oceanographic models would be analyzed to understand the effects of dumping wastes at sea.[101] But appealing to possible oceanographic research was an old trick, and it rang hollow; even some of the NEA countries who wished to dump were embarrassed by the IAEA levels, not wanting environmental politics at home to see such unrestrained use of the oceans sanctioned at the international level. They stressed that these were hypothetical figures and no one expected anyone actually to dump at such levels. Using an even older trick, they emphasized that these were simply threshold levels, below which no harm could come to the sea or humans.[102]

Conclusion

THE LONDON CONVENTION went into effect in 1975. Ten years later, the Commission of the European Communities inaugurated a study, Project Marina, to assess the radiation exposure to Europeans from radioactivity in the seas around northern Europe. Although most of the participants had to redirect their work in 1986 to assess the effects of the nuclear accident at Chernobyl, they did identify the primary pathways by which Europeans received radiation exposure from the sea. Windscale (at that point known as Sellafield) continued to be the primary source for annual exposure, exceeding the combined levels of exposure from natural radiation, all other nuclear sites, weapons testing, the Chernobyl accident, and packaged solid wastes. Of the other nuclear sites, La Hague (in France) and Dounreay (in Scotland) also put considerable amounts of radioactive material into the sea. At the time of Project Marina, as had been the case for decades, these chemical reprocessing facilities—rather than ships carrying packaged wastes—were the primary means of putting radioactive material into the oceans.[1]

Ocean disposal of radioactive waste did not stop with the London Convention, even in the case of packaged wastes dumped from ships. Initially, the convention only banned high-level waste. Through the rest of the 1970s, Europeans continued to use their Nuclear Energy Agency to conduct cooperative dumping operations, all in the same far-off site agreed upon by politicians. In the nearly half century between 1946 and the last reported operation in 1993, fourteen countries dumped radioactive waste into the sea, at about eighty different locations, mostly in the Atlantic, Arctic, and Pacific Oceans. Of these countries, very few used the ocean routinely. All the radioactive materials dumped into the Arctic have been attributed to the Soviet Union and (to a smaller extent) the Russian Federation, including the disposal near the island

Novaya Zemlya of six naval reactors with highly radioactive spent fuel and ten reactors without the fuel; these activities began around 1965. In the Pacific, Japan dumped about 1 percent of the total, New Zealand dumped an even tinier amount, and the rest is attributed to the United States (38.3 percent) and the Soviet Union (60.4 percent). In the 1970s, the Soviet Union dumped four reactors without spent fuel into the Sea of Japan. In the Atlantic, where the lion's share of waste was dumped, Britain was responsible for some 77.5 percent, while the United States, Belgium, Switzerland, France, the Netherlands, and other European countries dumped the rest. Worldwide, the Soviet Union (46.1 percent) and Britain (41.2 percent) were responsible for nearly 90 percent of the radioactive waste dumped from ships at sea between 1946 and 1993.[2]

Further international agreements in the 1970s put more curbs on radioactive waste disposal. Prohibitions on sea disposal went into effect in the Baltic Sea (1974), the Mediterranean Sea (1976), and the Black Sea (1992); the 1980s would see even more areas cordoned off as safe zones, free of drums of radioactive waste. When in 1978 the environmental group Greenpeace used its ship *Rainbow Warrior* to follow the ships on the NEA's annual dumping operation, politicians became even more nervous about the political fallout. Greenpeace showed films of their unsuccessful efforts to hinder the dumps, and in 1982 two drums of radioactive waste fell onto a boatful of Greenpeace protesters. These actions and the adverse media attention they brought to the governments sponsoring the dump, led the signatories of the London Convention to call a voluntary moratorium on ocean dumping in 1983. In the early 1990s, the dumping regime changed even further: instead of requiring scientific evidence to move a pollutant from the regulated "gray" list to the prohibited "black" list, the convention emphasized that preventive measures could be taken in the absence of conclusive evidence. Amidst the revelations in 1993 about Soviet dumping during the cold war, these countries—supported strongly by the United States—pushed for a ban on all dumping of radioactive waste, high or low level. During the 1993 vote, thirty-seven countries voted for the ban and none voted against. Britain, despite stalwart backing of sea disposal over the decades, abstained. Even without scientific consensus, environmental protection prevailed.[3]

Or so it seemed. This book has emphasized that the 1972 London Convention did not significantly change the practices of atomic energy establishments. Even in subsequent decades environmental legislation had only a small effect on what nuclear powers put into the sea and land. One could argue that the situation has changed since then; after all, international laws now ban the dumping of all radioactive waste, not just high-level wastes. This has proven impossible to accomplish, as it turns out. The rules became so stringent that scientists began to agonize over the concept of "de minimus"—which was used to define levels of radioactivity so small that materials should not be regulated. After all, when sediment is dredged from the sea floor for whatever reason—construction of a harbor, for example—its natural radioactivity immediately implicates it as radioactive waste; if the rules were strictly enforced, this material could not legally go back into the sea.

The fact that scientists have to identify these miniscule levels might imply a radical shift in thinking about the sea. However, as British health physicists pointed out in the early 1960s, the radioactivity from all the "dumping" from ships in the first decade or so of Britain's operations in the English Channel was equivalent to about a week's discharge from its pipeline at Windscale. At that time, they did not understand what merited such fuss about solid wastes being dumped at sea, but they did not often point out the irony. There has been a consistent and sharp disparity between the two kinds of operations in the amounts of radioactive waste introduced into the sea; yet in the early years it captured the imaginations of neither the public nor the diplomats who drafted the London Convention.[4] And because bans on dumping had often been achieved by regional agreements, what the regulations in the global convention amounted to—as one scholar has put it—"is flogging horses killed elsewhere."[5] This was absolutely the case with the United States, which turned away from the oceans long before even the first incarnation of the convention, and gave up almost nothing to ban the dumping of radioactive waste at sea. And it now seems clear that to another major nuclear power, the Soviet Union, the convention was simply ignored.

Especially because Britain, France, and others managed to preserve the lifeblood of their sea disposal operations, namely the bulk liquid discharges from pipelines, the ban on all radioactive dumping obscures the truth about radioactivity in the seas. As oceanographer Richard Fleming pointed out back in the 1950s, it probably does not matter how the radioactivity reaches the ocean. The avenue of introduction is irrelevant, whether it comes from solid wastes, bulk liquids, sunken reactors, or nuclear fallout. Activists positioning boats underneath drums being dumped at sea, fishermen finding radioactive plastic in their nets, or drums washing onto the shore—all these make headlines and they make frightening premises for novels, films, and doomsday scenarios. However, dumping operations from ships accounted for a tiny share of the total amount put into the sea, despite the high level of attention devoted to them. An environmentalist might rejoice at the London Convention, but a cynic might justifiably call it a legalistic chimera. It led nations to scrutinize even construction silt dumped far out at sea, but to stare unblinkingly at the continuous flow of waste emanating from nuclear reprocessing pipelines right offshore.

The purpose of this book, however, has not been to sit in judgment of environmentalists' aims and the empty promises of environmental treaties. Instead, it has been to illuminate some of the ways in which the uses of the ocean were conceptualized and negotiated in the years between World War II and the London Convention of 1972. These uses were shaped more by the relationships between the scientific communities and institutions of the cold war than by a sudden corrective made by the environmental movement. Yet the confrontation with environmentalism remains fascinating, and it is crucial in the historiography of science, technology, and the environment: how did these communities and institutions, shaped by the needs and opportunities of national security states, learn (or fail to learn) to operate in such a different political climate? In this book, the key communities were the marine scientists, the health physicists, the geneticists, and the variety of others who belonged to and/or carved out places for themselves in nuclear affairs from the 1940s through the 1960s. The institutions were the national academies, research institutions, radiation protection boards, and atomic energy establishments in the United States and Europe. To be sure, from the 1940s atomic energy establishments dumped their radioactive waste into nearby waters, and they worked closely with government bodies—such as the National Committee on Radiation Protection (United States) and the Medical Research Council (United Kingdom)—to establish health and safety guidelines for river and coastal discharges. And despite some qualms about genetic thresholds and the ocean's annual capacity to accept wastes, marine scientists accepted and promoted the view of the ocean as a global sewer. Oceanographers, particularly American ones, carved a niche for themselves in the world of atomic science during the 1950s, giving them unprecedented influence and opportunities.

The role of oceanographers in judging the uses of the oceans, however, often came at the expense of other experts, particularly the health physicists and sanitary engineers who designed policies from within the atomic energy establishments. All too often, these other experts felt, oceanographers emphasized the uncertainty of knowledge about the oceans, which undermined the scientific credibility of existing policies while pointing out the need for more funding. By the late 1950s, leading oceanographers were asking for millions of dollars annually in research funds, to help understand the fate of radioactive materials in the sea. Although the Atomic Energy Commission often complained, it also complied to a great extent, funding an array of subjects in the marine sciences in the interest of supporting and legitimizing its policies. But in Britain, the Atomic Energy Authority found these demands for patronage too opportunistic and too expensive, far beyond what knowledge was needed to develop sensible policies.

Although atomic energy officials did not call it blackmail, they knew quite well how important it was to please academic scientists. They disliked how oceanographers pandered to public anxieties and, later, to politicians wanting to mark themselves as environmentalists. But without the blessing of scientists outside the atomic energy establishments, waste disposal policies would be targeted by the lay public, egged on by scientists emphasizing how much was not yet known. This role of scientists was well known in government circles. When John Kennedy was elected president, his transition team was advised by outside consultants to be careful of scientists' opinions when trying to manage the Atomic Energy Commission. Not only were scientists capable of providing the foundation upon which technical decisions could be made, but also the scientific community was "a molder of public opinion." Much care had to be paid to the relations between the AEC and these scientists, the consultants asserted, for the scientists' public statements "will be much conditioned by the confidence members of the scientific community have in the integrity, competence, and sincerity of the Commission."[6]

This contest of scientific expertise between physicists, sanitary engineers, and health physicists working inside atomic energy establishments and oceanographers working outside of it was not a matter of one group wishing to put radioactive waste in the sea and the other wishing to ban it. It bears emphasizing that most oceanographers did not oppose dumping radioactive waste at sea. The conflict was about who could choose the thresholds. Even the major, media-intensive disputes about sea disposal—like the abortive attempt in 1960 to put waste in the Mediterranean—had more to do with oceanographers being left out of consultations and decision making than with opposing atomic energy policies. The oceanographers' aims—involvement in atomic energy affairs, influence in policy decisions, and patronage for ongoing research—all hinged on the assumption that the sea could in fact be used for radioactive waste disposal.

The spats between scientific bodies and atomic energy establishments fueled public controversies, and the relationship between the two groups changed with the evolving political landscape. The American decision to focus on land disposal took the wind out of the sails of oceanographers' grand aspirations for patronage; by dumping on land, the AEC also dumped a considerable economic and political liability. It allowed them to swat away the presumptuous grasp for authority by oceanographers, but this strategy also alienated academic scientists and intensified their criticism. Soon the AEC came into conflict with geologists about land disposal and with its own scientists about radiation exposure, leading to major cleavages in opinion about whether to trust the AEC's commitment to safety. Elsewhere in the 1960s,

French oceanographer Jacques-Yves Cousteau cast himself as a protector of the sea's living resources, which was easy to do given his conflict with the French atomic energy establishment at the start of the decade. Scientists inside and outside atomic energy establishments made choices in the 1960s as they confronted the era of environmentalism, with its new consciousness about protecting the living resources of the sea (not just the route to man) and its new opportunities for political allies. In the United States, almost the whole spectrum of politics was pro-environment by 1972, in form if not always in substance.

Though long a supporter of dumping radioactive waste at sea, the United States abandoned Britain to the proverbial hounds during a spate of national and international environmental lawmaking in the early 1970s. But its years of fighting off the "poison in the well" propaganda of the Soviet Union had already steeled Britain against the kind of political sniping it encountered due to environmental critiques of its policies. The absence of a powerful oceanographic community (and thus a stubborn unwillingness to support research that did not apply directly to policy formulation) and some shrewd environmental diplomacy gave Britain room to maneuver during the early 1970s. Britain's attitude was not static, however. Its atomic energy establishment adapted to political change through the 1960s with a well-developed public relations effort. Like the United States, it learned that patronage for science was an integral part of good public relations. This could take the form of patronage for laboratories (like the IAEA laboratory in Monaco), but more often it meant sandwiching dumping operations with science, such as by conducting research beforehand and afterward in order to legitimate them with "environmental monitoring." They were willing to include oceanographers in policy discussions, but despite some massaged egos this had no real effect. Moreover, they politely listened to ideas and objections of other countries, all in the cause of spreading out the responsibility for dumping at sea. And when the world turned against sea dumping, British atomic energy scientists did not despair. Instead, they acted behind the scenes, allowing agreements to come to pass without bitter objections and then fighting more significant battles—such as on the definition of "high-level" waste—in technical forums away from the gaze of politicians and the media.

From the American and British experiences, it is clear that the experts who formulated policy did not give up such roles easily, nor did they bear lightly challenges to their scientific integrity. Apparent concessions to new views by the 1970s, upon closer inspection, prove more illusory than real. It is true that the United States gave up ocean disposal, but doing so allowed the AEC to ignore the ax-grinding of oceanographers, the propaganda broadsides of the Soviet Union, and the grumblings of politicians at home and abroad. The fact that the Nixon administration later passed laws banning activities no longer practiced does not amount to much. If Britain fought a bit harder, it was because it had less land, less money to spend on enterprising oceanographers, and a far greater need to protect ocean disposal, with its nuclear facilities discharging directly into the sea. Although Britain too signed the London Convention, its health physicists did not alter their assumptions about the best policies; instead, they successfully argued for them in less visible ways. The scientists who built waste disposal policies in the 1970s and beyond were not a new breed; they traced their professional and intellectual lineages back to the health physicists of the 1940s.

As a final thought, the careful reader will have noticed the emphasis in this book upon patronage: strategies to promote it, opportunities to gain it, trends that promise it, and policies that presume its importance. It is not the goal of this book to characterize every academic scientist as opportunistic and greedy, nor is it to characterize everyone in atomic energy establishments as disingenuous, miserly, and careless of safety. A magazine editor who had published an exposé about the United States government's past dumping operations off Massachusetts once said to Congress, "In researching this story and in writing it, I have been struck by the comparative absence of villainy." He was surprised to find that, rather than a conspiracy, he had found "not only honorable but usually likable folk" doing their jobs, adhering to standards, and conducting legal businesses that satisfied government needs.[7] Perhaps the same might be said of the scientists inside and outside of governments who tackled the problem of waste disposal at sea. They pursued what they thought were the interests of themselves, their professional communities, and their countries. Nevertheless, thinking in terms of research funds is a helpful antidote to fast conclusions about the protagonists in environmental decision making. If Kennedy's transition team was correct about scientists' influence on the lay public, then it is worth remembering that scientists possess great stores of potential leverage when the political and financial stakes are high.

Moreover, one can make a career in uncertainty, exploring year after year the technical vicissitudes of a highly contentious policy without ever solving a practical problem. Atomic energy establishments sponsored a great deal of research, willingly or grudgingly; sometimes they did so in order to have a say in the outcome, as historians of science have suggested about the corrupting effects of patronage. But just as often, as this book has pointed out, they were compelled to spend money on what they considered superfluous projects because of the demands of alternative experts who might have made political trouble for them otherwise. And when this particular well dried up, marine scientists found sponsorship elsewhere because of persistent environmental concerns. In 1981, for example, the U.S. Food and Drug Administration, National Oceanic and Atmospheric Administration, and the Environmental Protection Agency agreed to sponsor public health monitoring programs in the oceans—finding drums of waste, sampling sediments and biota, and testing seafood. The purpose was not in support of any dumping operations, but to ensure that the activities that ceased two decades earlier were still having no discernable harmful effects on local people.[8]

It may be true that the idiom "He who pays the piper calls the tune" can be used to implicate all kinds of patrons, from atomic energy establishments to various military bodies, for corrupting science and its practice. And certainly this book has discussed atomic energy establishments using scientists, often disingenuously, to legitimize policies that no amount of scientific study would have changed. But let us acknowledge that scientists pursued their interests, too, and that calling into question existing policies could lead to research funding regardless of whether it contributed to the public's welfare. In the decades after World War II, atomic energy establishments complained that they could not get out of paying for science, for fear of being accused of negligence. That scientific patronage became a strategy for legitimizing policies also implies that some scientists were quite easily mollified by research funds. That their projects often had scant, if any, direct connection to waste disposal policies made their environmental concerns seem somewhat false. When mutual frustrations and resentments were vented, they fueled public relations fires already stoked by the Soviet Union's propaganda machine, which spoke of poison in the village well as if the ocean were a global well.

Looking back at the collaborative relationships between scientists and atomic energy establishments nationally and internationally in the 1950s, and contrasting them with the more testy relationships decades later, perhaps the well of nuclear-related funding from governments also had poison in it. Scientists' efforts to demonstrate the need for research typically implied uncertainty, hubris, and/or negligence in existing policies, and experts competed fiercely for the general public's approbation—essentially for the right to sit as authoritative experts about the uses of the ocean. The irony was that both groups of experts lost the battle for the right to speak for the sea and its appropriate role in the nuclear age. One side seemed to disintegrate: atomic energy establishments in the United States and Britain were reorganized in the 1970s and no longer exist as the powerful, authoritative bodies they once were, and the specter of public opinion has made all but the most die-hard countries chary of utilizing the oceans at all. Still, from the ashes of these institutions, health physicists moved to other national and international bodies, continuing to dominate radiation protection policies.

For their part, marine scientists helped to drive atomic energy establishments out of the sea. This was far from their original intention. They had seen in the future of atomic energy a clear role for the marine sciences, with bright prospects for cooperation and funding—all seemingly ensured by the major political and financial commitments to weapons and civilian nuclear power made during the 1950s. But in two short decades, political winds shifted and marine scientists followed the breeze. By the late 1970s and 1980s, they were trying to interest governments in radioactive monitoring programs and exploring dump sites with submersible crafts. But they had given up pointing out the importance of the oceans to the future of nuclear energy. Instead, they emphasized the uncertain effects of past actions and the continued threats posed by the artifacts of the atomic age lying on the ocean floor.

Notes

Abbreviations

AB	Records of the United Kingdom Atomic Energy Authority and Its Predecessors, UK National Archives, Kew, England.
AHSP	Alfred H. Sturtevant Papers, California Institute of Technology, Pasadena.
FO	Records Created and Inherited by the Foreign Office, UK National Archives, Kew, England.
GBP	George Beadle Papers, California Institute of Technology, Pasadena, California.
HCCEA	Fonds Haut-Commissaire, Commissariat à l'Énergie Atomique, Fontenay-aux-Roses, France.
LRDP	Lauren R. Donaldson Papers, University of Washington Libraries, Seattle.
NA326	Record Group 326, Atomic Energy Commission, National Archives and Records Administration, College Park, Md.
NA59	Record Group 059, Department of State, National Archives and Records Administration, College Park, Md.
NAA	National Academies Archives, Washington, D.C.
NIO	National Institute of Oceanography, Southampton, England.
ODWHOI	Records of the Office of the Director, Woods Hole Oceanographic Institution, Woods Hole, Mass.
RHFP	Richard H. Fleming Papers, Accession 4110–2, University of Washington Libraries, Seattle.
RRP6	Roger Randall Dougan Revelle Papers, MC 6, Archives of the Scripps Institution of Oceanography, La Jolla, Calif.
RRP6A	Roger Randall Dougan Revelle Papers, MC 6A, Archives of the Scripps Institution of Oceanography, La Jolla, Calif.
SIOS	Subject Files, Scripps Institution of Oceanography, La Jolla, Calif.
UNESCOR	Records of the United Nations Educational, Scientific and Cultural Organization, Paris.

Introduction

1. For Yablokov's report, see Government Commission on Matters Related to Radioactive Waste Disposal at Sea, *Facts and Problems Related to the Dumping of Radioactive Waste in the Seas Surrounding the Territory of the Russian Federation.*

For analyses of these revelations, see U.S. Congress Office of Technology Assessment, *Nuclear Wastes in the Arctic*, and Makhijani et al., *Nuclear Wastelands*, 285–392.

2. Peterson, *Troubled Lands*; Feschbach and Friendly, *Ecocide in the USSR*; Feshbach, *Ecological Disaster*. For a discussion of environmental impacts related particularly to the Soviet focus on production of nuclear power, see Josephson, *Red Atom*. Some scholars suggest that despite the apparently contrite attitudes of the government in the early years of the Russian Federation, many of the faults of the old regime have continued to flourish in the new regime. See Ziegler and Lyon, "The Politics of Nuclear Waste in Russia."
3. An outline of the history of sea dumping, giving details about each country's responsibility for introducing radionuclides into the oceans, can be found in Linsley, Sjöblom, and Cabianca, "Overview of Point Sources of Anthropogenic Radionuclides in the Oceans."
4. On fallout, see Divine, *Blowing on the Wind*; Kopp, "The Origins of the American Scientific Debate over Fallout Hazards." On reactors, see Mazuzan and Walker, *Controlling the Atom*; Carlisle, *Supplying the Nuclear Arsenal*. On the general scientific, production, and safety challenges faced by American and British atomic energy establishments, see Gowing and Arnold, *Independence and Deterrence*, and Hewlett and Duncan, *Atomic Shield*.
5. See Macfarlane, "Underlying Yucca Mountain."
6. See Lindee, *Suffering Made Real*; Walker, *Permissible Dose*.
7. On the cultural products elicited by the bomb, see Boyer, *By the Bomb's Early Light*; Weart, *Nuclear Fear*.
8. On early efforts to trace Strontium-90 in children's teeth, see Reiss, "Strontium-90 Absorption by Deciduous Teeth."
9. For an overview of this kind of exposure through ecological pathways, see Preston and Wood, "Monitoring the Marine Environment."
10. An overview of the reprocessing facilities built for weapons production during these decades, along with descriptions of waste disposal practices broken down by country, can be found in Makhijani et al., *Nuclear Wastelands*.
11. On the proliferation of radioisotopes after World War II, see Creager, "Tracing the Politics of Changing Postwar Research Practices"; Creager, "Nuclear Energy in the Service of Biomedicine."
12. See Hacker, *Elements of Controversy*; Mazuzan and Walker, *Controlling the Atom*; Walker, *Permissible Dose*.
13. The tendency of the AEC to focus on direct exposure rather than environmental exposure is discussed in Silverman, "No Immediate Risk." Daniel Grossman has argued that even when risks of environmental exposure were recognized, they were subordinated to production goals. See Grossman, "A Policy History of Hanford's Atmospheric Releases."
14. For an example, the process by which scientific truth should speak to power is outlined in Price, *The Scientific Estate*. For a discussion of the decline in public confidence in scientists to speak truth to power, see Jasanoff, "Science, Politics, and the Renegotiation of Expertise at EPA."
15. Peter Haas has employed the term "epistemic communities" to describe how knowledge-based experts arrive at a consensus and then help policy makers articulate and pursue their interests. This approach has generated lively discussion about how solutions to complex problems can be achieved at the international level. It is especially pertinent to environmental issues, because these touch on a variety of often-contradictory political interests and usually they entail a great deal of scientific uncertainty. See Haas, "Do Regimes Matter?" These ideas were

expanded in Haas, *Saving the Mediterranean*. See also the essays in *Knowledge, Power, and International Policy Coordination*, which is a special issue of *International Organization* 46, no. 1 (1992).

16. Lasse Ringius implores us to beware of tales that cast the formation of environmental regimes as products of power relations, self-interest, or scientific consensus. Where others see changes in scientific consensus and economic and political interests, he sees changes in environmental values and the mobilization of public opinion. Ringius leaves space for personal or institutional agency—for example, for a particularly active congressperson or scientist to be a "policy entrepreneur" mobilizing public views in favor of a particular point of view. See Ringius, *Radioactive Waste Disposal at Sea*.

Chapter 1 **Threshold Illusions**

1. Richard H. Fleming to James H. Jensen, 24 October 1952, folder "1949–1955," box 14, RHFP.
2. Richard H. Fleming to Allyn C. Vine, 20 March 1953, folder "1949–1955," box 14, RHFP.
3. Richard H. Fleming to Charles E. Renn, 17 June 1955, folder "1949–1955," box 14, RHFP.
4. Establishing acceptable fishing limits became the prime practical object of the International Council for the Exploration of the Sea, founded in 1902. See Rozwadowski, *The Sea Knows No Boundaries*.
5. Muller, "Artificial Transmutation of the Gene."
6. Muller, "The Production of Mutations."
7. "Science News."
8. Clark, *Radium Girls*.
9. Quinn, *Marie Curie*.
10. Welsome, *The Plutonium Files*.
11. Morgan, "The Responsibilities of Health-Physics," 93–94.
12. Ibid., 96.
13. Ibid.
14. James Forrestal to the President, 18 November 1946, folder 18.4, GBP.
15. See Beatty, "Genetics in the Atomic Age," and Lindee, *Suffering Made Real*.
16. Gregg's enthusiasm for genetics was linked to his view that human behavior had stronger genetic causes than social ones. See Paul, "The Rockefeller Foundation and the Origins of Behavior Genetics."
17. Draft Minutes, Advisory Committee for Biology and Medicine, 12 September 1947, folder 22.1, GBP.
18. C. J. Watson to Carroll Wilson, 21 January 1948, folder 22.1, GBP.
19. Draft Minutes, Advisory Committee for Biology and Medicine, 12 September 1947, folder 22.1, GBP.
20. Robert S. Stone to Alan Gregg, 4 November 1948, folder 22.2, GBP.
21. For this early criticism of the AEC, see Mazuzan and Walker, *Controlling the Atom*. For Lilienthal's comments, see page 18.
22. Draft Minutes, Advisory Committee for Biology and Medicine, 12 September 1947, folder 22.1, GBP.
23. Ibid.
24. Kevles, *The Physicists*, 66–67.
25. Lauriston S. Taylor, "The National Committee on Radiation Protection," October 1952, box 14, folder "US National Bureau of Standards," RHFP. See also Taylor, "Brief History of the National Committee on Radiation Protection and

Measurements (NCRP) Covering the Period 1929–1946." The early years of the NCRP are also discussed in Walker, *Permissible Dose*, 10–12.

26. Lauriston S. Taylor, "The National Committee on Radiation Protection," October 1952, box 14, folder "US National Bureau of Standards," RHFP.
27. Ibid.
28. Ibid. .
29. Draft minutes, Advisory Committee for Biology and Medicine, 12 September 1947, folder 22.1, GBP.
30. H. J. Muller to A. H. Sturtevant, 22 September 1954, folder 11.1, AHSP.
31. Such changes in terminology are emphasized in Walker, *Permissible Dose*, 10–12.
32. Hinton, "Weapons Using Radioactive Poison Pushed by Atomic Energy Board."
33. Grutzner, "Atomic Waste Pile Menace to Safety."
34. Press release, "U.S. Atomic Energy Commission Holds Seminar on Disposal of Radioactive Wastes," 30 January 1949, box 64, folder 723 (7–19–48), Entry 67A1, NA326.
35. Press release, "U.S. Atomic Energy Commission Holds Seminar on Disposal of Radioactive Wastes," 30 January 1949, box 64, folder 723 (7–19–48), Entry 67A1, NA326.
36. U.S. Atomic Energy Commission, "The Handling of Radioactive Waste Materials in the U.S. Atomic Energy Program," 22 September 1949, box 64, folder 723 (7–19–48), Entry 67A1, NA326.
37. "Press Conference on Radioactive Waste Disposal," Atomic Energy Commission, 28 January 1949, box 64, folder 723 (7–19–48), Entry 67A1, NA326.
38. E. I. Du Pont de Nemours & Co., Inc., "Summary of the Plutonium Project," 25 March 1949, box 47, folder 411.53, Entry 67A1, NA326.
39. This issue is addressed at greater length in Creager, "Nuclear Energy in the Service of Biomedicine."
40. "Radioactive Material 'Graveyards' Proposed." Details of human experimentation can be found in Welsome, *The Plutonium Files*.
41. See Westwick, *The National Labs*.
42. "Close Check Urged on Atomic Waste."
43. "Talks Scheduled on Disposal of Atomic Waste."
44. "Atomic 'Cemetery' Needed for Waste."
45. "Safe Disposal of Radioactive Wastes," "Dangers to Public in Radiation Noted."
46. Kaempffert, "Dangerous Radioactive Wastes, Which Now Are Buried, May Have Many Industrial Uses."
47. S. Allan Lough to various doctors, 19 July 1949, Tab F Early History, Briefing Book, Fifth Meeting, July 25–26, 1994, Washington, D.C., Advisory Committee on Human Radiation Experiments, National Security Archive, George Washington University.
48. "Guided Missile May Use Radioactive Waste."
49. Minutes of 27th AEC-MLC Meeting, 31 March 1948, NARA A015. Peter Westwick has pointed out that radiological warfare was one of the reasons for the expansion of biomedicine in the national laboratories. See Westwick, *The National Labs*, 241–242. For an analysis of the difficulties in adapting radioactive waste to radiological warfare, see Bruhèze, "Radiological Weapons and Radioactive Waste in the United States."
50. Atomic Energy Commission, Press Conference on the Eighth Semiannual Report of the Atomic Energy Commission, 28 July 1950, folder 22.6, GBP.
51. Hinton, "Weapons Using Radioactive Poison Pushed by Atomic Energy Board"; "Atomic Death Belt Urged for Korea"; "Atomic Belt Plan Held Not Feasible."

For the press conference, see Atomic Energy Commission, Press Conference on the Eighth Semiannual Report of the Atomic Energy Commission, 28 July 1950, folder 22.6, GBP.

52. "Atomic Death Belt Urged for Korea."
53. "Atomic Belt Plan Held Not Feasible."
54. Baldwin, "An Atomic Maginot Line."
55. For the early activities of Harwell, see Cockroft, "The Scientific Work of the Atomic Energy Research Establishment."
56. On the 1949 dumping operation, see "Radioactive Waste Disposal Planned" and "Atomic Wastes Dumped in Sea."
57. "Britain to Sink Atomic Wastes."
58. "Atomic Refuse Moon Dump Held Possible" and Kaempffert, "Atomic Ashes?"
59. Hinton, "Weapons Using Radioactive Poison Pushed by Atomic Energy Board"; Kaempffert, "How to Dispose of Deadly Radioactive Wastes Is a Difficult Problem for Atomic Plants." The transcript of the press conference is in Atomic Energy Commission, Press Conference on the Eighth Semiannual Report of the Atomic Energy Commission, 28 July 1950, folder 22.6, GBP.
60. "Clay 'Seal' Tested for Atom Wastes."
61. "Atomic Waste Made Safe."
62. These were National Bureau of Standards Handbooks 48 and 49, respectively. This and information on other subcommittees can be found in Lauriston S. Taylor, "The National Committee on Radiation Protection," October 1952, box 14, folder "US National Bureau of Standards," RHFP.
63. Ross Peavey to Lauriston S. Taylor, 18 July 1949, box 14, folder "1949–1955," RHFP.
64. Ibid.
65. Ibid.
66. Feitelberg is quoted in R. C. Peavey to Lauriston Taylor, 25 July 1949, box 14, folder "1949–1955," RHFP.
67. Sverdrup, Johnson, and Fleming, *The Oceans.*
68. Agenda for the meeting is attached to Lauriston S. Taylor to R. H. Fleming, 20 September 1949, box 14, folder "1949–1955," and a similar agenda describing Fleming's role is found in "Waste Disposal and Decontamination Committee Assignments," 3 November 1949, box 14, folder "US National Bureau of Standards National Committee on Radiation Protection," RHFP.
69. This summary is drawn from committee assignment reports, one dated 17 February 1950 and another undated but received by Fleming 10 November 1950, box 14, folder "US National Bureau of Standards National Committee on Radiation Protection," RHFP.
70. The complete references were J. N. C. Scott, "On the Proposed Investigations of the Dissipation of Effluent in the Sea," AERE T/M 1; "The Dissipation of Effluent in the Sea," Report on the Experiment at Sellafield, Harwell, AERE C/R-210; K. D. E. Johnson and J. Wilkinson, "The Absorption of Fission Products from Sea-Water, On Sea-Bed and Shore," AERE C/R-294. These citations are from G. W. Morgan to R. H. Fleming, 5 December 1949, box 14, folder "1949–1955," RHFP. The citation of Fuchs is K. Fuchs and J. N. C. Scott, "Investigation 'Wind Scale,'" T/N 17. This and citations of other authors such as J. S. Mitchell and H. J. Dunster are included in Walter H. Sullivan to James Jensen, 11 February 1952, box 14, folder "1949–1955," RHFP.
71. James H. Jensen to Walter Claus, 20 February 1950, box 14, folder "1949–1955," and Lauriston S. Taylor, General Communication No. 28, 10 September 1951, box 14, folder "US National Bureau of Standards," RHFP.

72. Minutes, Meeting of Subcommittee on Waste Disposal and Decontamination, National Committee on Radiation Protection, 31 January–1 February 1952, folder "1951–53," box 14, RHFP.
73. Ibid.
74. Ibid.
75. Lauriston S. Taylor to R. H. Fleming, 3 January 1952, box 14, folder "1949–1955," RHFP.
76. R. H. Fleming to James H. Jensen, 24 October 1952, box 14, folder "1949–1955," RHFP.
77. Ibid.
78. Minutes, Meeting of Subcommittee on Waste Disposal and Decontamination, 26–27 February 1953, folder "1951–53," box 14, RHFP.
79. Allyn C. Vine to Richard Fleming, 4 March 1953, and Richard H. Fleming to Allyn G. Vine, 20 March 1953, box 14, folder "1949–1955," RHFP.
80. Richard H. Fleming, "Factors Affecting the Disposal of Radioactive Wastes in the Ocean," Revised Draft, 28 April 1953, box 14, folder untitled, RHFP, pp. 1–2.
81. Ibid.
82. Ibid., 3–4.
83. Ibid., 3.
84. Ibid., 5–6.

Chapter 2 **Radiation Anxieties**

1. The meeting is described in Detlev Bronk to Harold Himsworth, 30 April 1956, folder "ORG: NAS: Coms on BEAR: Cooperation with other Organizations: Medical Research Council of Great Britain, 1955–1960," NAA.
2. Minutes of Meeting held at Dean Bradley House (Ministry of Works), 13 August 1946, AB 6/388.
3. Ibid.
4. See Green, "The Constitution and Functions of the United Kingdom Medical Research Council."
5. On this committee of Britain's Medical Research Council, see De Chadarevian, "Mice and the Reactor."
6. Minutes of Meeting held at Dean Bradley House (Ministry of Works), 13 August 1946, AB 6/388.
7. Proposals for Dealing with Effluent from Atomic Energy Research Establishment, Harwell, 24 October 1946, AB 6/388.
8. W. G. Marley to Director of AERE, 10 October 1946, AB 6/388.
9. Russian scientist A. Suslov, commenting on the ecological disaster zone around the Techa River, near the chemical reprocessing facility at Kyshtym, said that the "academicians of those times knew as much about the atom as ninth-graders do today." B. V. Nikipelov has pointed out that in the early years of chemical reprocessing, beginning around 1948, Soviet scientists had no practical knowledge or body of literature upon which to draw to ensure the safety of workers or the environment. See Donnay et al., "Russia and the Territories of the Former Soviet Union," 327.
10. H. Tongue to Lt. Col. A. J. Fisher, 8 August 1947, AB 6/388.
11. The public health concerns, including a description of the elaborate system developed by Tongue, are discussed in Gowing, *Independence and Deterrence*, 105–110.
12. H. Tongue to Lt. Col. A. J. Fisher, 11 April 1947, AB 6/181.

13. Initial plans for effluent pipeline at Windscale are discussed in several letters in the same file, including C. Hinton to Sir Charles G. Darwin, 7 June 1947, and C. C. Inglis to W. L. Owen, 16 July 1947, AB 16/500. The construction of the pipeline is discussed in Gowing, *Independence and Deterrence*, 107.
14. This overview is drawn from Dunster and Farmer, *A Summary of the Biological Investigations of the Discharges of Aqueous Radioactive Waste to the Sea from Windscale Works, Sellafield, Cumberland.* The early work is also discussed in Gowing, *Independence and Deterrence*, 106–107.
15. D. H. H. Peirson to W. Boyce, 27 May 1948, AB 16/637.
16. W. Boyce to D. E. H. Peirson, 3 June 1948, AB 16/637.
17. See J. D. Cockcroft to D. E. H. Peirson, 8 June 1948, AB 16/637; J. D. Cockcroft to D. E. H. Peirson, 30 July 1948, AB 16/637; J. D. Cockcroft to D. E. H. Peirson, 6 September 1948, AB 16/637; and other letters in the same file.
18. On Cockcroft's attempt to form a working party about ocean disposal, see J. D. Cockcroft to D. E. H. Peirson, 8 June 1948, AB 16/637; J. D. Cockcroft to D. E. H. Peirson, 30 July 1948, AB 16/637; J. D. Cockcroft to D. E. H. Peirson, 6 September 1948, AB 16/637; and other letters in the same file.
19. J. Diamond, Working Party on the Disposal of Radioactive Waste at Sea, Minutes, 16 February 1949, AB 6/181.
20. T. S. Leach to J. Diamond, 12 March 1949, AB 6/181.
21. R. M. Fishenden to D. E. Peirson, 21 March 1949, AB 16/637.
22. W. G. Marley to D. W. Cole, 8 March 1949, AB 6/181.
23. H. J. Blythe, "Summary of the First Disposal of Radio-active Waste at Sea," n.d., AB 6/604.
24. Ministry of Supply, Press Notice, 5 April 1949, AB 16/637.
25. H. J. Blythe, "Summary of the First Disposal of Radio-active Waste at Sea," n.d., AB 6/604.
26. L. C. Smyth to Secretary, Ministry of Supply, 6 April 1949, AB 6/181; W. G. Marley to General Manager, Western Union Telegraph Company, 20 April 1949, AB 6/181; W. G. Marley, "Working Party on the Disposal of Radioactive Waste at Sea, Report on Two Disposal Operations in 1949," n.d., AB 6/181.
27. On the offer of Admiralty ships, see D. E. H. Peirson to Secretary, Admiralty, 30 December 1949, AB 16/637.
28. On the firm's attitude toward its lucrative mail contracts, see W. C. Baxter to R. G. Elkington, [no date] January 1950, AB 16/637.
29. H. Tongue to A. H. Smales (and others), 29 August 1950, AB 6/741.
30. Minutes of meeting "to discuss arrangements for the disposal of effluent at Aldermaston," 16 March 1950, AB 6/180.
31. H. Tongue to G. E. Walker, 27 September 1950, AB 6/741.
32. "Suggested Final Form of Draft Letter," n.d., AB 6/874.
33. This attitude about secrecy and press notices is clear in much of the Ministry of Supply's correspondence and is spelled out explicitly in R. G. Elkington to R. E. France, 19 June 1951, AB 16/637.
34. K. D. Outteridge to W. G. Marley, 22 June 1951, AB 6/181.
35. H. Griffiths, "Report on the Disposal of Waste Material in Deep Water in the Atlantic on 29th June 1951," n.d., AB 6/181.
36. Minutes of meeting held in Chief Engineer's office, 17 July 1951, AB 6/388.
37. H. Griffiths, "Report on visit of the Chief Engineer to the Director of Armament Supply," date missing [July or August 1951], AB 6/388.
38. On the reaction of crews, see W. G. Marley, "Working Party on the Disposal of Radioactive Waste at Sea, Report on Two Disposal Operations in 1949," n.d., AB

6/181. On choosing the Hurd Deep, see R. G. Elkington to D. E. H. Peirson, 29 September 1950, AB 16/637.

39. G. W. Clare, "Appraisal of Sea Disposal Facilities open to Harwell and Aldermaston Establishments," n.d., AB 6/604.
40. W. G. Marley to R. H. Burns, 29 May 1952, AB 6/181.
41. R. H. Burns to Colonel N. Wilson, 23 July 1953, AB 6/741.
42. W. G. Marley, "Disposal of Radioactive Wastes for the Establishments of the United Kingdom Atomic Energy Authority," n.d., AB 16/1450. Britain's authorization procedures are discussed in Berkhout, *Radioactive Waste.*
43. Hewlett and Holl, *Atoms for Peace and War.*
44. Cutler was the president's Special Assistant for National Security Affairs. Memorandum of Discussion at the 236th Meeting of the National Security Council, Washington, 10 February 1955, in Glennon, *Foreign Relations of the United States*, 20–34.
45. See the president's comments to Secretary of the Treasury George Humphrey in Memorandum of Discussion at the 236th Meeting of the National Security Council, Washington, 10 February 1955, Glennon, *Foreign Relations of the United States*, 27.
46. Memorandum of Discussion at the 236th Meeting of the National Security Council, Washington, 10 February 1955, Glennon, *Foreign Relations of the United States*, 20–34.
47. Memorandum of Discussion at the 240th Meeting of the National Security Council, Washington, 10 March 1955, Glennon, *Foreign Relations of the United States*, 41–45; National Security Council Report, Glennon, *Foreign Relations of the United States*, 46–55.
48. Pfau, *No Sacrifice Too Great.*
49. See also Creager, "The Industrialization of Radioisotopes by the U.S. Atomic Energy Commission."
50. Divine, *Blowing on the Wind*, 3–8.
51. Ibid., 12–13.
52. Strauss is quoted in Sturtevant, "Social Implications of the Genetics of Man," 406.
53. "Hot Ashes" and "Scientist's Fears Answered."
54. R. A. Brink to A. H. Sturtevant, 28 June 1954, box 11.1, AHSP.
55. These letters can be found in box 11.1, AHSP.
56. H. J. Muller to A. H. Sturtevant, 22 September 1954, box 11.1, AHSP.
57. "Expert Belittles Peril in Radioactive Waste."
58. Ibid.
59. Lauriston S. Taylor to R. H. Fleming, 30 November 1954, box 14, folder "1949–1955," RHFP.
60. Robert Rugh to A. H. Sturtevant, 14 January 1955, folder 11.2, AHSP.
61. Floyd Mulkey (Howson Fellowship Bible Class, Woodlawn Methodist Church, Chicago) to A. H. Sturtevant, 17 January 1955, folder 11.2, AHSP.
62. H. J. Muller to Earl L. Green, 29 March 1955, and excerpt from quoted newspaper, folder 11.3, AHSP.
63. Earl L. Green to H. J. Muller, 4 April 1955, folder 11.3, AHSP.
64. A. H. Sturtevant to Earl L. Green, 25 April 1955, folder 11.3, AHSP.
65. H. J. Muller to Earl L. Green, 13 April 1955, folder 11.3, AHSP.
66. A. H. Sturtevant to Earl L. Green, 25 April 1955, folder 11.3, AHSP.
67. H. J. Muller to Bruce Wallace, 21 April 1955, folder 11.3, AHSP.
68. Department of State and Atomic Energy Commission, press release, 23 February 1955, "Plans Announced for United States Participation in Geneva Conference on Atomic Energy," folder 23.1, GBP.

69. The exclusion of H. J. Muller is discussed in greater detail in Kopp, "The Origins of the American Scientific Debate over Fallout Hazards."
70. M. G. Candau to Walter G. Whitman, 5 May 1955, and undated text attached entitled "The General Problems of Protection against Radiations from the Public Health Point of View," folder 620.992:539.16 "Atomic Radiations," UNESCOR.
71. The dye experiment was reported by H. Seligman in a paper entitled "The Discharge of Radioactive Waste Products into the Irish Sea, Part I: First Experiments for the Study of Movement and Dilution of Released Dye in the Sea." The permissible discharges were discussed in H. J. Dunster, "The Discharge of Radioactive Waste Products into the Irish Sea, Part II: The Preliminary Estimate of the Safe Daily Discharge of Radioactive Effluent." These are cited in Dunster and Farmer, *A Summary of the Biological Investigations of the Discharges of Aqueous Radioactive Waste to the Sea from Windscale Works, Sellafield, Cumberland.*
72. Memorandum of Discussion at the 261st Meeting of the National Security Council, Washington, 13 October 1955, Glennon, *Foreign Relations of the United States,* 211–212. This enthusiasm was tempered by the realization, gained largely from this 1955 conference, that the Soviet Union had progressed further in the field of atomic energy than expected. See Krige, "Atoms for Peace, Scientific Internationalism, and Scientific Intelligence."
73. Libby, "Radiochemistry," 121.
74. Cockcroft, "Future of Atomic Energy," 140, 141.
75. "Atom Power Curb Seen" and Laurence, "Atomic Residues Pose a Challenge."
76. Laurence, "Waste Held Peril in Atomic Power."
77. Plumb, "'Coming of Age' of Nuclear Technology Puts Emphasis on Dangers of Atomic Radiation."
78. "Atom Waste Put in Sea" and "Atomic Wastes."
79. Extract from AEA Meeting held on 4 November 1954, AB 16/1863.
80. Examples of coverage include "Atomic Waste Put in Sea."
81. E. H. Underwood to Mr. Forward, 2 February 1955, AB 16/1863.
82. E. N. Plowden to A. H. K. Slater, 2 February 1955, AB 16/1863.
83. A. H. K. Slater to Eric Underwood, 8 February 1955, handwritten note on Underwood's copy of A. H. K. Slater to Edwin Plowden, 4 February 1955, AB 16/1863.
84. J. D. Cockcroft to Edwin Plowden, 4 February 1955, AB 16/1863.
85. Lord Salisbury's comments are quoted in A. H. K. Slater to Edwin Plowden, 4 February 1955, AB 16/1863.
86. E. N. Plowden to John Cockcroft, 8 February 1955, AB 16/1863.
87. The new attitude and role for the public relations director is stated explicitly in D. C. V. Perrott to D. R. Willson, 22 February 1955, AB 16/1450.
88. An account of BEAR origins is given by Bronk in Transcript, Afternoon Session, Study Group on Disposal of Radioactive Wastes, 23 February 1956, folder "ORG: NAS: Coms on BEAR: Disposal and Dispersal of Radioactive Wastes: Meeting Transcript, February 1956," NAA. The quotation is taken from "NAS-Atom Rad Press Release, 8 April '55," folder "Pub Rel: General, 1955–1962," NAA.
89. Minutes of Second Meeting, Chicago, 5–6 February 1956, BEAR Genetics Panel, box 17.1, GBP.
90. Ibid. Beadle's opinions on this subject are also discussed in Berg and Singer, *George Beadle, An Uncommon Farmer,* chap. 14.
91. Minutes of Second Meeting, Chicago, 5–6 February 1956, BEAR Genetics Panel, box 17.1, GBP.
92. Ibid.

93. W. F. Libby to Detlev Bronk, 27 April 1955, folder "ORG: NAS: Coms on BEAR: Cooperation with other Organizations: Medical Research Council of Great Britain, 1955–1960," NAA.
94. Detlev Bronk to Harold Himsworth, 3 July 1956, folder "ORG: NAS: Coms on BEAR: Cooperation with other Organizations: Medical Research Council of Great Britain, 1955–1960," NAA.
95. Harold Himsworth to Detlev Bronk, 4 May 1955, folder "ORG: NAS: Coms on BEAR: Cooperation with other Organizations: Medical Research Council of Great Britain, 1955–1960," NAA.
96. Harold Himsworth to Detlev Bronk, 6 February 1956, and Detlev Bronk to Harold Himsworth, 18 February 1956, folder "ORG: NAS: Coms on BEAR: Cooperation with other Organizations: Medical Research Council of Great Britain, 1955–1960," NAA.
97. Warren Weaver to Harold Himsworth, 6 April 1956, folder "ORG: NAS: Coms on BEAR: Cooperation with other Organizations: Medical Research Council of Great Britain, 1955–1960," NAA.
98. Ibid.
99. Warren Weaver to Harold Himsworth, 9 April 1956, folder "ORG: NAS: Coms on BEAR: Cooperation with other Organizations: Medical Research Council of Great Britain, 1955–1960," NAA.
100. Detlev Bronk to Harold Himsworth, 30 April 1956, folder "ORG: NAS: Coms on BEAR: Cooperation with other Organizations: Medical Research Council of Great Britain, 1955–1960," NAA.
101. Harold Himsworth to Detlev Bronk, 10 May 1956 (a), folder "ORG: NAS: Coms on BEAR: Cooperation with other Organizations: Medical Research Council of Great Britain, 1955–1960," NAA.
102. Harold Himsworth to Detlev Bronk, 10 May 1956 (b), folder "ORG: NAS: Coms on BEAR: Cooperation with other Organizations: Medical Research Council of Great Britain, 1955–1960," NAA.
103. Harold Himsworth to Detlev Bronk, 13 June 1956, folder "ORG: NAS: Coms on BEAR: Cooperation with other Organizations: Medical Research Council of Great Britain, 1955–1960," NAA.
104. Ibid.
105. The MRC report did not explicitly set thresholds of safety, and indeed it emphasized the cumulative dangers from radiation while pointing out that the dangers from fallout were negligible compared to radiation from natural sources. For the impact of the MRC's report in Britain, see Chadarevian, "Mice and the Reactor." The ambiguous reception of geneticists' no-threshold views, including the objections of physicians, is analyzed in detail in Jolly, "Thresholds of Uncertainty."

Chapter 3 **The Other Atomic Scientists**

1. Roger Revelle, "The Scientist and the Politician," 1957 Charter Address, University of California, Riverside, 22 March 1957, box 29, folder 1, RRP6.
2. On Revelle during these years, see Ronald Rainger, "Patronage and Science."
3. Proceedings, Study Group on Oceanography and Fisheries, National Academy of Sciences, 3 March 1956, Princeton Inn, Princeton, N.J., NAA, p. 49.
4. On Donaldson's long experience studying salmon under AEC patronage, see Klingle, "Plying Atomic Waters."
5. "Project Chronology Chart and Summary of the Applied Fisheries Laboratory," n.d. [coverage ends in 1948], box 7, folder "Project Files," LRDP.

6. Lauren R. Donaldson to W. M. Chapman, 6 June 1947, box 2, folder "Chapman, Dr. W. M., 1947–1950," LRDP.
7. On the sardine fisheries, see McEvoy and Scheiber, "Scientists, Entrepreneurs, and the Policy Process."
8. Ronald Rainger makes this point more explicitly, saying that American oceanographers such as Revelle "embedded or incorporated their interests within a military framework." See Rainger, "Science at the Crossroads."
9. This thesis about oceanographers' patronage strategies is developed more fully in Hamblin, *Oceanographers and the Cold War.*
10. Richard H. Fleming to Charles E. Renn, 17 June 1955, box 14, RHFP.
11. Tentative agenda, conference entitled "Disposal of Radioactive Waste at Sea," held at Woods Hole Oceanographic Institution, 22–23 June 1955, box 14, RHFP.
12. Bostwick H. Ketchum to Edward Smith, 30 December 1955, folder "Inst: AEC," box 21, ODWHOI.
13. Alfred C. Redfield to Walter D. Claus, 21 December 1955; Bostwick H. Ketchum, "The Disposal of Fission Product Wastes at Sea: Hydrographic, Geochemical and Biological Problems in a Coastal Area," n.d., folder "Inst: AEC," box 21, ODWHOI.
14. Allyn Vine, "Memorandum on the Disposal of Radioactive Waste," 27 February 1956, folder "Pers: Vine, Allyn," box 24, ODWHOI.
15. Ibid.
16. On the promising use of radioisotopes, see Creager, "Nuclear Energy in the Service of Biomedicine;" and Santesmases, "Peace Propaganda and Biomedical Experimentation."
17. For an overview of the importance of radioactivity in shaping American oceanography, particularly with regard to Revelle's attitudes, see Rainger, "'A Wonderful Oceanographic Tool.'"
18. R. Revelle, T. R. Folsom, E. D. Goldberg, and J. D. Isaacs, "Nuclear Science and Oceanography," contribution from the Scripps Institution of Oceanography to the International Conference on the Peaceful Uses of Atomic Energy, 30 June 1955, box 13, RHFP.
19. Ibid.
20. Hillaby, "Sea May Become Atom Waste Site."
21. Kaempffert, "Problem of Radioactive Wastes" and Hillaby, "Seas May Become Atom Waste Site."
22. Everett, "Eugen Glueckauf."
23. The point about the conference's scientific blandness amid technological fanfare is made in Krige, "Atoms for Peace, Scientific Internationalism, and Scientific Intelligence."
24. Robert W. Hiatt, Howard Boroughs, Sidney J. Townsley, and Geraldine Kau, "Radioisotope Uptake in Marine Organisms with Special Reference to the Passage of Such Isotopes as Are Liberated from Atomic Weapons through Food Chains Leading to Organisms Utilized as Food by Man," USEAC Contract AT(04–3)-56, Annual Report 1954–55, 1 September 1955, box 13, folder "1955–59," RHFP.
25. Transcript of press conference, "The Biological Effects of Atomic Radiation, June 12, 1956," folder "ADM: Public Relations: Press Conferences: June 1956," NAA, p. 39.
26. Roger Revelle to Richard Fleming, 11 February 1956, box 13, folder "1956–59," RHFP.
27. Hacker, *Elements of Controversy*, 148–151. For the ABCC, see Lindee, *Suffering Made Real.*

28. Proceedings, Study Group on Oceanography and Fisheries, National Academy of Sciences, 4 March 1956, Princeton Inn, Princeton, N.J., NAA, p. 340.
29. Miyake, Sugimura, and Kameda, "On the Distribution of Radioactivity in the Sea Around Bikini Atoll in June 1954."
30. The experience during these years of one of Miyake's students, Katsuko Saruhashi, is told briefly in Normile, "A Prize of One's Own."
31. More details of TRANSPAC and NORPAC can be found in Hamblin, *Oceanographers and the Cold War.*
32. Diary of Japanese Trip by Lill and Revelle, 26 June 1955–10 July 1955, box 138, folder 15, RRP6A.
33. Roger Revelle, outline of presentation "Science and Natural Resources," n.d., box 138, folder 15, "Japanese Trip 6 & 7 1955, Revelle & Lill," RRP6A.
34. Roger Revelle, outline of speech, "Science and Natural Resources," n.d. (speech given July 1955), box 138, folder 15, RRP6A
35. Ibid.
36. Roger Revelle, "The Scientist and the Politician," 1957 Charter Address, University of California, Riverside, 22 Mar 1957, box 29, folder 1, RRP6.
37. Report of this telephone call is in R. G. Paquette to Richard Fleming, 19 July 1955, box 4, RHFP.
38. Warren S. Wooster to Kanji Suda, 28 June 1956, AC 6, box 7, folder 47, SIOS.
39. The political facets of NORPAC and EQUAPAC in relation to radioactivity are discussed briefly in Rainger, "'A Wonderful Oceanographic Tool.'"
40. Proceedings, Study Group on Oceanography and Fisheries, National Academy of Sciences, 4 March 1956, Princeton Inn, Princeton, N.J., NAA, p. 341–346.
41. Proceedings, Study Group on Oceanography and Fisheries, National Academy of Sciences, 4 March 1956, Princeton Inn, Princeton, N.J., NAA, p. 341–346.
42. Hacker, *Elements of Controversy*, 170–171.
43. Proceedings, Study Group on Oceanography and Fisheries, National Academy of Sciences, 4 March 1956, Princeton Inn, Princeton, N.J., NAA, p. 352–362. A discussion of Revelle's work with AEC can be found in Bruno, "The Bequest of the Nuclear Battlefield."
44. Proceedings, Study Group on Oceanography and Fisheries, National Academy of Sciences, 3 March 1956, Princeton Inn, Princeton, N.J., NAA, p. 30.
45. Ibid., p. 31–32.
46. Ibid.
47. Ibid., 33–37.
48. Ibid., 38.
49. Ibid., 39, 43.
50. Ibid., 44.
51. Ibid., 45.
52. Ibid., 48.
53. Proceedings, Study Group on Oceanography and Fisheries, National Academy of Sciences, 4 March 1956, Princeton Inn, Princeton, N.J., NAA, p. 363–365.
54. Ibid., 365–370.
55. Ibid., 373.
56. Proceedings, Study Group on Oceanography and Fisheries, National Academy of Sciences, 3 March 1956, Princeton Inn, Princeton, N.J., NAA, p. 49.
57. Ibid., 49.
58. Ibid., 50.
59. Proceedings, Study Group on Oceanography and Fisheries, National Academy of Sciences, 5 March 1956, Princeton Inn, Princeton, N.J., NAA, p. 456.
60. Ibid., 462.

61. Ibid., 458–460.
62. "General Summary," folder, "ADM: ORG: NAS: Coms on BEAR: Oceanography and Fisheries: Summary Reports: Drafts," NAA, p. 2–3.
63. Ibid., 5–6.
64. Ibid., 7.
65. Ibid. 12.
66. "Oceanography, Fisheries, and Atomic Radiation," 13.

Chapter 4 ***Forging an International Consensus***

1. Transcript of press conference, "The Biological Effects of Atomic Radiation, June 12, 1956," folder "ADM: Public Relations: Press Conferences: 1956 Jun," NAA. P. 31–36.
2. Ibid.
3. Department of State Memorandum of Conversation, Detlev Bronk, Douglas Cornell, George C. Spiegel, 25 October 1955, folder "ADM: ORG: NAS: Coms on BEAR: Cooperation with Other Organizations: Department of State," NAA.
4. Ibid.
5. Detlev Bronk to Gerard Smith, 11 January 1957, folder "Cooperation with Other Organizations: Department of State," NAA.
6. Ibid.
7. John Cockcroft to Roger Revelle, 10 July 1956, folder "Foreign Govts, 1956–1959," NAA.
8. J.A.V. Willis (British Embassy, Washington, D.C.) to Roger Revelle, 20 June 1956, box 13, folder "Conferences," RHFP
9. "Report of a Meeting of United Kingdom and United States Scientists on Biological Effects of Radiation in Oceanography and Fisheries," 31 October 1956, folder "ADM: ORG: NAS: Coms on BEAR: Oceanography and Fisheries: Meetings: Minutes: 1956 Sep," NAA, p. 2.
10. Ibid.
11. Ibid.
12. Ibid., 3–4.
13. Ibid., 3.
14. Ibid., 4–6.
15. Philip J. Farley, memorandum for Charles L. Dunham and Edward E. Gardner (both of the Atomic Energy Commission), 10 October 1956; "Report of a Meeting of United Kingdom and United States Scientists on Biological Effects of Radiation in Oceanography and Fisheries," 31 October 1956, folder "ADM: ORG: NAS: Coms on BEAR: Oceanography and Fisheries: Meetings: Minutes: 1956 Sep," NAA, p. 8.
16. By the late 1950s most radiation protection bodies, including the International Commission on Radiological Protection, identified internally absorbed radioisotopes as a significant subject to be distinguished from external radiation exposure. See "Recommendations of the International Commission on Radiological Protection."
17. Gunnar Randers to Sir (form letter to Member States), 9 April 1956, folder 620.992.539.16 "Atomic Radiations (incl. UN Scientific Committee on the Effects of Atomic Radiation)," UNESCOR.
18. W. A. Higinbotham to Detlev Bronk, 14 April 1955, folder "Pub Rel: General, 1955–1962," NAA.
19. Charles L. Dunham to George Beadle, 10 December 1956, box 17.5, GBP.
20. George Beadle to Charles L. Dunham, 17 December 1956, box 17.5, GBP.

21. "Excerpt from report by Paul Weiss on the 12th General Assembly of IUBS held in Rome, Italy, April 12–16, 1955," n.d., folder "IR: 1955–1960," NAA.
22. International Council of Scientific Unions, Seventh General Assembly (9–12 August 1955), Resolution, n.d., folder 620.992.539.16, UNESCOR. On ICSU, see Greenaway, *Science International.*
23. Strauss's "packed jury" comment can be found in Memorandum of a Conversation, United States Mission at the United Nations, New York, 20 May 1955, Glennon, *Foreign Relations of the United States*, 90–92.
24. On the State Department's views, see Telegram from the Department of State to the Embassy in the United Kingdom, 10 May 1955, Glennon, *Foreign Relations of the United States*, 75.
25. Sven Hörstadius, "Special Committee on the Biological Effects on Nuclear Radiation, Report of ad hoc Committee," 16 May 1956, folder 620.992.539.16 "Atomic Radiations (incl. UN Scientific Committee on the Effects of Atomic Radiation)," UNESCOR.
26. G. G. Brown, minute, 31 October 1956; Sir Pierson Dixon to Foreign Office, 30 October 1956, FO 371/123132.
27. Nestore B. Cacciapuoti to Director-General (Luther Evans), 7 November 1956, folder 620.992.539.16 "Atomic Radiations (incl. UN Scientific Committee on the Effects of Atomic Radiation)," UNESCOR.
28. These requests for information are repeated in H. Charnock to Director, Department of Natural Sciences, UNESCO, 15 October 1957, Folder 620.992:539 16 Part II, UNESCOR.
29. "Oceanic Disposal of Radioactive Wastes," Interim Report to United Nations Special Committee on the Effects of Atomic Radiation—April 1957," n.d., Folder 620.992:539 16 Part II, UNESCOR.
30. M. Yoshida to M. Fontaine, 15 March 1957; "Ocean Disposal of Radioactive Wastes: Interim Report to United Nations Special Committee on the Effects of Atomic Radiation—April 1957," n.d., folder 620.992.539.16 "Atomic Radiations (incl. UN Scientific Committee on the Effects of Atomic Radiation)," UNESCOR.
31. Henry Charnock to Director, UNESCO, 15 October 1957, Folder 620.992:539 16 Part II, UNESCOR.
32. Henry Charnock, draft, "Sea and Ocean Disposal of Radioactive Wastes," n.d., Folder 620.992:539 16 Part II, UNESCOR.
33. Ibid.
34. Ibid.
35. Henry Charnock to The Director, Department of Natural Sciences, UNESCO, 15 October 1957; Henry Charnock, "Sea and Ocean Disposal of Radioactive Wastes," n.d., folder 620.992.539.16 "Atomic Radiations (incl. UN Scientific Committee on the Effects of Atomic Radiation)," UNESCOR.
36. G. L. Kesteven to Henry Charnock, 8 March 1957, Folder 620.992:539 16 Part II, UNESCOR.
37. G. L. Kesteven to Henry Charnock, n.d., Folder 620.992:539 16 Part II, UNESCOR.
38. I. S. Eve to Masao Yoshida, 18 November 1957, Folder 620.992:539 16 Part II, UNESCOR.
39. G. L. Kesteven [FAO] to Henry Charnock, n.d.; I. S. Eve [WHO] to Henry Charnock, 18 November 1957; I. S. Eve to M. Yoshida, 18 November 1957, n.d., folder 620.992.539.16 "Atomic Radiations (incl. UN Scientific Committee on the Effects of Atomic Radiation)," UNESCOR.
40. Harold Himsworth to Detlev Bronk, 7 July 1958, folder "ADM: ORG: NAS: Coms on BEAR: Cooperation with Other Organizations: Medical Research

Council of Great Britain"; National Academy of Sciences—National Research Council, Press Release, 10 August 1958, folder "ADM: PUB Rel: Press Releases, 1955–1960," NAA.

41. Josephson, *Red Atom*, 20–21.
42. United States Official Representatives, "Supplementary Report on International Conference on the Peaceful Uses of Atomic Energy," n.d. [August 1955], Entry 67B1, box 69, folder "Organization and Management: International Scientific Congress (BP), NA326. On Lysenko's influence, see Joravsky, *The Lysenko Affair.*
43. United States Official Representatives, "Supplementary Report on International Conference on the Peaceful Uses of Atomic Energy," n.d. [August 1955], Entry 67B1, box 69, folder "Organization and Management: International Scientific Congress (BP), NA326.
44. An overview of the implications of this crisis can be found in Kunz, *The Economic Diplomacy of the Suez Crisis.*
45. Goldschmidt, *The Atomic Complex*, 264–272.
46. Ibid., 137.
47. Estimates about energy consumption, including British predictions, are in Hoffman, "The Role of Nuclear Power in Europe's Future Energy Balance."
48. "Treaty Establishing the European Atomic Energy Community (EURATOM)," 955.
49. "Statute of the International Atomic Energy Agency," 467.
50. Alexander, "The Peaceful Atom Organization."
51. These developments are discussed in greater detail in Fischer, *History of the International Atomic Energy Agency.*
52. "Radioactive Drum Sighted at Sea"; "Drifting Atomic Waste Can Sunk"; "Atom Waste Adrift; Shipping Menaced."
53. Reston, "A Peril of the Atomic Age."
54. Smith, "Detroit Reactor Called Best Bet."
55. "German Town Boycotts Vote."
56. Wittner, *Resisting the Bomb*, 107–124.
57. Arnold, *Windscale 1957*, 47–52.
58. Ibid., 55.
59. Ibid., 62.
60. Ibid., 80–83.
61. Ibid., 95.
62. Several documents in this folder explain the situation, most clearly in an unsigned (quoted) letter to Mr. Wickenden (Ministry of Housing and Local Government), n.d., PRO B298. The figure of 5 percent, and a comment about the need to avoid undermining the authorization procedure, is given in Wickenden to Williams, 18 July 1958, AB 6/2547.
63. Huw Howells to H. J. Dunster, 20 October 1958, AB 6/2547.
64. J. A. Dixon to H. J. Dunster, 5 November 1958, AB 6/2547.

Chapter 5 **No Atomic Graveyards**

1. For this episode and more details about the oceanographic work during the IGY, see Hamblin, *Oceanographers and the Cold War.*
2. "Atomic Refuse Leaves Pacific Unaffected."
3. "Current Is Found Far Down in Sea."
4. "Oceanographers Split." See also "'Vityaz' Again Investigates Dangers of Dumping Atomic Waste," translated extract from *Soviet News* of 11 November 1958, AB 16/2161.

5. "Sea Canyons Held Poor Atom Ash Cans."
6. Polikarpov, "Ability of Some Black Sea Organisms to Accumulate Fission Products."
7. U.S. Congress Office of Technology Assessment, *Nuclear Wastes in the Arctic*, 36–45.
8. Finney, "Atom Talks End on Hopeful Note." Also see Josephson, *Red Atom*, 175.
9. "Fusion Engine Seen."
10. Josephson, *Red Atom*, 175–176.
11. "Report by the United States Official Representatives to the Second United Nations International Conference on the Peaceful Uses of Atomic Energy, Geneva, Switzerland, September 1–13, 1958," n.d., AB 54/73.
12. A recent of study of Plowshare is Kirsch, *Proving Grounds*.
13. Finney, "Russian Charges U.S. Is Deceptive in Atom Proposal."
14. "Atomic Politics in Geneva."
15. Finney, "Soviet Discloses Biggest Reactor."
16. Finney, "Radiation Curbs Urged by Dutch."
17. "U.S. Offers to Pay for Atom Studies."
18. "Russia Assails U.N. Atomic Agency."
19. This paper by Kreps is found in Richard Fleming's papers. E. M. Kreps, "The Problem of Radioactive Contamination of Ocean and Marine Organisms," n.d., box 13, folder "1955–1959," RHFP.
20. E. M. Kreps, "The Problem of Radioactive Contamination of Ocean and Marine Organisms," n.d., box 13, folder "1955–1959," RHFP.
21. Ibid.
22. "Report by the United States Official Representatives to the Second United Nations International Conference on the Peaceful Uses of Atomic Energy, Geneva, Switzerland, September 1–13, 1958," n.d., AB 54/73.
23. For a then-contemporary discussion of UNCLOS, see Swan and Ueberhorst, "The Conference on the Law of the Sea: A Report." On the relevance of UNCLOS to the IAEA's recommendations, see Linsley, Sjöblom, and Cabianca, "Overview of Point Sources of Anthropogenic Radionuclides in the Oceans."
24. Harrison Brown to Detlev Bronk, 8 January 1958; Detlev Bronk to Harrison Brown, 3 February 1958, folder "ADM: ORG: NAS: Coms on BEAR: Oceanography and Fisheries: Cooperation with NRC Com on Oceanography," NAA.
25. Arnold B. Joseph to Richard C. Vetter, 19 June 1956, folder "ES: Com on Ocean: Subcommittee on Radioactive Waste Disposal into Atlantic and Gulf Coastal Waters, General, 1958," NAA. On *Savannah*'s waste control system, see Resner, Schmidt, and Van Turner, "Health Physics for the N.S. Savannah."
26. James M. Smith Jr., "Draft of Information for Working Group on Nuclear Ship Waste Disposal," 20 November 1958, folder "ES: Com on Ocean: Subc on Rad Waste Disp from Nuclear-Powered Ships: General, 1958," NAA.
27. Douglas L. Worf to Richard C. Vetter, 18 June 1956, folder "ES: Com on Ocean: Subc on Rad Waste Disp from Nuclear-Powered Ships: General, 1958," NAA.
28. Ibid.
29. Richard C. Vetter to Roger Revelle, 1 August 1958, folder "ADM: ORG: NAS: Coms on BEAR: Oceanography and Fisheries: Cooperation with NRC Com on Oceanography, 1958–1961," NAA.
30. U.S. National Academy of Sciences, *Considerations on the Disposal of Radioactive Wastes from Nuclear-Powered Ships into the Marine Environment*.
31. Ibid., emphasis in original.
32. Schaefer, "Some Fundamental Aspects of Marine Ecology in Relation to Radioactive Waste."

33. Pritchard, "Disposal of Radioactive Wastes in the Ocean."
34. Dunster and Farmer, *A Summary of the Biological Investigations of the Discharges of Aqueous Radioactive Waste to the Sea from Windscale Works, Sellafield, Cumberland.*
35. In the litany of international bodies typically used to give policy statements authoritative backing, the ICRP was the one used most frequently. However, this group did not differ significantly from the bodies in the United States and Britain. Through most of the 1950s, the membership of the main commission was dominated by American and British scientists (seven of thirteen members), with one Canadian and one Western European scientist making up the rest. The same figures who formulated policies in the United States, for example—National Committee on Radiation Protection chairman Lauriston S. Taylor, Oak Ridge health physicist Karl Z. Morgan, physician Robert S. Stone, radiobiologist Gioacchino Failla—sat on the commission. American and British health physicists visiting the various committees of the ICRP would have found many more familiar faces. In the 1950s, only one Soviet scientist, M. N. Pobedisnki, was a member of any of the commission's committees. The committee memberships from 1953 to 1959 are listed in "Recommendations of the International Commission on Radiological Protection."
36. Dunster and Farmer, *A Summary of the Biological Investigations of the Discharges of Aqueous Radioactive Waste to the Sea from Windscale Works, Sellafield, Cumberland.*
37. Claus, "What Is Health Physics?"
38. W. L. Templeton, "Report on a Visit to the U.S.A. and Canada: Biological Aspects of Waste Disposal," 5 May 1959, PRO B363. The role of the AEC in stimulating ecological research in the United States is discussed in Hagen, *An Entangled Bank*, and Bocking, *Ecologists and Environmental Politics*.
39. W. L. Templeton, "Report on a Visit to the U.S.A. and Canada: Biological Aspects of Waste Disposal," 5 May 1959, AB 7/8931.
40. Ibid.
41. Smith, "'Hot Cargo' Gets a One-Way Sea Trip."
42. "Fears 'Hot' Shellfish."
43. Arnold Joseph, Memorandum to Richard C. Vetter, 10 January 1958, folder "ES: Com on Ocean: Subcommittee on Radioactive Waste Disposal into Atlantic and Gulf Coastal Waters, General, 1958," NAA.
44. Ibid.
45. Ibid.
46. Ibid.
47. "Sea Areas Picked for Atom Wastes."
48. Thomas N. Downing to John A. McCone, 25 June 1959, folder "ES: Com on Ocean: Subc. On Rad Waste Disposal into Atlantic and Gulf Coastal Waters, 1959," NAA.
49. R. Bennett to Thomas N. Downing, 7 July 1959, folder "ES: Com on Ocean: Subc. On Rad Waste Disposal into Atlantic and Gulf Coastal Waters, 1959," NAA.
50. A. B. Metsger to Bob Casey, 1 July 1959, folder "ES: Com on Ocean: Subc. On Rad Waste Disposal into Atlantic and Gulf Coastal Waters, 1959," NAA.
51. On the dispute about the commercial license for the Gulf of Mexico, see Mazuzan and Walker, *Controlling the Atom*, 358–359.
52. Bob Casey to Richard C. Vetter, 22 July 1959, folder "ES: Com on Ocean: Subc. On Rad Waste Disposal into Atlantic and Gulf Coastal Waters, 1959," NAA.
53. M. Glenn to Hugh L. Dryden, 20 July 1959, folder "ES: Com on Ocean: Subc. On Rad Waste Disposal into Atlantic and Gulf Coastal Waters, 1959," NAA.
54. Radio Reports, Inc., "Interviews with Two Congressmen," 10 July 1959, John McCone Files, box 36, folder "MAT 12, Radioactive Waste and Waste Disposal, 1959 File," NA326.

55. Mazuzan and Walker, *Controlling the Atom*, 361.
56. Statement by A. R. Luedecke, General Manager, U.S. Atomic Energy Commission, before the Special Sub-Committee on Radiation of the Joint Committee on Atomic Energy of the Congress, 29 July 1959, John McCone Files, box 36, folder "MAT 12, Radioactive Waste and Waste Disposal, 1959 File," NA326.
57. Ibid.
58. Statement by A. R. Luedecke, General Manager, U.S. Atomic Energy Commission, before the Special Sub-Committee on Radiation of the Joint Committee on Atomic Energy of the Congress, 29 July 1959, John McCone Files, box 36, folder "MAT 12, Radioactive Waste and Waste Disposal, 1959 File," NA326.
59. Draft letter from Dr. McHugh to Mr. Miller, 23 July 1959, folder "ES: Com on Ocean: Subc. On Rad Waste Disposal into Atlantic and Gulf Coastal Waters, 1959," NAA.
60. Harrison Brown to Dayton Carritt, 25 July 1959; Detlev Bronk to S. Douglas Cornell, 31 July 1959, folder "ES: Com on Ocean: Subc. On Rad Waste Disposal into Atlantic and Gulf Coastal Waters, 1959," NAA.
61. Press release, from the Office of the Joint Committee on Atomic Energy, 3 September 1959, folder "CONGRESS: Committees: Atomic Energy: Joint Hearings: Summary Analysis, 1959," NAA.
62. Mazuzan and Walker, *Controlling the Atom*, 363.
63. Minutes, Meeting of Chairmen of BEAR Committees, 8 October 1959, folder "ADM: ORG: NAS: Coms on BEAR: Meetings: Chairmen of BEAR Coms," NAA.
64. Mazuzan and Walker, *Controlling the Atom*, 364.
65. On the NASCO report, see Hamblin, *Oceanographers and the Cold War*, chap. 5.
66. National Academy of Sciences—National Research Council, *Oceanography 1960 to 1970: A Report by the Committee on Oceanography*. Washington, D.C.: National Academy of Sciences—National Research Council, 1959.
67. Arnold B. Joseph to Richard C. Vetter, 19 June 1956, folder "ES: Com on Ocean: Subcommittee on Radioactive Waste Disposal into Atlantic and Gulf Coastal Waters, General, 1958," NAA.
68. Ibid.
69. John D. Isaacs to Pacific Coast Committee, 21 December 1959, folder "ES: Com on Ocean: Subc on Disposal of Low-Level Rad Waste into Pacific Coastal Waters: Report: Review, 1959," NAA.
70. John D. Isaacs to Members, Pacific Coast Committee on Sea Disposal of Radioactive Waste, 10 June 1960, folder "ES: Com on Ocean: Subc on Disposal of Low-Level Rad Waste into Pacific Coastal Waters: Rept: Review: General: 1960," NAA.
71. Arnold B. Joseph to John Isaacs, 22 September 1960, folder "ES: Com on Ocean: Subc on Disposal of Low-Level Rad Waste into Pacific Coastal Waters: Rept: Review: General: 1960," NAA.
72. John D. Isaacs to Ralph S. O'Leary, 15 March 1961, folder "ES: Com on Ocean: Subc on Disposal of Low-Level Rad Waste into Pacific Coastal Waters: Rept: Review: 1961," NAA.
73. Arnold B. Joseph to Roger Revelle, 20 July 1961, folder "ES: Com on Ocean: Subc on Disposal of Low-Level Rad Waste into Pacific Coastal Waters: Rept: Review: 1961," NAA.
74. Arnold B. Joseph, "Comments on Report of Pacific Coast Subcommittee on Sea Disposal of Low Level Radioactive Waste," revised April 1961, folder "ES: Com on Ocean: Subc on Disposal of Low-Level Rad Waste into Pacific Coastal Waters: Rept: Review: 1961," NAA.
75. Ibid.
76. Ibid.

77. Ibid.
78. R. R. Rubottom Jr., Assistant Secretary of State (signing for Secretary of State) to John A. McCone, Chairman, Atomic Energy Commission, 18 November 1959, AB 16/3002.
79. "Oral Report of Dr. Milton Eisenhower on His Trip to Mexico, August 12 to 27, 1959," 10 September 1959, Country Files, box 427, folder 21.61, NA59.
80. R. R. Rubottom Jr., Assistant Secretary of State (signing for Secretary of State) to John A. McCone, Chairman, Atomic Energy Commission, 18 November 1959, AB 16/3002.
81. Lowenstein, "Some Legal Considerations in the Ocean Disposal of Radioactive Wastes."
82. Ibid.
83. A. R. Luedecke to Chairman McCone [and others], 15 April 1960; and "Burial and Disposal of Low-level Radioactive Wastes," n.d., John McCone Files, box 36, folder "MAT 12, Radioactive Waste and Waste Disposal," NA326.
84. The decision to stop issuing new licenses is reported in Howard C. Brown Jr. to Chairman McCone, 20 June 1960, John McCone Files, box 36, folder "MAT 12, Radioactive Waste and Waste Disposal," NA326. Coastwise Marine also embarrassed the AEC that year with some questionable business practices and an explosion at its facilities. This and the decision to stop issuing new licenses are discussed briefly in Mazuzan and Walker, *Controlling the Atom*, 364.
85. On Mexico, see "A.E.C. Bars Waste Plan." On the claim that the AEC stopped dumping at sea after the furor over the academy's report, see U.S. Congress, *Ocean Dumping of Waste Materials*, 238.
86. Glenn T. Seaborg to Nabor Carrillo Flores, 2 November 1961, Country Files, box 427, folder 21.61, NA59.

Chapter 6 **The Environment as Cold War Terrain**

1. The "authorization" process in Britain is discussed, along with other pertinent waste disposal issues, in Berkhout, *Radioactive Waste*.
2. *Disposal of Low-Level Radioactive Waste into Pacific Coastal Waters.*
3. H. J. Dunster to Roger Revelle, 17 July 1961, AB 54/16.
4. A.H.K. Slater, Note for Record, 5 June 1959, AB 16/3000.
5. H. J. Dunster to J. M. Hartog, "International Agency Conference on Disposal of Nuclear Waste," 20 April 1959, AB 16/3000.
6. W. G. Marley to J. F. Jackson, "International Atomic Energy Conference on Disposal of Radioactive Wastes," 27 April 1959, AB 16/3000.
7. H. J. Dunster to J. M. Hartog, "International Agency Conference on Disposal of Nuclear Waste," 20 April 1959, AB 16/3000.
8. H. J. Dunster to A.H.K. Slater, 19 October 1959, AB 16/3001.
9. Sterling Cole to Secretary of State for Foreign Affairs, Foreign Office, 24 March 1959, AB 16/3000.
10. P. W. Ridley to F. J. Ward, 28 August 1959, AB 16/3000.
11. J. C. Walker to A.H.K. Slater, 12 October 1959, AB 16/3001.
12. H. J. Dunster to J. H. Hartog, 20 April 1959, AB 16/3000.
13. "Note by Sir John Cockcroft," 7 December 1959, AB 16/3001.
14. Ibid.
15. H. J. Dunster to John Cockcroft, 8 December 1959, AB 16/3001.
16. Stephen Coulter, "All at Sea on Atomic Waste," *Sunday Times* (22 November 1959), press clipping, AB 16/3001.
17. P. W. Ridley to H. J. Dunster, 9 December 1959, AB 16/3001.

18. A. H. K. Slater to A. S. McLean, 15 January 1960, AB 16/3001.
19. H. J. Dunster to A. H. K. Slater, 4 December 1959, AB 16/3001.
20. W. G. Marley to J. F. Jackson, 27 April 1959, AB 16/3000.
21. Office of the Minister of Science, Atomic Energy Division, "IAEA Conference on Waste Disposal, Monaco, November, 1959 . . . Notes for Use by the UK Delegation if Necessary," 12 November 1959, AB 16/3001.
22. "Russian Puzzles Atomic-Waste Parley," *New York Herald Tribune, Paris* (18 November 1959), clipping found in AB 16/3001.
23. "Soviet Criticism of Radioactive Waste," *(London) Times* (19 November 1959), clipping found in AB 16/3001.
24. Sir J. Bowker to Foreign Office, 3 September 1960, AB 16/3001.
25. Sir J. Bowker to Foreign Office, 14 September 1960, AB 16/3598.
26. Sir John Cockcroft, note for the record, 9 February 1960, AB 16/3596. On the IAEA's Scientific Advisory Committee, see Fischer, *History of the International Atomic Energy Agency*, 79.
27. V. Emelyanov to J. McCone, 15 March 1960, AB 16/3003.
28. G. M. P. Meyers to Mr. Williams, 22 April 1960, AB 16/3596.
29. G. M. P. Meyers to G. Smith, 15 September 1960, AB 16/3002.
30. Minutes, Working Party to Review the National Academy of Science's Report on Radioactive Waste Disposal from Nuclear Powered Ships, 14 December 1960, FO 371/157243.
31. On the importance of nuclear power in French politics and culture, see Hecht, *The Radiance of France*. For a French insider's view on the political and diplomatic issues related to nuclear programs, see Goldschmidt, *The Atomic Complex*. The origins of the CEA are discussed in Weart, *Scientists in Power*.
32. On the history of nuclear safety in France, see Foasso, "Histoire de la Sûreté de l'Énergie Nucléaire Civile en France, 1945–2000."
33. Most of the elements would have been Ce 144 and Pr 144 (about 350 curies), Sr 90 (about 40 curies), and Pu (about 14 curies). "Projet de Rejet en Mer en Méditerranée de Déchets Radioactifs," 12 May 1960, box F2/23–18 80–1217 (HC), folder "Effluents/Rejets en Méditerranée," HCCEA.
34. "Projet de Rejet en Mer en Méditerranée de Déchets Radioactifs," 12 May 1960, box F2/23–18 80–1217 (HC), folder "Effluents/Rejets en Méditerranée," HCCEA.
35. Jean Furnestin to Secrétaire Général de la Marine Marchande, 6 October 1960, box F2/23–18 80–1217 (HC), folder "Effluents/Rejets en Méditerranée," HCCEA.
36. V. Romanovsky to Haut Commissaire, CEA, 8 October 1960, box F2/23–18 80–1217 (HC), folder "Effluents/Rejets en Méditerranée," HCCEA.
37. Louis Fage to Le Délégué Général à la Recherche Scientifique et Technique, 13 October 1960, box F2/23–18 80–1217 (HC), folder "Effluents/Rejets en Méditerranée," HCCEA.
38. Ibid.
39. A brief biography of Cousteau is Madsen, *Cousteau*.
40. The CEA's records include various newspaper clippings that reprint Cousteau's remarks. Quoted here is an unpaginated excerpt from *Le Patriote*, a newspaper in Nice, of 11 October 1960, box F2/23–18 80–1217 (HC), folder "Effluents/Rejets en Méditerranée," HCCEA.
41. The letters of protest from towns on the coast are all collected in the same folder, near Jean Médecin to Francis Perrin, 12 October 1960, box F2/23–18 80–1217 (HC), folder "Effluents/Protestations," HCCEA.
42. The statements about Rainier and Cousteau in *Le Monde* are from clippings saved by the CEA dated 11 October 1960, box F2/23–18 80–1217 (HC), folder "Effluents/Rejets en Méditerranée," HCCEA.

43. "France to Delay Atomic Disposal."
44. Extract from interview of Cousteau by Pierre Ichaac, 26 October 1960, Fonds M. A. Gauvenet, folder "Déchets dans la Mer," HCCEA.
45. "Éléments d'Appréciation de l'Importance de l'Action de M. J.-Y. Cousteau dans la Campagne d'Opposition au Projet de Rejet Expérimental en Méditerranée," Department of External Relations, CEA, November 1960, box F2/23–18 80–1217 (HC), folder "Effluents/Rejets en Méditerranée," HCCEA.
46. Francis Perrin to Louis Fage, 14 October 1960, box F2/23–18 80–1217 (HC), folder "Effluents/Rejets en Méditerranée," HCCEA.
47. "Déchets Radioactifs," no author, 15 November 1960; Henri Baïssas to Jean Furnestin, n.d., box F2/23–18 80–1217 (HC), folder "Effluents/Rejets en Méditerranée," HCCEA.
48. "Déchets Radioactifs," no author, n.d., box F2/23–18 80–1217 (HC), folder "Effluents/Protestations," HCCEA.
49. Henri Baïssas, "Compte Rendu de ma Mission à Monaco," 15 December 1960, box F2/23–18 80–1217 (HC), folder "Effluents/Protestations," HCCEA.
50. Henri Baïssas, note, 6 February 1961, box F2/23–18 80–1217 (HC), folder "Effluents/Rejets en Méditerranée," HCCEA.
51. J. P. H. Trevor, "Panel on Disposal of Radioactive Waste into the Sea," 25 January 1961, and R. J. Garner, "Panel on Disposal of Radioactive Wastes into the Sea," 31 January 1961, FO 371/123132.
52. M. I. Michaels to H. C. Hainworth, 15 February 1961, FO 371/123132.
53. Unofficial Translation, "Poisoners of Wells and their Accomplices," *Pravda*, 19 January 1961, FO 371/123132.
54. Ibid. Brynielsson had been considered for being the first director general of the IAEA and had lost out to Sterling Cole, the American candidate. Brynielsson was being considered again around the time of the Brynielsson Report, but he withdrew his candidacy. For more on the appointments of IAEA directors general, see Fischer, *History of the International Atomic Energy Agency*.
55. M. I. Michaels to D. E. H. Peirson, 9 February 1961, AB 16/2161.
56. G. M. P. Myers to R. A. Thompson, 30 January 1961, FO 371/123132.
57. "Disposal of Waste by Sea Dumping: Answers to Possible Press Questions," n.d., FO 371/123132.
58. Ibid.
59. R. A. Thompson to G.M.P. Myers, 15 February 1961, FO 371/123132.
60. J. McAdam Clark to M. I. Michaels, 18 April 1961, FO 371/123132.
61. Ibid.
62. Anne E. Stoddart, IAFO Minute, 25 April 1961, FO 371/123132.
63. This incident is analyzed in greater detail in Beschloss, *Mayday*.
64. Joint Intelligence Bureau to M. I. Michaels, 25 August 1961, AB 16/4007.
65. Ibid.
66. Medvedev, "Nuclear Disaster in the Urals," and Medvedev, *Nuclear Disaster in the Urals*.
67. Donnay et al., "Russia and the Territories of the Former Soviet Union."
68. M. I. Michaels to A. Todd, Ministry of Defense, Joint Intelligence Bureau, 31 August 1961, AB 16/4007.
69. M. Phillips to Mr. Croome, 19 September 1961, AB 16/4007.
70. Donnay et al., "Russia and the Territories of the Former Soviet Union."
71. The French newspaper of 31 August 1961 is referred to by A. Todd to M. I. Michaels, 7 September 1961, FO 371/157244.
72. Anne E. Stoddart, IAFO Minute, 23 September 1961, FO 371/157244.
73. I.G.K. Williams to Mr. Myers, 12 October 1960, AB 54/16.

74. Ibid.
75. I.G.K. Williams to Dr. Gaunt, 24 July 1961, AB 54/16.
76. Overseas Relations Committee, Radioactive Effluent Disposal at the C.E.A. Centre de la Hague. Note by Authority Health and Safety Branch, 5 June 1962, AB 54/16.
77. D. E. H. Peirson to John Cockcroft, 4 November 1960, AB 16/3002. Quotation taken from I.G.K. Williams to J. C. Walker, "Collaboration between the IAEA and the Government of Monaco," 7 November 1960, AB 16/3585.
78. The brief for the United States delegation is entitled "Scientific Collaboration with the Government of Monaco on Research on Disposal of Radioactive Waste into the Sea," n.d., AB 16/3002.
79. "Record of Discussion of Anglo-US Collaboration on Waste Disposal Problems Prior to November Meeting of IAEA Committee," 14 October 1960, AB 16/3002.
80. Public relations liaison was to be handled by Jean Renou (France) and Eric Underwood (UK), while technical aspects would be handled by André Gauvenet (France) and Ian Williams (UK). See Bertrand Goldschmidt to D.E.H. Peirson, 20 December 1960, AB 16/3588.
81. "Notes of Talks with CEA 14th–15th December: Public Relations Aspects of Waste Disposal," n.d., AB 16/3588.
82. Eric H. Underwood, "Public Relations in Health and Safety Matters," AEA-CEA Discussions, 14–15 December 1960, AB 16/3588.
83. International Atomic Energy Agency, press release, "Finnish Scientist to Direct Major Research Program in Monaco," 8 June 1961, AB 16/3585.
84. United Kingdom Delegation to NATO to Atomic Energy and Disarmament Department, Foreign Office, 31 May 1961, FO 371/157244.
85. U.S. Atomic Energy Commission, press release, "Remarks Prepared by Joseph A. Lieberman, Chief, Environmental and Sanitary Engineering Branch, US Atomic Energy Commission, for Delivery to the Second Sanitary Engineering Conference on Radiological Aspects of Water Supplies, University of Illinois, Urbana, Ill., January 27, 1960," 27 January 1960, AB 16/3002.
86. Ibid.
87. Anne E. Stoddart, minute, 12 September 1961, FO 371/157244.
88. Emelyanov is quoted and discussed by J. C. Walker to I.G.K. Williams, "I.A.E.A. Annual General Conference, 1962, Radioactive Waste Disposal," 1 October 1962, AB 54/15.
89. "Radioactive Waste 'Polluting Sea.'"
90. "A.E.C. Rebuts Soviet on Pollution of Sea."
91. "Discharge Control at Windscale."
92. Beschloss, *The Crisis Years*.
93. H. J. Dunster, Note for the Record, Exchange Visit with the USSR on Waste Disposal, 9–18 April 1964, n.d., AB 54/46.
94. H. J. Dunster to I.G.K. Williams, 23 October 1963, AB 54/46.
95. I. G. K. Williams to M. I. Michaels, 4 May 1964, AB 54/46.
96. H. J. Dunster to J. M. Hill, 21 April 1964, AB 54/46.
97. H. J. Dunster to I.G.K. Williams, 23 October 1963, AB 54/46.
98. C. A. Mawson to R. J. Garner, 17 September 1964, AB 54/89.

Chapter 7 ***Purely for Political Reasons***

1. H. J. Dunster, "ENEA Seminar on the Experimental Disposal of Solid Radioactive Waste into the Atlantic Ocean," 20 April 1967, AB 54/91.

2. Ibid.
3. This is reported in J. Gaunt to I.G.K. Williams, 20 October 1959, AB 16/3002.
4. Cabinet Office London to BJSM Washington, D.C., 26 October 1959, AB 16/3002.
5. J. Gaunt to I.G.K. Williams, 27 October 1959, AB 16/3002.
6. Alexander, "Sea Water Monitoring Following Radioactive Waste Disposal Operations."
7. H. J. Dunster to F. A. Vick, "Radioactive Waste Disposal at Sea," 27 July 1960, AB 6/2087.
8. U.S. Congress, *Ocean Dumping of Waste Materials*, 238.
9. B. C. Peatey to M. Phillips, 18 February 1963, AB 54/15.
10. R. J. Garner to M. Phillips, 5 March 1963, AB 54/15.
11. P. Mehew to G.M.P. Myers, 28 June 1961, AB 6/2087.
12. F. J. Neary to Eric Underwood, 30 June 1961, AB 6/2087.
13. F. J. Neary to P. Mehew, 11 July 1961, AB 6/2087.
14. G. Smith to P. Mehew, 13 Jul 1961, AB 16/1654.
15. Cabinet Office London to BJSM Washington, 9 March 1960, PRO B541; J. M. Jones to A. Joseph, 14 March 1960, AB 16/3002.
16. W. A. langmead to A. Quinton, May 1960, AB 6/2541.
17. Report of a Research Group Board of Inquiry into the Circumstances Relating to the Recovery of Authority Radioactive Waste Containers from the Bay of Biscay on 29th May and 2nd June 1962, UKAEA Research Group, Atomic Energy Research Establishment, Harwell, November 1962, AB 54/5. The chairman of this Board of Inquiry was R. F. Jackson.
18. Ibid.
19. R. H. Burns, "Radioactive Solid Waste at AERE, Harwell: The Problems Associated with Land Burial and Sea Burial," March 1973, AB 15/7314.
20. R. H. Burns to Dr. F. A. Vick, 26 October 1962, AB 54/5.
21. R. F. Jackson to R. H. Burns, 22 October 1962, AB 54/5.
22. M. Phillips to R. Spence, 28 May 1964, AB 54/27.
23. M. Phillips to W. T. Potter, 9 June 1964, AB 54/15.
24. W. T. Potter to R. J. Garner, 18 June 1964, AB 54/15.
25. Ibid.
26. W. H. King, Report on Unusual Incidents during Atlantic Dumping Operation No. A.101, 28 Jul 1964, AB 54/31.
27. E. H. Bott to W. T. Potter, 6 July 1964, AB 54/31.
28. W. H. King, Report on Unusual Incidents during Atlantic Dumping Operation No. A.101, 28 July 1964, AB 54/31.
29. The substance of this narrative is taken from King's report of the incident. W. H. King, Report on Unusual Incidents during Atlantic Dumping Operation No. A.101, 28 July 1964, AB 54/31.
30. Extract of Minutes of the AEX (65) 13th Meeting, 31 July 1964, AB 54/31.
31. A.J.D. Winnifrith to Sir Alan Hitchman, 10 September 1964, AB 54/31.
32. E. A. Sharp to Sir John Winnifrith, 17 September 1964, AB 54/31.
33. A.J.D. Winnifrith to Sir Alan Hitchman, 10 September 1964, AB 54/31.
34. A.J.D. Winnifrith to Dame Evelyn Sharp, 2 October 1964, AB 54/31.
35. Extract from the Minutes of the AEA (64) 18th Meeting, 15 October 1964, AB 54/31.
36. F. A. Vick to A. S. McLean, 28 September 1964, AB 54/15.
37. R. H. Burns to F. A. Vick, 25 August 1964, AB 54/31.
38. R. H. Burns, "Radioactive Solid Waste at AERE, Harwell: The Problems Associated with Land Burial and Sea Burial," March 1973, AB 15/7314.

39. R. F. Jackson to F. A. Vick, 8 October 1964, AB 54/31.
40. Authority Health and Safety Branch, "Discharge of Radioactive Effluent from Cap de la Hague," 13 May 1965, AB 54/017.
41. H. J. Dunster and W. L. Templeton, "Notes on the Visit by the French CEA to Windscale to Discuss Marine Biological Programme, 17/18 June 1965," 12 July 1965, AB 54/017.
42. I.G.K. Williams, Note for the Record, 22 February 1966, AB 54/017.
43. Ibid.
44. I.G.K. Williams to Eric Underwood, 30 July 1965, AB 16/4282.
45. D. E. H. Peirson to D. G. Avery, 5 July 1961, AB 16/4008.
46. A brief for the AEA participants in a meeting with the CEA includes discussion of EURATOM meetings as well. See "AEA/CEA Exchanges on Radioactive Contamination of the Marine Environment, Meeting at Fontenay-aux-Roses, 19th December, 1961, Brief for Authority Participants," 13 December 1961, AB 16/3764.
47. B. Taylor to B. C. Peatey, 24 January 1964, AB 54/89.
48. R. J. Garner, note for the record, 11 May 1964, AB 54/89.
49. Ibid.
50. Comments on the lack of publicity on dumping exercises shared with the French are included in I.G.K. Williams to G. Bresson, 16 July 1964, AB 54/017.
51. M. Phillips to C. E. Coffin, 21 May 1964, AB 54/89.
52. Note by the French Delegation, "Proposal for the Setting Up of a Group of Experts on Sea Water Pollution," 22 April 1965, AB 54/89.
53. H. J. Dunster to E. C. Appleyard, 3 May 1965, AB 54/89.
54. The comments about ENEA taking over radioactive issues is found in a memorandum of conversation between I.G.K. Williams and ENEA director E. Wallauschek, 3 May 1965, AB 54/89.
55. M. Phillips to F. Morgan, 22 June 1964, AB 54/89.
56. R. J. Garner, note for the record, 14 July 1964, AB 54/89.
57. Extract from *Applied Atomics*, no. 461, 29 July 1964, AB 54/15.
58. Extract from *Applied Atomics*, no. 460, 22 July 1964, AB 54/15.
59. J. Mabile, "Projet OCDE de Rejets en Mer," 16 December 1965, Box: M3/07–61 (HC), "folder OCDE 1965–1967," HCCEA.
60. M. Phillips to R. H. Burns, 8 May 1964, AB 54/26.
61. Ibid.
62. R. H. Burns to M. Phillips, 14 May 1964, AB 54/26.
63. ENEA Steering Committee for Nuclear Energy: Health and Safety Sub-committee, Extract from Note of Meeting of the Restricted Working Group established by the Sub-committee to review the programme and prepare a schedule of work for 1966, 15–16 July 1965, AB 54/26.
64. Miss Rosemary Cargill, Scientific Relations Department Minute, 10 March 1966, FO 371/189438.
65. I.G.K. Williams to R. H. Burns, 11 March 1966, AB 54/26.
66. P. J. Kelly [Ministry of Technology] to J. McAdam Clark [Foreign Office], 21 January 1966, FO 371/189438.
67. I.G.K. Williams to R. H. Burns, 11 March 1966, AB 54/26.
68. Ibid.
69. F. A. Vick to Williams, 22 March 1966, AB 54/26.
70. G. W. Clare, "Meeting of Specialists on Sea Disposal Containers," 4 March 1966, AB 54/90.
71. Ibid.
72. R. Pilgrim to E. W. Jackson, 20 April 1966, PRO A197; R. Pilgrim to R. H. Burns and H. J. Dunster, 29 April 1966, AB 54/90.

73. L.D.G. Richings to McAdam Clark, 11 October 1966, FO 371/189438.
74. H. J. Dunster to C. A. Mawson, 11 May 1966, AB 54/90.
75. H. J. Dunster to E. Wallauschek, 24 May 1966, AB 54/90.
76. H. J. Dunster to E. Wallauschek, 2 June 1966, AB 54/90.
77. I.G.K. Williams to H. J. Dunster, 13 June 1966, AB 54/90.
78. H. J. Dunster to I. G. K. Williams, 18 October 1966, AB 54/90.
79. H. J. Dunster to I. G. K. Williams, 26 October 1966, AB 54/90.
80. L.D.G. Richings to H. J. Dunster, 26 October 1966, AB 54/90.
81. John Swallow to V. Romanovsky, 15 December 1966, AB 54/90.
82. H. J. Dunster, "ENEA Seminar on the Experimental Disposal of Solid Radioactive Waste into the Atlantic Ocean," 20 April 1967, AB 54/91.
83. "Compte Rendu de la Mission de Mm. Avargues et Rodier," n.d., box M3/07–61 (HC), folder "OCDE 1965–1967," HCCEA.
84. L.D.G. Richings to P. J. Kelly, 18 April 1967, AB 54/26.
85. Ibid.
86. Ibid.
87. Ibid.
88. H. J. Dunster to W. G. Belter, 26 May 1967, AB 54/91.
89. E. R. Greening to Chairman, Atomic Energy Authority, 12 March 1967, AB 54/26.
90. L.D.G. Richings to R. H. Burns, 25 April 1967, AB 54/26.
91. Eric Underwood to R. M. Fishenden, 8 May 1967, AB 54/26.
92. For an overview of the operation, see European Nuclear Energy Agency, *Radioactive Waste Disposal Operation into the Atlantic, 1967.*
93. ENEA Steering Committee for Nuclear Energy, "Progress Report, June 1967," 8 June 1967, AB 54/91.
94. M. de Rouville to various, 12 March 1968, box M3/07–61 (HC), folder "Rejets de Dechets Radioactifs dans l'Océans Atlantique," HCCEA.

Chapter 8 ***Confronting Environmentalism***

1. Hunt, *Radioactivity Studies in Lowestoft.*
2. Polikarpov, *Radioecology of Aquatic Organisms.*
3. Preston, "Artificial Radioactivity in Freshwater and Estuarine Systems."
4. L.D.G. Richings, note for the record, 4 June 1969, AB 45/85.
5. Ibid.
6. A. Preston, "A Report on Visits to IAEA Laboratory at Monaco, the CEA Centre for Nuclear Studies at Cadarache and the CNEN Laboratory for the Study of Radioactive Contamination of the Sea at Fiascherino, 2–16 September 1969," 24 September 1969, AB 54/93.
7. Novick, *The Careless Atom.*
8. Kotin, "Letter to the Editor."
9. Petrow, *In the Wake of Torrey Canyon.*
10. Steinhart and Steinhart, *Blowout*, and Sollen, *An Ocean of Oil.*
11. Acting Secretary of the Interior (Train) to Elliot L. Richardson, Under Secretary of State, 17 November 1969, in Holly and McAllister, *Foreign Relations, 1969–1976*, document 288.
12. An overview of these environmental activities, particularly focused on the State Department, is in "Significant Activities of the Department of State in the Environmental Field," transmitted by Acting Deputy Director of the Office of Environmental Affairs (Salmon) to Secretary to Council on Environmental Quality (Gibbons), 14 July 1970, in Holly and McAllister, *Foreign Relations, 1969–1976*,

document 298. On Nixon's environmental agenda, see Flippen, *Nixon and the Environment*.

13. Telegram 171300 from Department of State to All Embassies in Europe, 14 October 1970, in Holly and McAllister, *Foreign Relations, 1969–1976*, document 301.
14. Report by Task Force III of the Committee on International Environmental Affairs, "U.S. Priority Interests in the Environmental Activities of International Organizations," December 1970, in Holly and McAllister, *Foreign Relations, 1969–1976*, document 303.
15. Klein, "The Careless Atom."
16. Smith, "A.E.C. Scored on Storing Waste."
17. This summary is based on a letter, J. Hoover Mackin to Frederick Seitz, 14 June 1967, folder "ES: Com on Geologic Aspects of Radioactive Waste Disposal: Background Info," NAA. On Hess's criticisms, see Walker, *Controlling the Atom*, 370–371.
18. Frederick Seitz to Glenn T. Seaborg, 30 August 1965, folder "ES: Group on Radioactive Waste Disposal: Ad hoc: Meeting, 1965," NAA.
19. Glenn T. Seaborg to Frederick Seitz, 1 November 1965, folder "ES: Group on Radioactive Waste Disposal: Ad hoc: Meeting, 1965," NAA.
20. "Management of Radioactive Wastes from the Nuclear Power Industry," n.d., received at NAS 1 November 1965, folder "ES: Group on Radioactive Waste Disposal: Ad hoc: Meeting, 1965," NAA.
21. Milner B. Shaefer to Earl F. Cook, 6 December 1965; Gordon Fair to Earl F. Cook, 7 December 1965, folder "ES: Group on Radioactive Waste Disposal: Ad hoc: Meeting, 1965," NAA.
22. M. King Hubbert to Abel Wolman, 29 December 1965, folder "ES: Group on Radioactive Waste Disposal: Ad hoc: Meeting, 1965," NAA.
23. Ibid.
24. John E. Galley to Abel Wolman, 11 December 1965, folder "ES: Group on Radioactive Waste Disposal: Ad hoc: Meeting, 1965," NAA.
25. E. F. Cook to Abel Wolman, 24 May 1966, folder "ES: Group on Radioactive Waste Disposal: Ad hoc," NAA; Walter H. Bailey to J. Hoover Mackin, 26 August 1966; Walter H. Bailey to Members, Ad hoc Group on Radioactive Waste Disposal, 26 August 1966, folder "ES: Com on Geologic Aspects of Radioactive Waste Disposal: Background Info," NAA. Bailey addressed the letter about the stickers to the wrong committee (the ad hoc committee was Wolman's, whereas the standing committee was Galley's).
26. "Comments on NAS Report," n.d., folder "ES: Com on Geologic Aspects of Radioactive Waste Disposal, 1966," NAA.
27. Milton Shaw to Frederick Seitz, 7 November 1966, folder "ES: Com on Geologic Aspects of Radioactive Waste Disposal, 1966," NAA.
28. John E. Galley to J. Hoover Mackin, 20 May 1967, reply to AEC comments, folder "ES: Com on Geologic Aspects of Radioactive Waste Disposal: Background Info," NAA.
29. Milton Shaw to Frederick Seitz, 25 May 1967, folder "ES: Com on Geologic Aspects of Radioactive Waste Disposal: Background Info," NAA.
30. Frederick Seitz to Milton Shaw, 5 June 1967, folder "ES: Com on Geologic Aspects of Radioactive Waste Disposal: Background Info," NAA.
31. J. Hoover Mackin to Frederick Seitz, 14 June 1967, folder "ES: Com on Geologic Aspects of Radioactive Waste Disposal: Background Info," NAA.
32. Ibid.
33. Ibid.
34. On this controversy, see Walker, "The Atomic Energy Commission and the Politics of Radiation Protection, 1967–1971."

35. Lawson, "Nuclear Split."
36. Pace, "Control of Radioactive Wastes in Sea Urged at Experts' Parley."
37. See Lyons, "Army Will Transport Nerve Gas across South for Disposal at Sea," and "Trains Loaded for Nerve Gas Shipment."
38. Halloran, "Still Lots of Deadly Gas Awaiting Disposal."
39. "Fascell Asks Dumping Ban."
40. Smith, "Panel Urges Curbs on Ocean Dumping; Nixon Hails Report."
41. Ibid.
42. The claim about Vietnam has been made persuasively in Flippen, *Nixon and the Environment.*
43. Kenworthy, "Nixon Asks Great Lakes Dumping Ban."
44. Curtis and Fisher, "The Seven Wonders of the Polluted World."
45. Harwood, "We Are Killing the Sea Around Us."
46. "Senate Backs Curb on Ocean Dumping," and "Bill to Limit Dumping of Wastes in Oceans Is Passed by House."
47. Semple, "Vexed Nixon Prods Congress on Ecology."
48. R. H. Burns, "Notes on a Meeting Held in Paris (ENEA Offices) on 17th and 18th October 1968," 28 October 1968, AB 54/72.
49. See E. Reynolds, "Report on a Voyage in the Deutsches Hydrographisches Institut's Research Vessel Meteor, Irish Sea, November 1968," 16 December 1968, AB 54/92.
50. G. W. Clare to L.D.G. Richings, 21 March 1969, AB 54/92.
51. Y. Sousselier, " Opération de Rejet en Mer . . ." 11 March 1969, box M3/07–61 (HC), folder "Rejets de Dechets Radioactifs dans l'Océans Atlantique," HC-CEA.
52. G. W. Clare to L.D.G. Richings, 21 March 1969, AB 54/92.
53. Ibid.
54. L.D.G. Richings to H. J. Dunster, 26 March 1969, AB 54/92.
55. A. Preston to L.D.G. Richings, 29 April 1969, AB 54/92.
56. A. G. Perrin to L.D.G. Richings, 22 April 1970, AB 45/85.
57. Ibid.
58. The details of this reorganization are explained briefly on the Web site of the United Kingdom Atomic Energy Authority. See http://www.ukaea.org.uk.
59. L.D.G. Richings to J. B. Lewis, 23 April 1970, AB 45/85.
60. J. H. Axford to L. D. G. Richings, 27 October 1970, AB 54/93.
61. A. W. Kenny to J. H. Axford, 30 October 1970, AB 54/93.
62. A. Preston to J. H. Axford, 2 November 1970, AB 54/93.
63. L.D.G. Richings to J. H. Axford, 3 November 1970, AB 54/93.
64. Preston and Wood, "Monitoring the Marine Environment."
65. Ibid. 454.
66. Ibid.
67. Ibid.
68. M. I. Michaels to L.D.G. Richings, 4 November 1970, AB 45/148.
69. J. B. Lewis, "Radioactive Waste Disposal into the Atlantic: First Meeting of the Operations Executive Group of ENEA," 14 December 1970, AB 54/53.
70. J. B. Lewis, "ENEA Radioactive Waste Dump 1971," 12 February 1971, AB 54/53.
71. J. B. Lewis, "Radioactive Waste Disposal into the Atlantic: First Meeting of the Operations Executive Group of ENEA," 14 December 1970, AB 54/53.
72. P. J. Kelly to R. Press, 12 January 1971, AB 54/93.
73. Clippings of these press articles were collected by the CEA and dated 19 August 1971, box F2/23–04 80–1203 (HC), folder "Déchets/Evacuation des déchets dans l'Atlantique," HCCEA.

74. L.D.G. Richings, note for the record, 14 March 1972, AB 54/53.
75. Intelligence Note, Bureau of Intelligence and Research, "The Soviets in the International Environment: After Stockholm," 11 August 1972, in Holly and McAllister, *Foreign Relations, 1969–1976*, document 328.
76. Miller N. Hudson Jr., Scientific Attaché, U.S. Embassy, Rio de Janeiro, to Christian A. Herter, Office of Environmental Affairs, 12 February 1971, in Holly and McAllister, *Foreign Relations, 1969–1976*, document 306.
77. Christian A. Herter, "Classified Report of the United Nations Conference on the Human Environment, Stockholm, Sweden, June 5–16, 1972," 28 July 1972, in Holly and McAllister, *Foreign Relations, 1969–1976*, document 325.
78. Memorandum from the Chairman of the Council on Environmental Quality (Train) to President Nixon, 19 Jun 1972, in Holly and McAllister, *Foreign Relations, 1969–1976*, document 324.
79. OECD Nuclear Energy Agency, "Radioactive Waste Disposal into the Atlantic (June-July 1972)," 2 August 1972, AB 54/73.
80. J. B. Lewis, "Radioactive Waste for Sea Dumping in 1973," 10 October 1972, AB 54/73.
81. J. B. Lewis to J. Williams, 13 October 1972, AB 54/53, emphasis in original.
82. Ibid.
83. B. Verkerk to J. B. Lewis, 8 November 1972, AB 54/53.
84. J. B. Lewis, "Radioactive Waste Sea Dump 1973," 15 November 1972, AB 54/73.
85. J. A. H. Broughton, note, "NEA Steering Committee Meeting, 3 May 1973," 4 May 1973, AB 54/53.
86. A. Preston to L.D.G. Richings, 6 November 1972, AB 54/53.
87. Ibid.
88. J. B. Lewis to J. Williams, 16 January 1973, AB 54/73.
89. L.D.G. Richings to J. B. Lewis, 22 January 1973, AB 54/53.
90. Ibid.
91. J. H. Clarke, "Meeting at DTI to Discuss the 'Panel on IAEA Responsibilities under the Convention on the Prevention of Marine Pollution by Dumping of Waste & Other Matter, Vienna 4–8 June 1973," n.d., AB 54/73.
92. A. Preston and H. J. Dunster, note for the record, 11 June 1973, AB 45/85.
93. On Britain's hopes for these IAEA negotiations, see J. H. Clarke, "Meeting at DTI to Discuss the 'Panel on IAEA Responsibilities under the Convention on the Prevention of Marine Pollution by Dumping of Waste & Other Matter, Vienna 4–8 June 1973," n.d., AB 54/73.
94. These definitions were subject to periodic review. The fifty-year half-life for alpha activity and the restrictions on strontium-90 were specified in the definition submitted by the IAEA to governments in late 1974. See Sigvard Eklund to F. H. Jackson, 6 December 1974, AB 54/125.
95. A. Preston and H. J. Dunster, note for the record, 11 June 1973, AB 54/125.
96. R. A. Vining to J.A.H. Broughton, 24 November 1972, AB 54/73.
97. A. Preston and H. J. Dunster, note for the record, 11 June 1973, AB 45/85.
98. J. B. Lewis, brief of meeting, Operations Executive Group, NEA, meeting held 5 November 1973, 12 November 1973, AB 54/53.
99. J.K.L. Thompson, "The Topaz Story—A Saga of the Atlantic Waste (the NEA Sea-dump)," 17 July 1973, AB 54/124.
100. F. S. Feates and J. B. Lewis, note for the record, 3 September 1973, AB 54/124.
101. Sigvard Eklund to F. H. Jackson, 6 December 1974, AB 54/125.
102. Some wanted to have the figures officially lowered (Sweden), while others emphasized that the figures were really for health physics calculations (France) and not intended as actual disposal guidelines. The sources for these reflections is J.

B. Lewis, note for the record, "Meeting of the NEA Operations Executive Group Held in Paris on Thursday, 24 October 1974," 31 October 1974, AB 54/73.

Conclusion

1. Commission of the European Communities, *The Radiological Exposure of the Population of the European Community from Radioactivity in North European Marine Waters, Project "Marina."*
2. The North East Atlantic received 42.3 PBq (2.3 MCi), as compared to 38.4 PBq in the Arctic, 2.9 PBq in the Northwest Atlantic, and 1.5 PBq in all of the Pacific. These figures, and all of the data from this paragraph, are drawn from Linsley, Sjöblom, and Cabianca, "Overview of Point Sources of Anthropogenic Radionuclides in the Oceans."
3. These post-1972 developments are discussed in much greater detail in Ringius, *Radioactive Waste Disposal at Sea*, chap. 8.
4. M. Phillips, "Note of a Meeting held on the 17th of April, 1963 at Harwell, to discuss the future of disposal of solid radioactive waste into the sea," 1 May 1963, AB 6/2574.
5. Stokke, "Beyond Dumping," 42.
6. McKinney and Company, Management Consultants, "Study of the 1960–61 Presidential Transition, Section X: Chairing and Managing Atomic Energy Commission," November 1960, John McCone Files, box 39, folder "O&M 1, General Policy," NA326.
7. U.S. Congress, *Waste Dumping*, 390. The quotation is from Michael Pogodzinski, editor of *New England Outdoors.*
8. U.S. Congress, *Waste Dumping*, 409.

Bibliography

"A.E.C. Bars Waste Plan." *New York Times*, 22 April 1962.

"A.E.C. Rebuts Soviet on Pollution of Sea." *New York Times*, 16 November 1962.

Alexander, Holmes. "The Peaceful Atom Organization." *Los Angeles Times*, 18 October 1957.

Alexander, R. E. "Sea Water Monitoring Following Radioactive Waste Disposal Operations." *Health Physics* 7 (1961–1962), 106–113.

Arnold, Lorna. *Windscale 1957: Anatomy of a Nuclear Accident*. New York: St. Martin's Press, 1992.

"Atom Power Curb Seen." *New York Times*, 6 April 1955.

"Atom Waste Adrift; Shipping Menaced." *New York Times*, 15 July 1957.

"Atom Waste Put in Sea." *New York Times*, 29 January 1955.

"Atomic 'Cemetery' Needed for Waste." *New York Times*, 29 July 1950.

"Atomic Belt Plan Held Not Feasible." *New York Times*, 18 April 1951.

"Atomic Death Belt Urged for Korea." *New York Times*, 17 April 1951.

"Atomic Politics in Geneva." *New York Times*, 5 September 1958.

"Atomic Refuse Leaves Pacific Unaffected." *Los Angeles Times*, 3 December 1957.

"Atomic Refuse Moon Dump Held Possible." *Los Angeles Times*, 27 June 1950.

"Atomic Waste Made Safe." *New York Times*, 11 January 1953.

"Atomic Wastes Dumped in Sea." *New York Times*, 22 May 1949.

"Atomic Wastes." *New York Times*, 18 September 1955.

Baldwin, Hanson W. "An Atomic Maginot Line." *New York Times*, 20 May 1954.

Beatty, John. "Genetics in the Atomic Age: The Atomic Bomb Casualty Commission, 1947–1956." In *The Expansion of American Biology*, ed. Keith R. Benson, Jane Maienschein, and Ronald Rainger, 284–324. New Brunswick: Rutgers University Press, 1991.

Berg, Paul, and Maxine Singer. *George Beadle, An Uncommon Farmer: The Emergence of Genetics in the 20th Century*. Cold Spring Harbor, N.Y.: Cold Spring Harbor Laboratory Press, 2003.

Berkhout, Frans. *Radioactive Waste: Politics and Technology*. London: Routledge, 1991.

Beschloss, Michael R. *The Crisis Years: Kennedy and Khrushchev, 1960–1963*. New York: HarperCollins, 1991.

———. *Mayday: Eisenhower, Khrushchev, and the U-2 Affair*. New York: Harper and Row, 1986.

"Bill to Limit Dumping of Wastes in Oceans Is Passed by House." *Wall Street Journal*, 10 September 1971.

Bleise, A., P. R. Danesi, and W. Burkart. "Properties, Use and Health Effects of Depleted Uranium (DU): A General Overview." *Journal of Environmental Radioactivity* 64 (2003): 93–112.

Bocking, Stephen. *Ecologists and Environmental Politics: A History of Contemporary Ecology*. New Haven: Yale University Press, 1997.

Boyer, Paul S. *By the Bomb's Early Light: American Thought and Culture at the Dawn of the Atomic Age*. New York: Pantheon, 1985.

"Britain to Sink Atomic Wastes." *New York Times*, 4 October 1949.

Bruhèze, Adri August Albert de la. "Political Construction of Technology: Nuclear Waste Disposal in the United States, 1945–1972." Ph.D. diss., University of Twente, The Netherlands, 1992.

———. "Radiological Weapons and Radioactive Waste in the United States: Insiders' and Outsiders' Views, 1941–1955." *British Journal for the History of Science* 25 (1992): 207–227.

Bruno, Laura A. "The Bequest of the Nuclear Battlefield: Science, Nature, and the Atom during the First Decade of the Cold War." *Historical Studies in the Physical and Biological Sciences* 33, no. 2 (2003): 237–260.

Carlisle, Rodney P., with Joan M. Zenzen. *Supplying the Nuclear Arsenal: American Production Reactors, 1942–1992*. Baltimore: Johns Hopkins University Press, 1994.

Carson, Rachel. *The Sea Around Us*. Rev. ed. New York: Signet, 1961.

Chadarevian, Soraya de. "Mice and the Reactor: The 'Genetics Experiment' in 1950s Britain." *Journal of the History of Biology* 39 (2006): 707–735.

Clark, Claudia. *Radium Girls: Women and Industrial Health Reform, 1910–1935*. Chapel Hill: University of North Carolina Press, 1997.

Claus, W. D. "What Is Health Physics?" *Health Physics* 1 (1958): 56–61.

"Clay 'Seal' Tested for Atom Wastes." *New York Times*, 26 October 1954.

"Close Check Urged on Atomic Waste." *New York Times*, 13 January 1950.

Cockcroft, John. "Future of Atomic Energy." *Scientific Monthly* 82, no. 3 (March 1956): 136–141.

———. "The Scientific Work of the Atomic Energy Research Establishment." *Proceedings of the Royal Society of London, Series B. Biological Sciences* 139 (24 April 1952): 300–313.

Commission of the European Communities. *The Radiological Exposure of the Population of the European Community from Radioactivity in North European Marine Waters, Project "Marina."* Luxembourg: Office for Official Publications of the European Communities, 1990.

Cousteau, Jacques-Yves, with Frédéric Dumas. *The Silent World*. New York: Harper & Row, 1953.

Creager, Angela N. H. "Nuclear Energy in the Service of Biomedicine: The U.S. Atomic Energy Commission's Radioisotope Program, 1946–1950." *Journal of the History of Biology* 39 (2006): 649–684.

———. "The Industrialization of Radioisotopes by the U.S. Atomic Energy Commission." In *The Science-Industry Nexus: History, Policy, Implications*, proceedings of Nobel Symposium 123, ed. Karl Grandin, Nina Wormbs, and Sven Widmalm, 143–167. Watson Publishing, 2004.

———. "Tracing the Politics of Changing Postwar Research Practices: The Export of 'American' radioisotopes to European biologists." *Studies in History and Philosophy of Biological and Biomedical Sciences* 33 (2002): 393–415.

"Current Is Found Far Down in Sea." *New York Times*, 29 June 1958.

Curtis, Richard, and Dave Fisher. "The Seven Wonders of the Polluted World." *New York Times*, 26 September 1971.

"Dangers to Public in Radiation Noted." *New York Times*, 4 September 1951.

"Discharge Control at Windscale." *Times*, 16 November 1962.

Divine, Robert A. *Blowing on the Wind: The Nuclear Test Ban Debate, 1954–1960*. New York: Oxford University Press, 1978.

Donnay, Albert, Martin Cherniack, Arjun Makhijani, and Amy Hopkins. "Russia and the Territories of the Former Soviet Union." In *Nuclear Wastelands: A Global Guide to Nuclear Weapons Production and Its Health and Environmental Effects*, ed. Arjun Makhijani, Howard Hu, and Katherine Yih, 285–392. Cambridge: MIT Press, 2000.

"Drifting Atomic Waste Can Sunk." *Los Angeles Times*, 17 July 1957.

Dunster, H. J., and F. R. Farmer. *A Summary of the Biological Investigations of the Discharges of Aqueous Radioactive Waste to the Sea from Windscale Works, Sellafield, Cumberland*. Risley: UKAEA Safety Branch, 1958.

Edgerton, David. *Warfare State: Britain, 1920–1970*. Cambridge: Cambridge University Press, 2006.

Emery, William J. "The *Meteor* Expedition, an Ocean Survey." In *Oceanography: The Past*, ed. Mary Sears and Daniel Merriman, 690–702. New York: Springer Verlag, 1980.

European Nuclear Energy Agency. *Radioactive Waste Disposal Operation into the Atlantic, 1967*. Paris: OECD, 1968.

Everett, D. H. "Eugen Glueckauf." *Biographical Memoirs of Fellows of the Royal Society* 30 (1984): 192–224.

"Expert Belittles Peril in Radioactive Waste." *Los Angeles Times*, 7 December 1954.

"Fascell Asks Dumping Ban." *New York Times*, 16 August 1970.

"Fears 'Hot' Shellfish." *New York Times*, 19 December 1957.

Feshbach, Murray. *Ecological Disaster: Cleaning Up the Hidden Legacy of the Soviet Empire*. New York: Twentieth Century Fund Press, 1995.

Feshbach, Murray, and Arthur Friendly. *Ecocide in the USSR: Health and Nature under Siege*. New York: Basic Books, 1992.

Finney, John W. "Radiation Curbs Urged by Dutch." *New York Times*, 13 September 1958.

Finney, John W. "Russian Charges U.S. Is Deceptive in Atom Proposal." *New York Times*, 4 September 1958.

———. "Soviet Discloses Biggest Reactor." *New York Times*, 9 September 1958.

———. "Atom Talks End on Hopeful Note." *New York Times*, 14 September 1958.

Fischer, David. *History of the International Atomic Energy Agency: The First Forty Years*. Vienna: IAEA, 1997.

Flippen, J. Brooks. *Nixon and the Environment*. Albuquerque: University of New Mexico Press, 2000.

Foasso, M. Cyrille. "Histoire de la Sûreté de l'Énergie Nucléaire Civile en France, 1945–2000." Ph.D. diss., Université Lumière, Lyon II, 2003.

Forman, Paul. «Behind Quantum Electronics: National Security as Basis for Physical Research in the United States, 1940–1960.» *Historical Studies in the Physical and Biological Sciences* 18 (1985): 149–229.

"France to Delay Atomic Disposal." *New York Times*, 13 October 1960.

"Fusion Engine Seen." *New York Times*, 2 September 1958.

Gerber, Michele Stenehjem. *On the Home Front: The Cold War Legacy of the Hanford Nuclear Site*. Lincoln: University of Nebraska Press, 2002.

"German Town Boycotts Vote." *New York Times*, 16 September 1957.

Glennon, John P., ed. *Foreign Relations of the United States, 1955–1957, vol. XX: Regulation of Armaments; Atomic Energy*. Washington, D.C.: Government Printing Office, 1990.

Goldschmidt, Bertrand. *The Atomic Complex: A Worldwide Political History of Nuclear Energy*. La Grange Park, Ill.: American Nuclear Society, 1982.

Government Commission on Matters Related to Radioactive Waste Disposal at Sea. *Facts and Problems Related to the Dumping of Radioactive Waste in the Seas Surrounding the Territory of the Russian Federation*. Moscow: Office of the President of the Russian Federation, 1993.

Gowing, Margaret. *Independence and Deterrence: Britain and Atomic Energy, 1945–1952*. New York: St. Martin's Press, 1974.

Gowing, Margaret, and Lorna Arnold. *Independence and Deterrence: Britain and Atomic Energy, 1945–1952*. New York: St. Martin's Press, 1974.

Green, Francis H. K. "The Constitution and Functions of the United Kingdom Medical Research Council." *Science*, n.s., 116 (1 August 1952): 99–105.

Greenaway, Frank. *Science International: A History of the International Council of Scientific Unions*. Cambridge: Cambridge University Press, 1996.

Grossman, Daniel P. "A Policy History of Hanford's Atmospheric Releases." Ph.D. diss., Massachusetts Institute of Technology, 1994.

Grutzner, Charles. "Atomic Waste Pile Menace to Safety." *New York Times*, 15 April 1948.

"Guided Missile May Use Radioactive Waste." *Los Angeles Times*, 2 November 1947.

Haas, Peter M. "Do Regimes Matter? Epistemic Communities and Mediterranean Pollution Control." *International Organization* 43, no. 3 (1989): 377–403.

———. *Saving the Mediterranean: The Politics of International Environmental Cooperation*. New York: Columbia University Press, 1990.

———, ed. *Knowledge, Power, and International Policy Coordination*. Special issue of *International Organization* 46, no. 1 (1992).

Hacker, Barton C. *Elements of Controversy: The Atomic Energy Commission and Radiation Safety in Nuclear Weapons Testing, 1947–1974*. Berkeley: University of California Press, 1994.

Hagen, Joel B. *An Entangled Bank: The Origins of Ecosystem Ecology*. New Brunswick: Rutgers University Press, 1992.

Halloran, Richard. "Still Lots of Deadly Gas Awaiting Disposal." *New York Times*, 23 August 1970.

Hamblin, Jacob Darwin. *Oceanographers and the Cold War: Disciples of Marine Science*. Seattle: University of Washington Press, 2005.

Harwood, Michael. "We Are Killing the Sea Around Us." *New York Times*, 24 October 1971.

Hecht, Gabrielle. *The Radiance of France: Nuclear Power and National Identity after World War II*. Cambridge: MIT Press, 1988.

Hewlett, Richard G., and Francis Duncan. *Atomic Shield: A History of the United States Atomic Energy Commission, Volume 2, 1947–1952*. University Park: Pennsylvania State University Press, 1969.

Hewlett, Richard G., and Jack M. Holl. *Atoms for Peace and War, 1953–1961: Eisenhower and the Atomic Energy Commission*. Berkeley: University of California Press, 1989.

Hillaby, John. "Sea May Become Atom Waste Site." *New York Times*, 19 August 1955, 5.

Hinton, Harold B. "Weapons Using Radioactive Poison Pushed by Atomic Energy Board." *New York Times*, 1 August 1950.

Hoffman, George W. "The Role of Nuclear Power in Europe's Future Energy Balance." *Annals of the Association of American Geographers* 47, no. 1 (1957): 15–40.

Holly, Susan K., and Warren B. McAllister. *Foreign Relations, 1969–1976, vol. E-1: Documents on Global Issues, 1969–1972*. Washington, D.C.: Department of State, 2005.

"Hot Ashes." *New York Times*, 27 June 1954.

Hounshell, David A. "Rethinking the Cold War; Rethinking Science and Technology in the Cold War; Rethinking the Social Study of Science and Technology." *Social Studies of Science* 31 (2001): 289–297.

Hunt, G. J. *Radioactivity Studies in Lowestoft: The First Fifty Years.* Science Series Technical Report Number 105. Lowestoft: Centre for Environment, Fisheries and Aquaculture Science, 1997.

Jasanoff, Sheila. "Science, Politics, and the Renegotiation of Expertise at EPA." *Osiris* 7 (1992): 194–217.

Jolly, J. Christopher. "Thresholds of Uncertainty: Radiation and Responsibility in the Fallout Controversy." Ph.D. diss., Oregon State University, 2004.

Joravsky, David. *The Lysenko Affair.* Cambridge: Harvard University Press, 1970.

Josephson, Paul R. *Red Atom: Russia's Nuclear Power Program from Stalin to Today.* New York: W. H. Freeman, 1999.

Kaempffert, Waldemar. "Atomic Ashes? Dump Them on Mars." *New York Times,* 18 April 1954.

———. "Dangerous Radioactive Wastes, Which Now Are Buried, May Have Many Industrial Uses." *New York Times,* 11 February 1951.

———. "How to Dispose of Deadly Radioactive Wastes Is a Difficult Problem for Atomic Plants." *New York Times,* 17 September 1950.

———. "Problem of Radioactive Wastes." *New York Times,* 26 June 1955.

Kenworthy, E. W. "Nixon Asks Great Lakes Dumping Ban." *New York Times,* 16 April 1970.

Kevles, Daniel J. *The Physicists: The History of a Scientific Community in Modern America.* Cambridge: Harvard University Press, 1995.

Kirsch, Scott. *Proving Grounds: Project Plowshare and the Unrealized Dream of Nuclear Earthmoving.* New Brunswick: Rutgers University Press, 2005.

Klein, Stanley. "The Careless Atom." *New York Times,* 9 March 1969.

Klingle, Matthew W. "Plying Atomic Waters: Lauren Donaldon and the 'Fern Lake Concept' of Fisheries Management." *Journal of the History of Biology* 31 (1998): 1–32.

Kopp, Carolyn. "The Origins of the American Scientific Debate over Fallout Hazards." *Social Studies of Science* 9 (1979): 403–422.

Kotin, Albert. "Letter to the Editor." *New York Times,* 26 June 1969.

Krige, John. "Atoms for Peace, Scientific Internationalism, and Scientific Intelligence." *Osiris* 21 (2006): 161–181.

Kunz, Diane B. *The Economic Diplomacy of the Suez Crisis.* Chapel Hill: University of North Carolina Press, 1991.

Laurence, William L. "Atomic Residues Pose a Challenge." *New York Times,* 21 April 1955.

———. "Waste Held Peril in Atomic Power." *New York Times,* 17 December 1955.

Lawson, Herbert G. "Nuclear Split." *Wall Street Journal,* 20 May 1970.

Libby, Willard F. "Radiochemistry." *Scientific Monthly* 83, no. 3 (September 1956): 115–121.

Lindee, M. Susan. *Suffering Made Real: American Science and the Survivors at Hiroshima.* Chicago: University of Chicago Press, 1994.

Linsley, G., K.-L. Sjöblom, and T. Cabianca. "Overview of Point Sources of Anthropogenic Radionuclides in the Oceans." In *Marine Radioactivity,* ed. Hugh D. Livingston, 109–138. Amsterdam: Elsevier, 2004.

Lowenstein, Robert. "Some Legal Considerations in the Ocean Disposal of Radioactive Wastes." *Health Physics* (1961): 110–113.

Lyons, Richard D. "Army Will Transport Nerve Gas across South for Disposal at Sea." *New York Times,* 30 July 1970.

Macfarlane, Allison. "Underlying Yucca Mountain: The Interplay of Geology and Policy in Nuclear Waste Disposal." *Social Studies of Science* 33 (2003): 783–807.
Madsen, Axel. *Cousteau: An Unauthorized Biography*. New York: Beaufort Books, 1986.
Makhijani, Arjun, Howard Hu, and Katherine Yih, eds. *Nuclear Wastelands: A Global Guide to Nuclear Weapons Production and Its Health and Environmental Effects*. Cambridge: MIT Press, 1995.
Mazuzan, George T., and J. Samuel Walker. *Controlling the Atom: The Beginnings of Nuclear Regulation, 1946–1962*. Berkeley: University of California Press, 1984.
McEvoy, Arthur F., and Harry N. Scheiber. "Scientists, Entrepreneurs, and the Policy Process: A Study of the Post-1945 California Sardine Depletion." *Journal of Economic History* 44, no. 2 (1984): 393–406.
Medvedev, Zhores A. *Nuclear Disaster in the Urals*. New York: Norton, 1979.
Mills, Eric L. "'Physische Meereskunde': From Geography to Physical Oceanography in the Institut für Meereskunde, Berlin, 1900–1935." *Historisch-Meereskundliches Jahrbuch* 4 (1997): 45–70.
Miyake, Y., Y. Sugimura, and K. Kameda. "On the Distribution of Radioactivity in the Sea around Bikini Atoll in June 1954." *Papers in Meteorology and Geophysics* 5 (1955): 420–427.
Morgan, Karl Z. "The Responsibilities of Health-Physics." *Scientific Monthly* 63, no. 2 (1946): 93–100.
Muller, H. J. "Artificial Transmutation of the Gene." *Science*, n.s., 66, no. 1699 (1927): 84–87.
———. "The Production of Mutations." Nobel Lecture, 12 December 1946. *Nobel Lectures: Physiology or Medicine, 1942–1962*. Amsterdam: Elsevier, 1964.
National Academy of Sciences—National Research Council, *Oceanography 1960 to 1970: A Report by the Committee on Oceanography*. Washington, D.C.: National Academy of Sciences—National Research Council, 1959.
Needell, Allan. *Science, Cold War, and the American State: Lloyd V. Berkner and the Balance of Professional Ideals*. Amsterdam: Harwood, 2000.
Normile, Dennis. "A Prize of One's Own." *Science*, n.s., 260, no. 5106 (1993): 424.
Novick, Sheldon. *The Careless Atom*. Boston: Houghton Mifflin, 1969.
"Oceanographers Split." *New York Times*, 22 November 1958.
"Oceanography, Fisheries, and Atomic Radiation." *Science*, n.s., 124 (6 July 1956): 13–16.
Pace, Eric. "Control of Radioactive Wastes in Sea Urged at Experts' Parley." *New York Times*, 1 July 1970.
Paul, Diane B. "The Rockefeller Foundation and the Origins of Behavior Genetics." In *The Expansion of American Biology*, ed. Keith R. Benson, Jane Maienschein, and Ronald Rainger, 262–283. New Brunswick: Rutgers University Press, 1991.
Peterson, D. J. *Troubled Lands: The Legacy of Soviet Environmental Destruction*. Boulder, Colo.: Westview Press, 1993.
Petrow, Richard. *In the Wake of Torrey Canyon*. New York: David McKay Co., 1968.
Pfau, Richard. *No Sacrifice Too Great: The Life of Lewis L. Strauss*. Charlottesville: University Press of Virginia, 1985.
Plumb, Robert K. "'Coming of Age' of Nuclear Technology Puts Emphasis on Dangers of Atomic Radiation." *New York Times*, 18 December 1955.
Polikarpov, G. G. "Ability of Some Black Sea Organisms to Accumulate Fission Products." *Science*, n.s., 133, no. 3459 (1961): 1127–1128.
———. *Radioecology of Aquatic Organisms*. Amsterdam: North-Holland, 1966.
Preston, A. "Artificial Radioactivity in Freshwater and Estuarine Systems." *Proceedings of the Royal Society of London, Series B, Biological Sciences* 180, no. 1061 (1972): 421–436.

Preston, A., and P. C. Wood. "Monitoring the Marine Environment." *Proceedings of the Royal Society of London, Series B, Biological Sciences* 177 (1971): 451–462.
Price, Don K. *The Scientific Estate*. Cambridge: Belknap, 1969.
Pritchard, Donald W. "Disposal of Radioactive Wastes in the Ocean." *Health Physics* 6 (1961): 103–109.
Quinn, Susan. *Marie Curie: A Life*. Reading, Mass.: Addison-Wesley, 1995.
"Radioactive Drum Sighted at Sea." *Los Angeles Times*, 16 July 1957.
"Radioactive Material 'Graveyards' Proposed." *Los Angeles Times*, 11 October 1949.
"Radioactive Waste Disposal Planned." *Los Angeles Times*, 6 April 1949.
"Radioactive Waste 'Polluting Sea.'" *Times* (London), 16 November 1962.
Rainger, Ronald. "Patronage and Science: Roger Revelle, the U.S. Navy, and Oceanography at the Scripps Institution." *Earth Sciences History* 19, no. 1 (2000): 58–89.
———. "Science at the Crossroads: The Navy, Bikini Atoll, and American Oceanography in the 1940s." *Historical Studies in the Physical and Biological Sciences* 30, no. 4 (2000): 349–371.
———. "'A Wonderful Oceanographic Tool': The Atomic Bomb, Radioactivity and the Development of American Oceanography." In *The Machine in Neptune's Garden: Historical Perspectives on Technology and the Marine Environment*, ed. Helen M. Rozwadowski and David K. Van Keuren, 93–131. Sagamore Beach, Mass.: Science History Publications, 2004.
Recommendations of the International Commission on Radiological Protection." *Health Physics* 2 (1959): 1–20.
Reiss, Louise Zibold. "Strontium-90 Absorption by Deciduous Teeth." *Science*, n.s., 134, no. 3491 (1961): 1669–1673.
Resner, E. P., G. D. Schmidt, and R. J. van Turner. "Health Physics for the N.S. Savannah." *Health Physics* 7 (1961): 114–119.
Reston, James. "A Peril of the Atomic Age." *New York Times*, 16 July 1957.
Ringius, Lasse. *Radioactive Waste Disposal at Sea: Public Ideas, Transnational Policy Entrepreneurs, and Environmental Regimes*. Cambridge: MIT Press, 2001.
Rozwadowski, Helen M. *The Sea Knows No Boundaries: A Century of Marine Science under ICES*. Seattle: University of Washington Press, 2002.
"Russia Assails U.N. Atomic Agency." *New York Times*, 27 September 1958.
"Safe Disposal of Radioactive Wastes." *New York Times*, 6 May 1951.
Santesmases, María Jesús. Peace Propaganda and Biomedical Experimentation: Influential Uses of Radioisotopes in Endocrinology and Molecular Genetics in Spain, 1950 1971. *Journal of the History of Biology* 39 (2006): 765–794.
Schaefer, Milner B. "Some Fundamental Aspects of Marine Ecology in Relation to Radioactive Waste." *Health Physics* 6 (1961): 97–102.
"Science News." *Science*, n.s., 91, no. 2366 (May 1940): 8, 10–11.
"Scientist's Fears Answered." *New York Times*, 26 June 1954.
"Sea Areas Picked for Atom Wastes." *New York Times*, 21 June 1959.
"Sea Canyons Held Poor Atom Ash Cans." *New York Times*, 3 August 1958.
Semple, Robert B., Jr. "Vexed Nixon Prods Congress on Ecology." *New York Times*, 6 September 1972.
"Senate Backs Curb on Ocean Dumping." *New York Times*, 25 November 1971.
Silverman, Michael Joshua. "No Immediate Risk: Environmental Safety in Nuclear Weapons Production, 1942–1985." Ph.D. diss., Carnegie Mellon University, 2000.
Smith, Gene. "Detroit Reactor Called Best Bet." *New York Times*, 20 July 1957.
———. "'Hot Cargo' Gets a One-Way Sea Trip." *New York Times*, 5 September 1957.
Smith, Robert M. "A.E.C. Scored on Storing Waste." *New York Times*, 7 March 1970.
———. "Panel Urges Curbs on Ocean Dumping; Nixon Hails Report." *New York Times*, 8 October 1970.

Sollen, Robert. *An Ocean of Oil: A Century of Political Struggle over Petroleum off the California Coast*. Juneau: Denali Press, 1998.
Statute of the International Atomic Energy Agency." *American Journal of International Law* 51, no. 2 (1957): 466–485.
Steinhart, Carol E., and John S. Steinhart. *Blowout: A Case Study of the Santa Barbara Oil Spill*. Belmont, Calif.: Wadsworth, 1972.
Stokke, Olav Schramm. "Beyond Dumping? The Effectiveness of the London Convention." In *Yearbook of International Co-operation on Environment and Development 1998/99*, ed. Helge Ole Bergesen, Georg Parmann and Oystein B. Thommessen, 39–49. London, Earthscan, 1998.
Sturtevant, A. H. "Social Implications of the Genetics of Man." *Science* 120 (10 September 1954): 405–407.
Sverdrup, H. U., Martin W. Johnson, and Richard H. Fleming. *The Oceans: Their Physics, Chemistry, and General Biology*. New York: Prentice-Hall, 1942.
Swan, Charles, and James Ueberhorst. "The Conference on the Law of the Sea: A Report." *Michigan Law Review* 56, no. 7 (1958): 1132–1141.
"Talks Scheduled on Disposal of Atomic Waste." *Los Angeles Times*, 5 September 1950.
Tarr, Joel A. *The Search for the Ultimate Sink: Urban Pollution in Historical Perspective*. Akron: University of Akron Press, 1996.
Taylor, L. S. "Brief History of the National Committee on Radiation Protection and Measurements (NCRP) Covering the Period 1929–1946." *Health Physics* 1 (1958): 3–10.
"Trains Loaded for Nerve Gas Shipment." *New York Times*, 8 August 1970.
"Treaty Establishing the European Atomic Energy Community (EURATOM)." *American Journal of International Law* 51, no. 4 (1957): 955–1000.
U.S. Congress Office of Technology Assessment. *Nuclear Wastes in the Arctic: An Analysis of Arctic and Other Regional Impacts from Soviet Nuclear Contamination*. Washington, D.C.: Government Printing Office, 1995.
U.S. Congress. House. Committee on Merchant Marine and Fisheries, Subcommittee on Oceanography. *Waste Dumping: Hearings before the Subcommittee on Oceanography and the Subcommittee on Fisheries and Wildlife Conservation and the Environment of the Committee on Merchant Marine and Fisheries, House of Representatives, Ninety-seventh Congress*. Washington: Government Printing Office, 1982.
———. Subcommittee on Fisheries and Wildlife Conservation and the Subcommittee on Oceanography of the Committee on Merchant Marine and Fisheries. *Ocean Dumping of Waste Materials Hearings*. 92d Cong., 1st sess., 5–7 April 1971.
U.S. National Academy of Sciences. *Considerations on the Disposal of Radioactive Wastes from Nuclear-Powered Ships into the Marine Environment*. Washington, D.C.: National Academy of Sciences, 1959.
———. *Disposal of Low-Level Radioactive Waste into Pacific Coastal Waters*. Washington, D.C.: National Academy of Sciences, 1962.
"U.S. Offers to Pay for Atom Studies." *New York Times*, 26 September 1958.
Walker, J. Samuel. "The Atomic Energy Commission and the Politics of Radiation Protection, 1967–1971." *Isis* 85, no. 1 (1994): 57–78.
———. *Permissible Dose: A History of Radiation Protection in the Twentieth Century*. Berkeley: University of California Press, 2000.
Weart, Spencer R. *Scientists in Power*. Cambridge: Harvard University Press, 1979.
———. *Nuclear Fear: A History of Images*. Cambridge: Harvard University Press, 1988.
Welsome, Eileen. *The Plutonium Files: America's Secret Medical Experiments in the Cold War*. New York: Delta, 1999.

Westwick, Peter J. *The National Labs: Science in an American System, 1947–1974*. Cambridge: Harvard University Press, 2003.
Wittner, Lawrence S. *Resisting the Bomb: A History of the World Nuclear Disarmament Movement, 1954–1970*. Stanford: Stanford University Press, 1997.
Ziegler, Charles E., and Henry B. Lyon. "The Politics of Nuclear Waste in Russia." *Problems of Post-Communism* 49 (2002): 33–42.

Index

About the Author

JACOB DARWIN HAMBLIN is an assistant professor of history at Clemson University. He is a former postdoctoral fellow of the Centre Alexandre Koyré in Paris and taught history for four years at California State University, Long Beach. He has published extensively on the international dimensions of science, technology, and the environment during the cold war era. His first book was *Oceanographers and the Cold War: Disciples of Marine Science* (2005).